How To Rebuild BIG-BLOCK FORD ENGINES

By Steve Christ

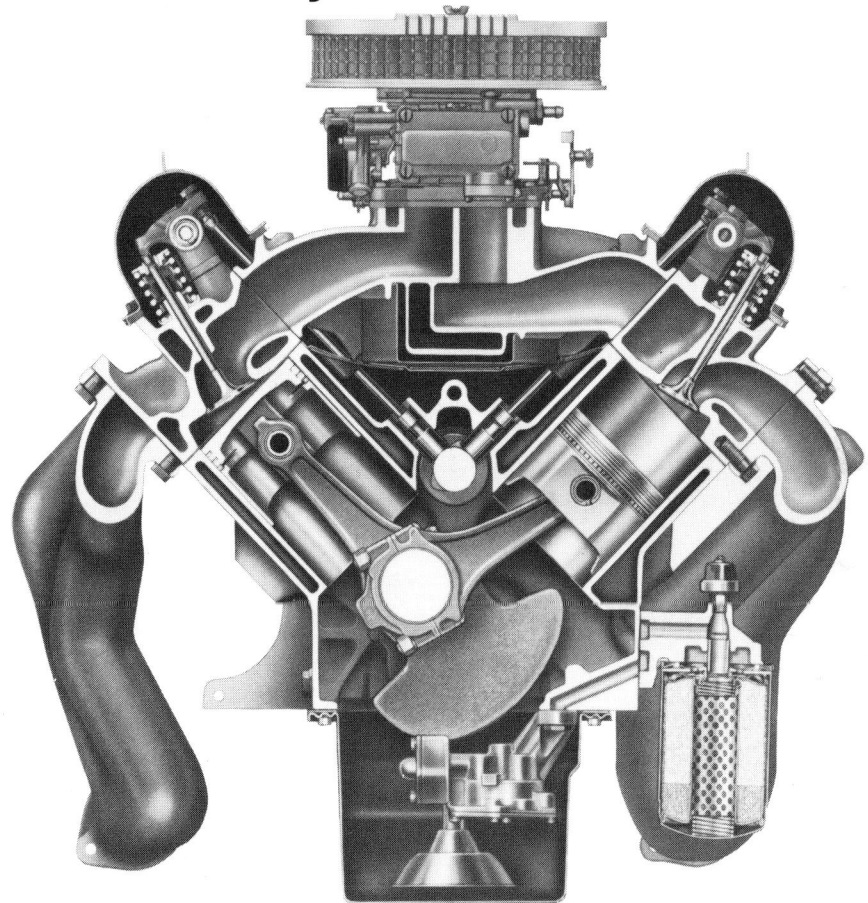

Published by HPBooks
a division of Price Stern Sloan, Inc.
11150 Olympic Boulevard, Los Angeles, California 90064
ISBN: 0-89586-070-8   Library of Congress Catalog Card Number 82-84702
© 1983 Price Stern Sloan, Inc.
Printed in the U.S.A.

18 17 16 15 14 13 12 11 10 9

Cover Photo by Michael Lufty

All rights reserved. No part of this publication may be reproduced, stored in a retrieval system or transmitted, in any form or by any means, electronic, mechanical, photocopying, recording or otherwise, without the prior written permission of the publisher.

NOTICE: The information contained in this book is true and complete to the best of our knowledge. All recommendations on parts and procedures are made without any guarantees on the part of the author or Price Stern Sloan. Tampering with or altering any emissions-control device is in violation of federal regulations. Author and publisher disclaim all liability incurred in connection with the use of this information.

CONTENTS

INTRODUCTION 3

Chapter 1
DO YOU NEED TO REBUILD? 4

Chapter 2
ENGINE REMOVAL 13

Chapter 3
PARTS IDENTIFICATION and INTERCHANGE 24

Chapter 4
TEARDOWN 58

Chapter 5
SHORT-BLOCK RECONDITIONING 70

Chapter 6
CYLINDER-HEAD RECONDITIONING 98

Chapter 7
REASSEMBLY 117

Chapter 8
INSTALLATION 145

Chapter 9
TUNEUP 157

INDEX 158

THANKS

Writing this book was made easier through the help and support of many high-caliber individuals. With great risk of leaving someone out, I would like to thank those who helped to ensure accuracy and completeness. They also made the job of authoring this book enjoyable and a reward in itself.

Wally and Perry Wagamon of Wagamon Brothers Engine Rebuilding gave me access to their engine-rebuild shop, equipment and employees. They were a constant source of technical advice and humor. Thanks to them and their employees for staying past closing time and interrupting their work on many occasions to help.

John Emmert, George Trainor and Earl Miller of Ford Motor Company kept my enthusiasm high with confidence and support. Howard Aula, Bill Barr and Don Sullivan of Ford dug facts out of their memories—as far back as 1957. I was surprised and pleased by their extra effort.

Bill Buffa, Walt Clark, Linda Lee, Al Rominsky, Ray Sobocinski and many others from Ford helped provide much needed information and material.

Thanks also to: John Havel of Clevite Bearings; Bob Bub of Cloyes Gear & Products; Myron Cottrell of Cottrell Racing Engines; Robert Morris, Irene Jeremize and Rich Livinair of Fel Pro; Hollander Publishing; Gary Kohn of M.A.S. (Racing Unlimited); Cal DeBruin of Sealed Power; Larry Davis of the Thunderbolt Club of America, Hueytown, Alabama; John Vermeersch of Total Performance; and Jerry Nyland, Mark Christ and Bob Stubbs. And to those unnamed individuals who helped—especially those Ford engineers who asked to remain anonymous.

Special thanks to Jim Miller for tuning my hieroglyphics into readable English, and to Tom Monroe for supporting my waxing writing talent.

INTRODUCTION

Big-block Ford was introduced in 1958 in 332 and 352 cubic-inch displacements (CID). Produced over 20 years, it was one of Ford's most-successful engines. Photo courtesy of Ford Motor Co.

Standard engine in Thunderbirds for years, the big-block later gained reputation as a workhorse in pickup, medium-duty and heavy-duty trucks. Photo courtesy of Ford Motor Co.

Designed in the mid-'50s, the big-block Ford was introduced in 1958 Ford, Edsel and Thunderbird cars. Although it later became a very successful racing engine, the big-block Ford was originally designed to handle the increasing weight of passenger cars. Highways were being improved so the speed and distance a car could travel was increasing. Reliability, higher performance and lower maintenance were also being sought by the consumer.

This book covers both *FE* and *FT* engines. The FE designation refers to passenger-car and light-truck engines. FE engines include: 332, 352, 352HP, 360, 361, 390, 390HP, 406, 410, 427, 428, 428CJ and 428SCJ. FT engines were installed in medium- and heavy-duty trucks. FT engines include: 330MD, 330HD, 359, 361, 389 and 391.

Both engine families share external dimensions and some parts. I'll refer to both as *big-blocks,* but where there is reason to make a distinction between them, I'll use the FE or FT designation. Besides covering the rebuild, the book also lists many parts and specifications for these engines.

In passenger cars, FE production started in 1958 and ended in 1971. FE engines were installed in pickup trucks from 1965 through 1976. Medium- and heavy-duty trucks used the FT, which was produced from 1964 until 1978. There were also many marine and industrial applications throughout this time, using both FE and FT engines.

The big-block Ford is a *Y-block* design—the cylinder block extends below the crankshaft center line. The engine is not usually referred to as a *Y-block* because the 239/312 series— predecessors of the big-block—are commonly known as *Y-blocks*. They were also offered during the big-blocks' first years.

On the big-block, the cylinder block extends 2-5/8-in. below the crank center line, more than 1-in. below the bottom of the crank journals. This makes the cylinder block very rigid.

Many big-block Fords are in use all over the world in their original vehicles. The big-block has been in nearly every type of race and every type of vehicle. It has won at stock-car tracks, drag strips, boat races, tractor and truck pulls, hill climbs and in sports-car and endurance racing such as Daytona and LeMans. And the engine is still competing and winning. Few engines have competed and won in such a wide variety of races.

Besides competition, the big-block does well in fuel economy—better than many engines of equal or smaller displacement. Full-size LTDs with 390s commonly obtain 18 or more miles per gallon (mpg) on the highway. The big-block is also a workhorse, used in trucks with payloads measured in tons.

As you read this book, you will note the durability of this engine. If you don't already, you will learn to appreciate the big-block Ford. When you rebuild your engine, put as much into it as you expect to get out. This will guarantee that you'll be satisfied with your rebuilt engine.

CHAPTER 1
Do you need to rebuild?

If you are reading this chapter, you probably have questions as to the condition of the engine. It may have high oil consumption, be low on power or have an undetermined noise. You may just want to improve its performance.

Many big-block Fords have gone well past the 100,000-mile mark without a rebuild or major repair. These engines give excellent service with regular oil and filter changes. I have seen some 352s and 390s on the third revolution of the odometer without a rebuild or valve job.

Any engine past the 100,000-mile mark will probably benefit from a rebuild, but one may not be absolutely necessary. At the same time, there are some engines with less than half that many miles that need a rebuild! Briefly read this chapter, then come back to the areas that apply to your engine.

HIGH OIL CONSUMPTION

If the engine uses a quart of oil every 1000 miles or less, it is using too much. Unless the engine is worked extremely hard or is operated at high rpm, the condition can be corrected. High-performance engines with compression ratios of 11:1 or more and larger clearances will naturally use more oil.

Before condemning the engine as a oil burner, check whether the oil is leaking. A bad rear-main seal or some gasket may be causing the excessive oil loss. A small amount of engine oil on the driveway usually means a lot is out on the road. If the engine doesn't leak oil, check some other possibilities.

Piston Rings—Three piston rings are used in all FE and most FT engines. The top two rings—three on four-ring FT pistons—share the job of sealing the combustion chamber. The bottom

High-performance 406, shown with three 2-barrel carburetors, was superseded by the 427 in less than two years. Short life of the 406 was due to horsepower race of '60s. Photo courtesy of Ford Motor Co.

ring handles oil control. The oil ring removes *most* of the oil sprayed or splashed onto the cylinder walls. It doesn't remove all the oil because the compression rings require some for lubrication and sealing.

Although the rings may not be worn, oil can get into the combustion chamber when excess oil is splashed on the cylinder walls. The excess is caused by too-large bearing-to-journal clearances on the main or rod journals. The higher volume of oil may be too much for the oil rings to control.

Just as oil can get into the combustion chamber past the oil rings, combustion gases can blow back past the compression rings into the crankcase. This is referred to as *blowby*. Crankcase gases or vapors have to be vented or pressure will build up in the crankcase.

Early-model big-blocks had a *road-draft tube* to vent these vapors to the atmosphere. Later big-block Fords use positive crankcase ventilation (PCV) to vent the vapors into the intake manifold. Vapors are then drawn into the combustion chambers and burned.

If the blowby is too great, the PCV system won't be able to cope. The result is smoke out the exhaust pipe or breather cap, provided the breather cap is not connected to the air cleaner. If the engine smokes, there will be carbon and oil deposits in the combustion chambers. The valves, spark-plugs, piston tops and piston-ring lands will also be coated. The inside of the intake manifold and exhaust system will have the deposits, too.

Before condemning the rings, check the PCV system, particularly

If engine has conventional point-type ignition, check point gap. Insufficient point gap causes the spark to retard ignition. This seriously cuts performance and fuel economy.

Most performance troubles are caused by worn or incorrectly adjusted parts. Complete, high-quality tuneup kit can go a long way in helping diagnose—and repair—many performance problems. Cost is quickly recovered with today's high fuel costs.

the valve. The valve should rattle when you shake it. Check the PCV hose to the intake manifold to make sure it is not plugged. Any restriction in the PCV system will cause smoke.

Valve Guides & Seals—Valve guides and stems have built-in clearance to allow some oil to pass for lubrication. As the clearance grows, excess oil goes down the valve stems. What happens next depends on whether the valve is an intake or exhaust. A worn intake-valve guide allows oil to enter the combustion chamber. Wear on the exhaust side allows oil into the exhaust port.

Oil loss down the valve guides is also aggravated by blowby. Crankcase pressure forces more oil down through the guides. Excess blowby and worn valve guides indicate that the engine should be rebuilt.

Valve-stem seals can be worn, cracked or missing. This causes excess oil to pass down the valve guide, regardless of stem or guide condition. Replacing valve-stem seals can bring oil consumption down to an acceptable level if the rest of the engine is in good shape. Valve-stem seals can be replaced without removing the cylinder heads.

Cracked Oil Passage—There is an oil passage for each rocker-arm assembly in the cylinder block and a head. The right passage runs from number-4 cam-bearing journal to number-3 rocker stand. The left passage runs from number-2 cam-bearing journal to number-2 rocker stand.

Occasionally these passages crack near the cam journal. Oil is then forced into the water jacket and mixes with the engine coolant. This is an extremely difficult condition to diagnose—fortunately it is rare.

The cure for this crack is to drill the passage out to the OD of a long small-diameter tube—a pushrod with the ends cut off does nicely. Cover the tube with sealant, insert it and trim the top flush with the block surface. I have never experienced this problem on any of my big-blocks, although some machine shops do the repair on all big-block rebuilds as a preventive measure. If you think the engine has this problem, have the machine shop do the repair after you tear the engine down. By the time you buy the drill bit and round up the tubing they can have the job done.

LOSS OF PERFORMANCE

If fuel economy isn't what it should be or power seems low, there are a few checks to make. Numerous problems can cause poor performance and not require a rebuild.

First, make sure the engine is in tune. Too much or too little fuel, or incorrect ignition operation can hurt performance. Also, excessive drag from the brakes or drive train reduces power at the rear wheels.

A good way to check power is to take your vehicle to a shop with a *chassis dynamometer*. A chassis dynamometer can check the power at the rear wheels—*road horsepower*. The operator can monitor the various components of the engine, such as the carburetor and distributor, while the car is "going down the road." A chassis dyno also allows comparing horsepower before and after changes are made.

Have the engine tuned and checked on a dyno only if you suspect the engine is in good condition, or you have the money and want to see how much horsepower the engine gained from a rebuild. If the engine hasn't had a tuneup recently, its performance should improve from having one.

Before getting into engine-problem diagnosis, let's take a look at other causes of performance loss.

Non-Engine-Related Problems—Before blaming the engine, be sure the rest of the drive train is in good condition. Make sure the clutch is not slipping in the case of a manual transmission. Be sure an automatic transmission shifts correctly and does not slip. Find a place where you can put the transmisssion in neutral and coast to make sure the brakes don't drag.

Incorrect Fuel Supply—A restricted fuel line or filter, a faulty fuel pump or bad carburetor adjustment can account for poor performance. Check that the carburetor is clean and the float level is set correctly. Make sure the throttle plate or plates are opening and closing all the way. Check the intake system for air leaks. These will lean out the air/fuel mixture and hurt engine performance.

Faulty Ignition—Check the ignition system. Make sure the engine is getting a good spark at the correct time. Incorrect sparkplugs, faulty wires or coil, cracked or worn distributor cap and rotor or incorrect timing can cause a huge loss in performance.

Having the engine checked on an analyzer will greatly simplify this task. If you check the system and still feel something is amiss, have the distributor's vacuum- and mechanical-advance mechanisms checked. If you have your vehicle checked on an engine analyzer, the technician should be able to do this for you. If not, remove the distributor and take it to an ignition-specialty shop.

If you have made it this far and still haven't located the problem, check for the following: blown head gasket, restricted exhaust, worn timing chain and sprockets, worn camshaft and lifters, burned valves and carbon deposits.

Blown Head Gasket—Often there will be an immediate loss in power and an increase in engine noise with a blown head gasket. If the leak occurs to the outside of the engine you'll hear it. The noise will sound like a rapid off-and-on release of high-pressure air. The rate increases with engine rpm.

Some head-gasket leaks will enter the cooling system, showing up as multi-colored oil spots in the coolant. These leaks are relatively easy to diagnose. Large white deposits will appear on the sparkplugs. There may also be white "smoke"—steam—from the exhaust as the engine draws coolant into the combustion chamber. The engine will eventually overheat as combustion gasses overpressurize the cooling system and force the coolant out.

A leak into an adjacent cylinder is harder to detect. The cylinders at fault should show up during a *compression* or *leak-down* test. A compression test should be done when the engine is warm because some head gaskets leak only when the engine is warm or hot.

Plugged Exhaust—A restricted exhaust system has perplexed many a mechanic. It doesn't happen often, but when it does it is found only after going through *everything* else. A plugged exhaust will usually allow an engine to start. But the engine will either die soon after or not reach high rpm, depending on how badly the exhaust is plugged.

Likely causes for a plugged exhaust are a stuck heat riser or a collapsed or crushed exhaust pipe or muffler. Another possibility is a clogged catalytic converter on '75 or '76 pickup trucks. It's not uncommon for a converter to plug, particularly if leaded gas has been used.

Check the heat riser first. It is usually on the right exhaust-manifold outlet. The heat riser is positioned by a spring. Grab the counterweight or shaft of the heat riser and try to rotate it 90°. The shaft should spring back to its resting position when you release it. A stuck heat riser can be freed by a special penetrating oil—Ford calls it Rust Inhibitor. Spray the ends of the shaft and then rotate the shaft back and forth until it works smoothly and the spring returns it to position.

To test the engine to see if the rest of the exhaust system is plugged, hook a vacuum gage to the intake manifold. If vacuum is normal—about 15 inches of mercury—when the engine is cranked and drops low or to zero once it starts, it is likely the exhaust system is plugged. If the exhaust is unrestricted, you should *feel* a definite pulse from the pipe/s. If you have a trained ear you can recognize a plugged exhaust system choking at the exhaust-pipe outlet.

If you suspect a restriction, look underneath for a crimped or collapsed pipe or muffler. If the pipe doesn't appear to be collapsed, don't assume the exhaust system is OK. The inner pipe or the catalytic converter may be collapsed or plugged, or a muffler baffle may have broken loose. Any of these can cause the restriction.

As a final check, the exhaust system can be disconnected behind the Y- or H-pipe, as these don't usually collapse. Run the engine to see if the condition improves. Make sure you do this where the neighbors won't mind the noise and only long enough to make the check.

Worn Timing Set—If the timing chain and sprockets are badly worn, valve timing will be off. This can affect performance severely, even to the point of the engine not running.

A worn timing chain and sprockets will sometimes show up while timing the engine and working the throttle. The timing mark on the vibration damper will appear to move erratically as you slowly open and close the throttle. This can also be caused by a worn distributor, so check it also if you find this condition.

Nylon chunks in the drain oil indicate broken teeth from a nylon-toothed cam sprocket. This means time for replacement.

A simple way to check timing-chain and sprocket wear is to use a 15/16-in. socket on the vibration-damper bolt. Remove the distributor cap and turn the crank back and forth about 1/2-in. using the damper and timing pointer as reference. If there is hesitation between the movement of the crank and the distributor rotor, the timing chain and sprockets are worn.

If wear is severe the chain can jump a tooth. Generally, if the timing chain jumps one tooth the engine will barely run. If the chain jumps two teeth the engine may not run at all.

Suspect a jumped timing chain if the engine won't start after you shut it off at high rpm or if it *ran on*—dieseled—after you shut it off. Wait a bit to make sure the lifters have not pumped up before blaming the timing chain and sprockets. Either condition will let the engine spin fast like it has no compression while cranking.

To double-check for a jumped timing chain, remove the air filter. If the chain has jumped, there will usually be a steady stream of air coming *out* of the carburetor as the engine is cranked.

Worn Cam Lobes & Lifters—This condition isn't very common in big-block Fords, but happens occasionally. If one lobe or lifter wears more than the others, the engine will run roughly, especially under load. If all the lobes or lifters are worn excessively, engine power will be severely reduced.

If an intake valve doesn't open enough, there won't be sufficient air/fuel mixture to supply that cylinder. If an exhaust valve doesn't open enough, the incoming mixture will be diluted with exhaust gases. Either way, the engine loses power.

Burned Valves—A burned exhaust valve can't seal the combustion chamber. This results in lower compression and power. A burned valve can be caused by carbon holding the valve open, retarded timing, lean mixture or by the valve not seating fully. Sometimes, it's a combination of these. A burned valve can be found during a compression or leak-down test, pages 10 and 11.

Carbon Deposits—Brownish or grey-black smoke pouring from the exhaust pipe of a car at wide-open throttle is carbon deposits being burned or knocked off.

These deposits form in the combustion chambers of an engine that is in a poor state of tune or condition, or is not being driven properly. Carbon deposits have a number of causes: rich fuel mixture, high oil consumption, cold operating temperature or extended low-speed driving or idling.

Carbon deposits can cause a number of problems: First, the carbon accumulates on the piston and cylinder head, taking up room in the combustion chamber. This raises the compression ratio.

If the compression ratio gets too high, the engine *detonates,* or pings. During normal combustion the combustion-chamber pressure builds to the point that the air/fuel mixture in the cylinder explodes, or *detonates,* instead of burning.

If the engine has a problem with detonation, don't assume it's due to carbon buildup. Detonation can also be caused by using fuel with an octane rating too low for the engine's compression ratio—a common occurrence with today's low-octane gas. Other causes may be overheating, over-advanced timing or the engine inhaling too much oil because of worn valve guides or rings.

Before enough carbon accumulates to cause detonation, it usually causes *preignition.* The carbon particles glow hot enough to ignite the air/fuel mixture *before* the sparkplug does—like the glow plug of a diesel or model-airplane engine.

Both conditions are hard on the engine. They put stress on the pistons, rods, crank, cylinder heads and finally, your wallet.

Carbon also causes problems at the valves. As carbon builds up around a valve seat it can prevent the valve from closing all the way. Then the hot exhaust gases act like an acetylene torch and burn, or cut, the edge of the valve as they escape into the exhaust port. Once the edge is burned the valve will not seal and the cylinder will lose compression. Meanwhile, the valve continues to burn.

Carbon deposits on valves and ports also reduce air/fuel-mixture flow into the combustion chambers. This robs an engine of power and cuts fuel efficency.

A number of "tuneup-in-a-can" elixirs are sold to remove carbon buildup. But additives often remove the carbon in big chunks. This is like throwing grit into a combustion chamber. The carbon increases ring wear and may lodge between a valve and seat.

Additives are OK if the carbon buildup is light, but judging how much buildup is on the piston and cylinder head is difficult with an assembled engine. Rather than using an additive, cure the problem causing the carbon buildup. If you cure the problem the carbon will eventually burn away.

Valve Adjustment—If the engine is equipped with solid lifters, check valve clearance. A few extra thousandths of an inch of clearance, or lash, can make all the difference in noise and lost power. Hydraulic lifters also make noise if the "clearance" is too great—they should have zero clearance.

Hydraulic lifters can also *pump up,* causing poor performance. The lifter fails to bleed down and holds the valve open more than it should—sometimes preventing the valve from closing. This problem can also be caused by weak valve springs. In either case, the engine will idle roughly, stall or not be capable of reaching high rpm.

DIAGNOSIS

With the foregoing problems in mind, you can now do the following checks to determine the condition of the engine. The most-common problems have been listed. The problems with your engine may be unusual, or the engine may have created an entirely unique one.

NOISES

Any moving part can make a noise, but locating that noise can be difficult. The first step in finding the source of the noise is to narrow your choices.

First determine if the noise is at *engine speed* or *half engine speed.* This will give you some clue as to where the problem lies. Noises at half engine speed are usually in the valve train. Two exceptions are piston slap and fuel-pump noises—they also occur at half speed. Noises at engine speed are normally crankshaft related, or in the bottom end of the engine.

Use a timing light to determine the

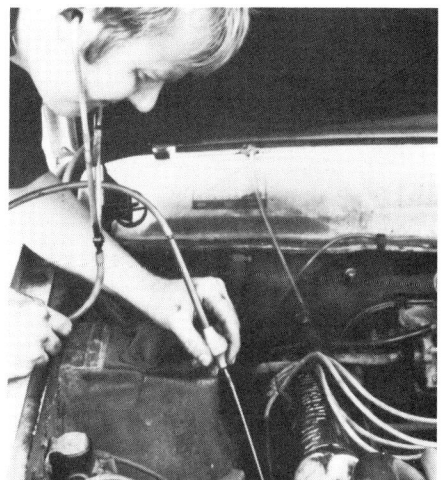

Stethoscope used to pinpoint engine noise. Be sure to listen to fuel pump on big-blocks—faulty fuel pump is often misdiagnosed as more-serious problem.

speed of the noise. Hook the light to any plug wire and listen. If the noise occurs once every time the light flashes, the noise is at half engine speed. If the noise occurs twice every time the light flashes, it is at engine speed.

Once you have determined the speed of the noise, you may need a listening device to pinpoint it. A mechanic's *stethoscope* works best; it is flexible and picks up sound well. You can make do with a screwdriver, piece of wood or garden hose.

A 3—4-ft section of hose works similar to the stethoscope. Use a plastic hose if possible—it transmits sound better than rubber. A screwdriver or wooden rod transmits the sound well, but is not flexible.

Unfortunately, sounds are also carried by the cylinder block, so you may have trouble pinpointing the noise. If you use a piece of wood or screwdriver, the blade or one end has to be placed against the engine. Press the other end against your ear or touch it to your head behind your ear.

Move the end around on the engine until you find where the noise is loudest. Place the probe next to a sparkplug. This is an excellent way to locate noise from that cylinder—especially from the piston-and-rod assembly.

If the noise seems to be coming from the *upper end* of the engine—the valve train—move the probe until the sound is the loudest from one of the cylinders. Then alternate between intake and exhaust ports of that cylinder to pinpoint the source. Don't forget the fuel pump. It also operates at half engine speed. Read on for this problem.

Valve-Train Noise—Noise from the valve train is normally a clicking sound, caused by excessive valve clearance. If the noise is in this area, remove the valve covers. Run the engine to help locate the faulty part. If the noise occurs at low rpm, reduce idle speed so oil doesn't splash all over.

With the valve covers off and the engine running, check that the pushrods rotate and the valves open and close. If a pushrod isn't rotating or a valve isn't opening as much as the rest, check for a bent pushrod, or a badly worn lifter or cam lobe.

As I previously stated, worn camshafts are not usually a problem with big-block Fords. If the engine has hydraulic lifters and a noisy valve train, a lifter is the probable culprit. Insert a feeler gage between the rocker arm and valve-stem tip. This should quiet the noise. If possible, adjust the valve clearance, page 140.

If the engine has hydraulic lifters, remove the pushrod and check that it is straight, page 84. If the pushrod is OK, the problem is probably a malfunctioning lifter.

Lifters can be removed with the intake manifold in place, so replacing them is not nearly as difficult as it is on most V8 engines. It's "simply" a matter of removing the rocker-arm shaft, pages 61 and 62, and fishing the lifter out of its bore.

Piston Slap—Piston slap makes a hollow or dull noise. If you were to tap a piston's thrust surface—the right side of the piston below the rings—with a plastic mallet it would make a similar sound. The noise from the engine is muffled by cast iron and coolant.

Piston slap is caused by excessive piston-to-bore clearance. It's usually the result of piston wear or a collapsed or broken piston skirt. These are indications of high mileage, overheating or poor cylinder-wall lubrication.

Piston slap is loudest when an engine is cold because piston-to-bore clearance is maximum. If the noise goes away after the engine warms up, the condition isn't serious.

To determine if the noise is piston slap, loosen the distributor hold-down bolt. Retard the ignition by slowly rotating the distributor counterclockwise. Retarded timing reduces the load placed on the piston by combustion. The noise should become quieter or disappear if it's piston slap. Be sure to reset the timing for further tests.

A hole in a piston makes a noise similar to piston slap. It can be detected by removing the dipstick and listening to the dipstick tube. You will be able to hear the noise when piston reaches TDC and feel the resulting pulse of air from combustion if there's a *holed* piston.

Main Bearing—A main bearing with excess clearance can knock at half speed. The noise will sound far away—muffled—in the engine. You can sometimes feel it through the accelerator pedal. The noise will usually occur when the engine is first started, hot or cold, before oil pressure builds up. It can also be detected under hard acceleration. Disabling the cylinders closest to the problem main bearing should quiet the noise and help you to locate which one is at fault.

Don't be fooled by detonation. It usually shows up under hard acceleration, but won't be as forceful or low-pitched sounding as that caused by excess main-bearing clearance. Detonation has a tinny, rattling sound.

Connecting Rod—A connecting rod will knock or pound if oil pressure is low or connecting-rod bearing-to-journal clearance is excessive. The noise will be loudest just after you let up on the accelerator after maintaining a constant speed.

Wrist Pin—Wrist-pin noise has a double click to it and is quite pronounced. This sound is normally heard while the engine is idling or at low speeds. If there is a common problem with the big-block Ford—other than valve-train noise—wrist-pin wear is it. Incorrect clearance during assembly is the usual cause. The pin can also come loose in the piston or the bushing in the rod can crack, flake or wear out.

Piston Rings—Bad piston rings have a chattering sound that is most noticeable during acceleration. This is usually caused by broken rings, but it may be the result of the rings losing their *tension* or springiness. This condition should show up as low compression during a compression test.

Fuel Pump—Before you tear the engine down because of a bottom-end noise, check the fuel pump. Fuel-pump noise will normally be a fairly loud, double click at *half engine speed*. It's often mistaken for a bad rod.

If you want to be sure the fuel pump is at fault, disconnect the fuel lines, plug them and remove the pump. Run the engine—there'll be enough fuel in the carburetor for a couple of minutes of idling. If the noise ceases, chances are the fuel pump is bad or the fuel-pump eccentric is loose on the camshaft.

PERFORMANCE PROBLEMS

Read the Plugs—Sparkplugs give you a chance to "see" what is going on in each cylinder. Remove the plugs, but keep them in order for future reference.

The two things to look for are a wet, shiny-black insulator or a plug with blister marks on the insulator or shell. Wet, black deposits indicate excessive oil in the combustion chamber. Blistering on the insulator indicates excessive heat.

If the plug insulators have a dry gray or black coating on them, check for over-rich mixture or weak ignition.

BE CAREFUL

Use *CAUTION* when working on a running engine. Make sure you or the probe don't get caught in the fan or pulleys and belts. Keep neckties and jewelry in your dresser drawer! Stay out of the plane of the fan. *Wear safety glasses.* Be very safety-conscious—accidents happen.

A few years ago I went to investigate a Talladega Torino that supposedly had a 427 in it. While the engine was running, I looked underneath the car into the engine compartment to check for the cross-bolted mains.

I stood up with what I thought was a piece of dirt in my eye. I had my wife check when I got back into the car, but she couldn't find anything. That night, at my wife's insistence, I went to the hospital emergency room. They used a scapel to pull a sliver of rusty metal from my eye. The piece of metal was thrown by the fan as the engine idled!

The following Monday I went innocently to the eye doctor. He put my head in a *vise* and proceeded to *drill* some remaining particles out.

The full impact of the ordeal didn't hit me until I left his office. I now wear goggles *often*. To add insult to injury, the engine wasn't a 427.

A   **B**   **C**   **D**

Reading sparkplugs is a good diagnostic technique. Worn-out plugs, A, are easily recognized by eroded electrodes and pitted insulator. Replace them and the engine's performance will immediately improve. Oil-coated and fouled plugs, B, indicate internal wear of piston rings, cylinder bores and valve guides. Carbon-fouled plugs, C, have dry, black and fluffy deposits, and indicate weak ignition, carburetor problems or poor driving habits. Normal plugs, D, have a brown to greyish-tan appearance with some electrode wear indicated by rounded edge on center electrode. When checking plugs, record the cylinder number as you remove each plug. Photos courtesy of Champion Spark Plug Company.

Extensive low-speed driving or excessive idling can also cause this condition. Other problems may be a high float level, stuck choke or a sunken float.

If the engine has a Holley carburetor there's another possibility. After sitting for a long time, the gasket between the main carburetor body and metering block may dry out and crack. Fuel then runs into the engine, causing an excessively rich mixture.

If you suspect this problem, remove the float bowl/s and metering block/s. Remove the gaskets and check them for cracks or tears. Replace them if they are bad. Keep all of the gaskets soaking in gasoline while you check them to keep them from drying out. If they are good, you don't want to ruin them.

Vacuum Test—An internal-combustion engine is basically an air pump. How well it pumps air is a good indication of the engine's overall condition. When running, an engine produces a vacuum in the intake manifold. If each cylinder doesn't contribute its share of power, a vacuum gage should show it.

The first test should be done with the engine running. Hook the vacuum gage to a fitting on the intake manifold and start the engine. At idle the reading should be about 16—18 inches of mercury (''Hg). Altitude, ignition timing and cam design all can affect manifold vacuum. It will be less at high altitudes and on engines with a high-overlap—performance—cam.

A reading of 15''Hg or less indicates either incorrect ignition timing or a worn engine. Set the timing and recheck vacuum. If the needle *floats* —moves slowly back and forth—the mixture is probably over-rich. Turn the idle-mixture screws in or increase engine speed to 2000—2500 rpm to see if this corrects the problem. If the reading is low—12''Hg or less—the engine may have a blown head gasket or an air leak.

Next, accelerate the engine rapidly and then release the throttle. When the engine is accelerating, the reading should drop but remain steady. If the reading fluctuates, the valve springs may be weak.

When you release the throttle, the reading should jump to about 5''Hg above the reading at idle and then settle back to the original idle reading. If the reading does not go that high, the pistons and rings are not sealing well.

If you get a normal reading that soon drops back to zero when you first start the engine, suspect a plugged exhaust system.

Once the engine is warm, make *cranking vacuum tests*. First, disable the ignition system. If the engine has a conventional point-type distributor, remove the points-to-coil lead. With electronic ignition, disconnect the distributor-to-amplifier lead.

Now with either a helping hand or a remote starter, crank the engine. The vacuum-gage needle should remain fairly steady while the engine is cranked. If the needle fluctuates, one of the cylinders is not doing its share. The cause of this could be: incorrect valve adjustment, a worn camshaft lobe, collapsed lifter, a leaky valve, worn cylinder bore or piston rings, broken piston rings, a hole in the piston or a leaky head gasket. The following tests will locate which cylinder or cylinders is at fault.

Power-Balance Test—This is a good test for locating a problem cylinder.

Vacuum gage: a simple but seldom-used tool. Unless engine has a "wild" camshaft, this gage will give you more information on engine condition than any other gage.

Aircraft mechanics use a test similar to this for one of their run-up tests on reciprocating engines.

If each cylinder is contributing an equal amount of power, eliminating any one will give the same power or rpm reduction. Most good engine analyzers with an oscilloscope have a power-balance-test feature. With this feature any combination of cylinders can be disabled to perform the test.

During the power-balance test, disable the cylinders one at a time as the engine is running. Note the power or rpm drop with the cylinder disabled. Most of us don't have a dynamometer at our disposal so I'll use rpm as the reference.

For the test, idle the engine at about 1000 rpm. Disable one cylinder and allow engine rpm to stabilize. The rpm drop is noted as each cylinder is disabled. The greater the rpm drop, the more power that cylinder was

9

> **INDUCED FIRING**
>
> If rpm rises when one cylinder is disabled, that plug wire may be connected to the wrong sparkplug. If the wiring is correct, the sparkplug was probably being fired by *induced current*.
>
> Induced current, or *inductance*, occurs when the magnetic field around one wire *induces* a current in an adjacent wire. This causes two sparkplugs to fire simultaneously. In most cases, induced firing causes no problems, because there is insufficient air/fuel mixture to burn in the cylinder. Problems arise when the firing is induced in a cylinder where there is enough *compressed* air/fuel mixture to burn.
>
> The two wires usually involved where induced firing is a problem are number-7 and -8. These cylinders normally fire in order, -7 followed by -8. If the two wires are run close to one another, number-8 cylinder can fire 90° too soon. This is hard on the engine as it tries to turn the crankshaft backwards. For this reason be sure number-7 and -8 sparkplug wires don't run next to each other for any distance.

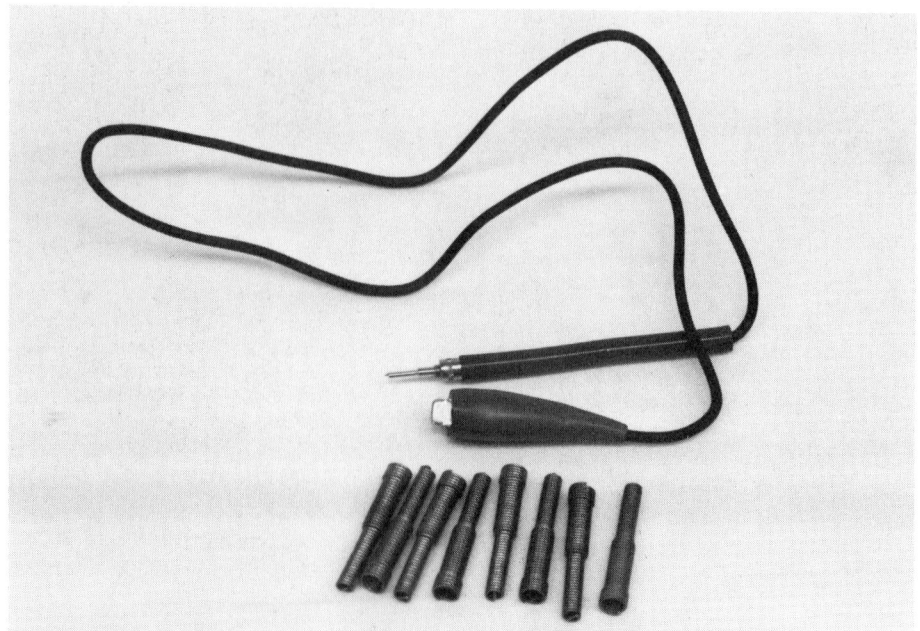

Using this type of test kit is simplest and safest way to check cylinder power balance. Jumpers are connected to each sparkplug so individual cylinders can be shorted without risking ignition-system damage. Photo by Tom Monroe.

producing. Conversely, the less rpm drop, the less that cylinder was producing.

Most engine analyzers have a built-in power-balance tester, which simplifies this test. The home method is not so difficult, but be careful. The problem is avoiding a 20,000—60,000-volt shock or damaging an electronic-ignition unit.

One method is to pull each plug wire off, then place a 4-in.-long piece of solid wire into the boot so it touches the lead. Bend the wire at the bottom of the boot, then place the boot back over the sparkplug. Finally, bend the wire as far away from the surrounding metal as possible. By using a ground wire or jumper cable, you can ground each plug individually to disable that cylinder.

This same method can also be used at the distributor cap, taking care that you have no arcing between cap towers. If you use this method keep track which cylinder you have disabled. Use the distributor-cap method if you have a Mustang, Cougar or '66—'67 Comet or Fairlane, particularly if it has air-injection. There is just no access to the plugs.

A number of manufacturers make power-balance test kits. The kits consist of a jumper wire and eight springs. The springs fit between the sparkplugs and the plug-wire boots. By touching the exposed spring with the grounded wire, each plug can be shorted out without risking damage to the ignition or your nervous system.

Another method is to use insulated pliers to pull the wires off while the engine is running. I seem to get shocks anyway, but the method does work. This method is OK with conventional, point-type ignition systems, but don't do it with an electronic ignition. The resulting high-voltage surge will damage the ignition.

Compression Test—This is the easiest test to determine the sealing quality of a cylinder. Compression testing should be done with the engine warm to give the most-accurate indication of engine condition. Run the engine to bring it up to operating temperature, then remove the sparkplugs.

Disable the ignition system. With an electronic ignition, unplug the distributor-to-amplifier lead. With conventional ignition, remove the points-to-coil lead. Block the throttle wide open. Make sure the choke is open.

Install the compression tester in the number-1 sparkplug hole. Crank the engine over at least three times and record the highest reading. Check the rest of the cylinders the same way and compare the readings. The lowest reading should be at least 75% of the highest.

For example, if the highest reading is 140 psi, the lowest should be at least 105 psi. Consider 75% the *absolute* lowest percentage. An engine being considered for performance should be in the high-90% range.

Caution: When doing a compression test, an explosive air/fuel mixture is being shot out the sparkplug hole of each cylinder while the engine is cranking. A spark or flame could cause a small explosion. So make sure the immediate area is free of any open flames—*don't smoke*.

What readings should you see? The range of pressures should be 100—250 psi. They should definitely be above 100 psi, but there can be a large difference between engines. An older 352 in excellent condition will show about 180 psi per cylinder. With 75% considered acceptable, the lowest reading of any other cylinder should be 135 psi. An engine with this wide range of readings is not in excellent condition.

If all the cylinders except one are good, squirt about a *teaspoon* of 20W or 30W oil through the sparkplug hole. Too much oil will raise the compression ratio and give you a false reading. Wait a minute, then check the cylinder again.

If the pressure comes up near the

 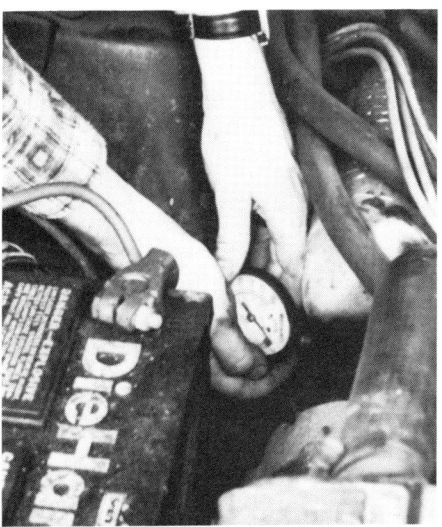

Installing compression gage on big-block in small engine compartment is difficult. Either gage works well, but screw-in type prevents accidental pressure loss. Lowest pressure for any cylinder should be within 75% of highest. Even that percentage is questionable.

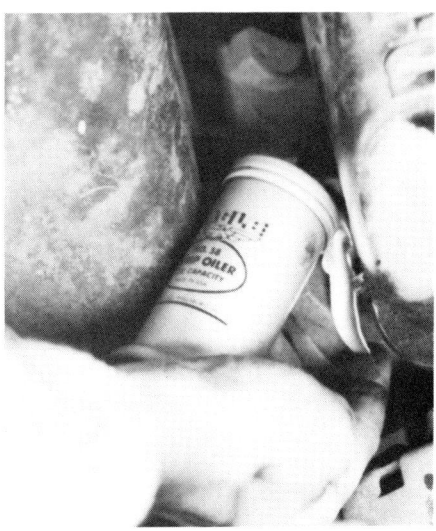

If cylinder reads on low side, do a *wet test*. Squirt small amount of oil into cylinder and repeat compression test. If pressure increases substantially, suspect bad rings or cylinder bore. Otherwise suspect bad valves or seats.

others, the rings are not sealing. If the pressure stays about the same, suspect that a valve is not sealing or the head gasket is blown.

Leak-Down Test—This is similar to a compression test with a little more sophistication, more answers and more money. The major difference between the two is that a leak-down test uses externally applied pressure. The rate at which the pressure leaks indicates the cylinder's sealing ability.

A leak-down tester won't be found in the typical toolbox. But except for the *percent-leakage gage,* the test can be done with a simple air tank and a plug adapter.

The leak-down tester requires a compressed-air source. If you already have this, you can perform all the tests without the leak detector—except for determining percentage of loss through leakage.

A leak-down test is performed with the piston at top dead center (TDC) on its *compression stroke*—both valves are closed. Air is supplied to the cylinder via a special fitting that adapts the air hose to the sparkplug hole. The cylinder *should* do a good job of holding the pressure. If it doesn't, the test can give you some indication as to what's wrong.

Caution: Be sure your hand isn't near the fan, belts or pulleys while doing a leak-down test. As the cylinder is pressurized the engine will turn over with a short burst if the piston is not on TDC.

While testing, have the breather and radiator cap off and the carburetor blocked wide open. This way, you can tell what is leaking if a leak shows up.

To set each piston at TDC, mark the distributor housing directly in line with each sparkplug tower. Remove the distributor cap and crank the engine to align the rotor with the corresponding mark for the cylinder you want to test. As you finish testing one cylinder, turn the crankshaft clockwise 90° to align the rotor with the next mark. Test the next cylinder in the firing order: 15426378.

Connect the air hose and pressurize the cylinder. This is when the leakdown gage comes in handy. The gage tells you what percentage of air supplied to the cylinder leaks out. A small amount of air leakage through the breather is common on a worn engine. Normal is 5—10%; 20% or more is poor.

If you don't have the gage, listen for air leaking at the carburetor, exhaust pipes, radiator and breather, oil filler and dipstick tube.

If you hear air escaping through the carburetor, the intake valve is leaking. If you can hear air escaping at the exhaust pipe, the exhaust valve is leaking. If either valve is leaking, be sure the piston is still at TDC on the compression stroke.

A head-gasket leak will show up as air leaking out an adjacent cylinder or through the radiator-filler neck. A cracked cylinder wall or head will also show up as air leaking out the filler neck. If you hear leakage at the breather cap or dipstick tube, the pistons and rings are not sealing in the cylinder bore.

Head Gasket: If pressure is down on two adjacent cylinders, chances are the head gasket has blown between them. A compression test of the two cylinders will show equally low pressure. During a leak-down test you should hear air escaping out one sparkplug hole when the other cylinder is pressurized and vice versa.

A head gasket can also blow between a cylinder and the cooling system. This will let combustion-chamber pressure escape into the cooling system and coolant into the combustion chamber. If the problem is severe, you can find it by pressurizing the cooling system. You'll be able to hear air or coolant escaping into that cylinder through the sparkplug hole.

A blown head gasket is also indicated by the sparkplugs. If one or two plugs show white fluffy deposits and the rest look OK, suspect a blown head gasket.

Valves: As I noted, a leak-down test will indicate a bad valve by allowing air to escape out the carburetor or tailpipe. If you don't have a leak-down tester, you'll have to depend on a compression check and further testing. If oil didn't correct a low-compression reading, chances are the valves are not sealing.

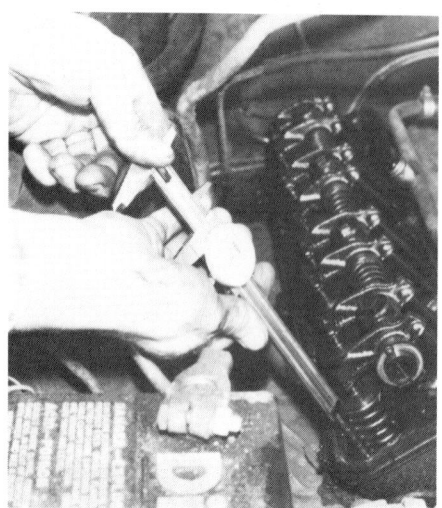

If engine has solid lifters, this method works fine for measuring valve lift. Just be sure to measure from same points for both open and closed readings. Hydraulic lifters usually bleed down too quickly for accurate readings or comparisons. With hydraulic lifters, remove rocker shaft to check cam-lobe lift at the pushrod.

This method is best for measuring valve lift. If engine is equipped with ball-end pushrods, use tape around pushrod end to hold indicator plunger tip on pushrod, or use cup-end—solid-lifter type—pushrod as shown.

The first thing to do is to make sure the valve is closing all the way. Remove the valve cover. Rotate the crankshaft until the piston for the bad cylinder is at TDC on the compression stroke—the distributor rotor will be pointing at that cylinder's distributor-cap tower. With the piston in this position you should be able to rotate both pushrods easily with your fingers. They shouldn't be loose.

If there is too much clearance—the pushrod is loose—an easy check is to remove the rocker-arm assembly. Refer to pages 66 and 140 for doing this. Place a straightedge across all the valve-stem tips or valve-spring retainers and see if any valves are at a different height.

There will be some variance, but it should be less than about 0.015 in. from one valve to another. A valve "taller" than the others may be worn or burned. A "short" one is probably bent or has carbon holding it open.

CAMSHAFT-LOBE WEAR

Cam-lobe wear is not a common problem. If a valve is not opening to specification, I would suspect a hydraulic lifter as the culprit. Although an occasional camshaft lobe in a big-block Ford may go away, I couldn't find a big-block Ford cam bad enough to show. The cam pictured on page 84 is from a non-Ford engine. It has less than 50,000 miles on it.

Checking Valve Lift—Cam-lobe wear will show up directly as reduced valve lift. If the engine has solid lifters, it's a simple matter to measure valve lift with the following technique. If the engine has hydraulic lifters, skip this section and measure the camshaft-lobe lift directly. The pressure of the valve springs will cause hydraulic lifters to bleed down—particularly on a high-mileage engine. This makes it difficult to get an accurate valve-lift reading.

If you have a vernier caliper or dial indicator, use it. If not, measure as closely as you can with a divider and ruler or just a ruler. To measure valve lift you must start with each lifter on the base of its cam lobe—piston at TDC on its compression stroke.

The best place to measure with a vernier caliper is from the machined spring seat, but there isn't enough room to measure from that point. Instead, take the measurements between each cylinder's intake and exhaust valves. If you measure all the cylinders at this point, the readings should be comparable.

Record the distance from the head to the top of the retainer with the valve closed. Turn the crankshaft until the valve is fully open—the spring is most compressed. The *difference* between the two measurements is valve lift.

If you are using a dial indicator, zero the indicator with the valve closed. Leave the full travel of the indicator for the open position. Crank the engine until the valve is fully open and read valve lift directly.

Compare the readings after you've finished the last cylinder, then check the charts, page 85. Recheck any valves that are not to specification. To get approximate lobe lift, divide valve lift by the rocker-arm ratio: 1.76 for adjustable—solid lifters—and 1.73 for non-adjustable—hydraulic lifters.

Checking Lobe Lift—To check the camshaft, remove the rocker-arm assembly. Measure lobe lift at the top end of each pushrod. This gives the most-accurate measurement.

Most of the pushrods are guided by holes in the intake manifold. There shouldn't be any problem measuring lobe lift with these pushrods. A couple of the holes do not completely surround the pushrods. Use your hand or a piece of rubber to hold these against the rounded part of the intake manifold.

A pushrod with a socket or cup-type rocker-arm end—such as those used with solid lifters—will make measurements a lot easier. Just move it from hole to hole for the measurements. Be careful when pulling the pushrod out of the lifter—don't pull the lifter out of its bore. If you don't have one of these pushrods, wrap a piece of tape around the end of the pushrod.

Measure the highest and lowest readings for each lobe as you rotate the crankshaft. The difference is lobe lift. Compare this to the chart, page 85. All lobes should be within 0.005-in. lift.

Pay particular attention to any pushrods that didn't rotate when the engine was running. The taper on the lobe has probably worn away—the first sign of a bad lobe and lifter.

Also note: If any of the lobes don't measure up to specification, but the engine has good compression and oil consumption is OK, you can just replace the camshaft *and lifters*. Never mix old lifters with a new camshaft. The old lifters will quickly destroy the lobes of a new cam.

CHAPTER 2
Engine removal

1959 352 with Ford carburetor, combination vacuum/fuel pump for vacuum wipers and early-style expansion tank. Photo courtesy of Ford Motor Co.

After the decision to rebuild comes the task—removing the engine. Engine removal may not seem difficult for the experienced mechanic, but you may be a little apprehensive if you've never done it.

Armed with the proper equipment—tools and knowledge—plus a little time and money, engine removal is relatively easy. The task is more enjoyable if you keep in mind the fresh engine you're going to put back in.

Safety—Keep safety foremost in your mind when removing an engine. The big-block Ford weighs more than 700 lb fully dressed. That much weight deserves respect. Make sure the equipment you use is strong enough for the job. Use it correctly. You should have someone help you. Make sure that person is safety conscious too. As the saying goes, "Safety is no accident."

EQUIPMENT NEEDED

Besides common hand tools you need a hoist to lift the engine, a jack to lift the vehicle and ramps or stands to support it. You also need a place to set the engine once it's out. You should have some boxes or cans for storing parts, a couple of drain pans and a pair of fender protectors.

Hoist—The engine hoist can be the block-and-tackle type, chain fall or a *cherry picker*. A cherry picker rolls on the floor and has an arm with a hook or a short chain hanging from it. This arm is raised hydraulically.

Although all work well, the cherry picker is my favorite. Available at most rental shops, it offers greater mobility when removing or installing the engine. If you use a block and tackle, you normally will have to move the car before you can lower the engine. With the cherry picker you move the engine, not the car.

The block-and-tackle or chain fall will work well if you find someplace suitable to hang it. The problem is most garages use 2x4 or 2x6 rafter ties or joists. These are suitable for lifting starter motors, not big-block Fords.

Although the cherry picker is considered a safe hoist, I once had a 390 drop less than a foot from me when a weld broke. After my heartbeat returned to normal, I towed the hoist back to the rental shop. I carefully inspected the next one. A word of caution here: Be careful when the engine is supported only by the hoist—*never* get under it.

Ramps and Stands—If you have a 4X4 F250 pickup, you definitely won't need jack stands or ramps. If you have a '66 Thunderbird like that shown, you not only need them; you need strong ones. The T-bird weighs nearly 5000 lb. And more than half of that is on the front wheels!

Ramps or jack stands should be selected carefully. They are all you have between you and the vehicle when you are underneath it. I have seen many poorly designed or low-quality ramps and stands. Stands should be heavy-gage steel and ramps cross braced.

Generally speaking, I think ramps are safer than jack stands. Ramps won't tip as easily if you get a little exuberant when removing the 700-lb engine.

Ramps aren't perfect; they take up more room than stands. If you are purchasing a set of ramps, look for the kind with the removable ramp section. With the ramp section removed there is extra working room.

Ramps also support the vehicle by the suspension, not the frame or body. The problem comes when lifting the engine. As engine weight is lifted off the suspension, the suspension unloads and the vehicle will raise 4 in. or more. If the engine hoist has limited travel, the engine may not clear the body.

13

Typical big-block-powered vehicle is heavy—such as this '66 Thunderbird. Support it well, or you may never live to rebuild your engine.

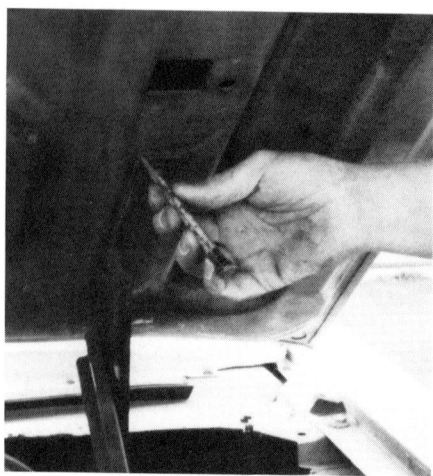

To restore hood alignment, reference hood hinges to hood. Matchmarks can be made with grease pen, chalk, punch or awl. Avoid scratching down to bare metal.

When looking for jack stands, look for those with wide, heavy bases. I prefer flat, square bases to triangular ones.

Stands should go under the frame or side rails so they *feel* more solid—that doesn't mean they're more stable. Stands also require more consideration of the surface they're on. Stands should be used on solid, level surfaces—anything other than concrete may require some assistance. Jack stands will sink in dirt, asphalt or tar—especially on a hot day—when you least expect it. Avoid this problem by placing a piece of plywood at least 3/8-in. thick under each stand.

When using either ramps or jack stands—particularly stands—I use a backup support. This can be a wheel or two or a big piece of wood under the frame or body. It is better to appear overcautious and be safe than to look brave and be dead. No matter how you support the front of the vehicle, block the rear wheels and apply the parking brake.

Jack—The only jack you should consider is a floor jack. Others may work but none work as well. Besides lifting the vehicle, the floor jack is indispensable as an adjustable support for the transmission when removing the engine. The same precautions apply here: Use a good-quality jack and don't overload it.

Tools and Containers—Common hand tools, including a 3/8- and 1/2-in.-drive rachet and socket set, a set of combination wrenches and a few screwdrivers should be all the tools needed for the task. Also gather some large cans or small boxes to store accessories and bolts as you remove them.

Label the containers: radiator parts, alternator parts, carburetor parts, and so on. This will speed up the job when you install the rebuilt engine. Use large boxes to protect the carburetor and other components.

PREPARATION

Fender Protectors—Place a pair of fender protectors over the fenders of the vehicle—LN600 owners ignore this—to protect the finish. These protectors also help to hold tools in place. A pair of old rugs, slipcovers, bedspreads, heavy rags or old coats work well—just make sure there are no snaps or zippers and such to scratch the finish.

Human-Body Protectors—On *unibody*—unitized construction—vehicles, there are a pair of nasty shock studs sticking through the shock towers in the engine compartment. If you haven't encountered these before, you will. A couple of small rubber balls with holes drilled in them or some fuel line will cover the studs.

Hood Removal—First prepare a place to store the hood where it won't be damaged. The roof of the car is OK if the car is in a garage. Place something on the roof of the car to prevent scratches and lay the hood upside down on the roof. If you are planning to stand the hood up in the garage, place a block of wood under each corner. This will protect the edges so they won't be bent or the paint chipped.

Before you loosen the hinges, trace around them so you can align the

Use large container to catch coolant—be sure it is wide or deep enough to contain splashing coolant.

hood when you reinstall it. Use a felt-tip pen, scriber or a grease pencil. Anything is good as long as it won't fade, wash off or wear off.

Place a short 2x4 between each rear corner of the hood and the cowl. These will prevent the hood from sliding back into the cowl or windshield. Remove the two front hood bolts and then get a friend to help. The hood is more awkward than heavy, but don't try to handle it alone. With a friend supporting the hood on the other side, remove the remaining hood bolts.

If you don't have someone to help, hook the engine hoist to the front of the hood. Remove the hood and swing it around and lean it against the hoist.

Exhaust Manifolds—Except for some factory cast-iron headers, exhaust manifolds need not be removed

You need room to work! If vehicle is equipped with fender braces, remove them now—don't wait until you have to.

Unless you plan to restart engine, remove battery for safety and extra elbow room.

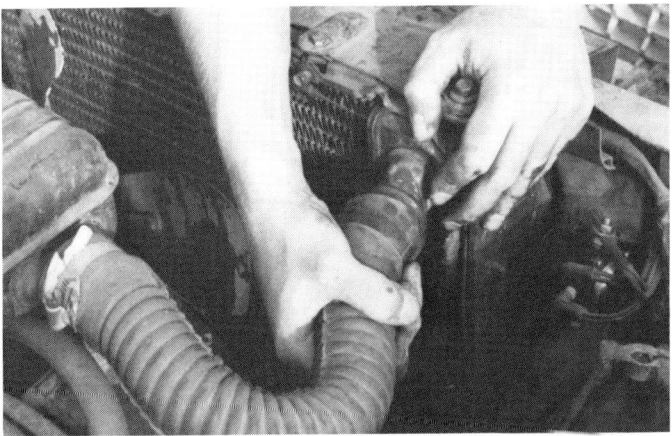
Once coolant is drained, remove radiator hoses. You may find it easier to remove lower radiator hose after you have more room.

Viscous-drive fan can be removed before radiator because bolts are on engine side of fan. Fixed-hub fan has nuts or bolt heads on front of fan, so remove radiator first. This prevents fins and cooling tubes from being bent—or bloodstained from skinned knuckles.

until engine teardown. But there is a good reason to loosen the bolts before you pull the engine. It is very easy to break the upper bolts.

On most big-blocks, the top bolt holes for the exhaust manifold are drilled completely through the exhaust-port flange. To make matters worse, the holes point up, so water or snow puddles in the holes at the end of the bolt or stud. After a few years the bolts get pretty rusty. When you try to remove them they simply twist off.

Soak the bolts or studs with penetrating oil before you try to break them loose. Then, if the engine will still run, start it and run it until the exhaust manifolds get hot. Loosen all the exhaust-manifold bolts you can reach. Remove the top ones completely.

I have let an engine run while I removed these bolts—I'll do anything to keep from breaking them. Although it's very easy to get burned using this method, it's worth the risk. I have yet to break an exhaust-manifold bolt using this method. If you do break a bolt, keep working. Fix it later when the cylinder-head work is being done.

If the engine has the factory cast-iron headers, disconnect them before removing the engine. You have to remove the right one anyway to remove the starter motor.

Make a Note—If you have a 1960 Ford you shouldn't have much trouble remembering where parts go and which vacuum hose goes where. But if your vehicle has Ram Air, an air pump (AIR), Improved-Combustion System (IMPCO), air conditioning or any of a dozen accessories, take good notes. This may not seem important now, but the notes will save you a lot of head scratching at installation time.

One way is to identify each component and mark where it connects. Wrap masking tape around each part and write the information on the tape with a ball-point pen. Another method is to take pictures of the engine compartment. Photograph both sides and the front with the air cleaner on. Remove the air cleaner and repeat the photo sequence. This will help greatly during installation.

Air Cleaner—When you remove the air-cleaner housing, try to leave all hoses and electrical connections attached to it. Disconnect them at the other end. This makes storing and installation easier.

Fender Braces—If the car has unitized construction, it probably has braces from the fire wall to the spring towers. You can work around these supports

15

Carefully remove radiator and store in safe place—like the trunk. This radiator was leaking from previous poor repair. If cooling system was dirty, full of rust, scales or sludge, have radiator flushed or recored before reinstalling. New engine demands—and deserves—good cooling system.

Removing radiator and fan gives more working room for accessory removal.

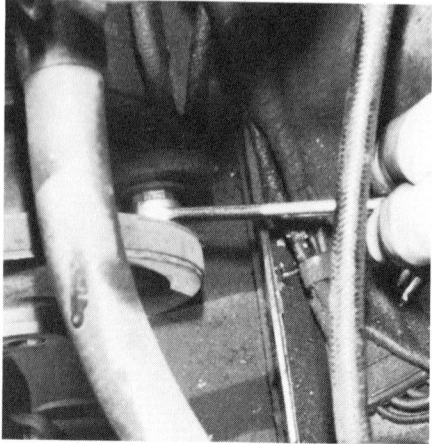
If at all possible, use box-end wrench or sockets to remove bolt and nuts.

Return line for hydraulic windshield-wiper motor. This type motor was used only on '63—'68 Thunderbird.

When last bolt is removed, set alternator aside. If you already removed battery, leave alternator wires attached.

but you'll save time by removing them. Put the bolts back in the braces for safekeeping.

Battery—Elbow room is short in the Fairlane/Comet or Mustang/Cougar, so remove the battery for more room. If you intend to drive the vehicle up a set of ramps, do it now before you remove the battery.

If the engine has an automatic transmission you may want to reinstall the battery later. That way you can crank the engine to gain access to the torque-converter bolts.

Remove the negative cable first to lessen the chance of a spark. Remove the battery hold-down and get a good grip on the battery. Be careful not to drop or tip it and spill the acid. I always wear glasses of some sort when removing a battery—even sunglasses are better than nothing. Store the battery off the ground in a cool area and away from small curiosity seekers.

Fan, Shroud & Radiator—If you plan to save the coolant, drain it into containers for storage. The containers should be able to handle two or more gallons. Open the petcock and remove the radiator cap to let the radiator *breathe*. While the coolant drains, turn your attention to the radiator and fan.

If the vehicle has a clutch-, or *viscous-drive,* fan, remove the nuts and fan. Be careful not to bump the radiator with either your hands or the fan. The fan will damage the radiator and the radiator will damage your hands. You can protect the radiator and your hands by loosening the shroud and slipping a piece of cardboard over the radiator. Tape the cardboard to the radiator. Once the fan is off, remove the shroud.

If the engine has the standard fan, remove the fan shroud first. Lay it back over the fan, remove the radiator and then the fan.

If the vehicle has an automatic transmission, the transmission-cooler lines must be removed before the radiator. Cooling lines are often bent, restricted or completely twisted off by someone not using the correct tools. Hold the nut next to the radiator with an open-end wrench. Use a tubing or flare-nut wrench to break the fitting loose.

If the line begins to twist, work the tubing wrench back and forth slowly. Give the tubing a chance to break loose from the radiator—sometimes penetrating oil or a *little* heat will help. If the line refuses to break loose, *don't twist it off.* Cut it with a tubing cutter 2—3 in. from the nut. You can always splice it back together with hose clamps and a piece of 5/16-in.-ID fuel

Tight spots may require use of an open-end wrench.

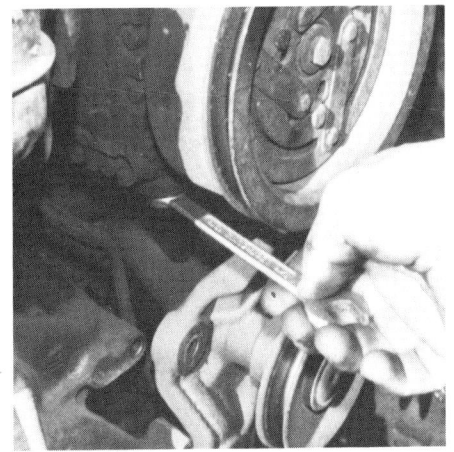

Before removing compressor from support bracket, know where you are going to put it. Most vehicles have enough room to lay compressor on front of wheel well or shock tower, but it should be wired in place.

Condenser wired to hood hinge for security. Condenser was removed with compressor because compressor-to-condenser line was too short to wire compressor to fender. Besides, there were only two bolts holding condenser in place. Old rug protects fender.

hose.

Some cooler lines are connected by rubber hoses from the factory. With these just loosen the clamps and slide the hose off the end. After disconnecting the cooler lines, connect the two lines with a hose to prevent further loss of transmission fluid.

Remove the upper and lower radiator hoses. The radiator is held in place either by two bolts on each side, or by a pair of saddles on the bottom with one or two hold-downs on top. Support the radiator while you loosen it.

Lift the radiator out—don't bump it into anything—and store it in a safe place. It can go in the trunk, but check the radiator first. If it has any rust streaks or loose fins, send it to the radiator shop to be repaired.

Pulleys, Belts & Accessories—With the radiator and fan out of the way you should have some working room at the front of the engine. If the engine is equipped with a lot of belt-driven accessories, take good notes or more photos.

Pay particular attention to the mounting-bracket installations. There is usually only one way everything fits back together. Finding that way can take a lot of time.

Remove the water-pump pulley. It should be held in place only by fan-belt tension. Next loosen the belts as you come to them—the power-steering belt is usually last. Remove the idler pulley if the engine has one and then the rest of the accessories.

Expansion Tank—Most big-blocks came with an expansion tank fastened to the intake manifold. This bolts in place of the thermostat housing. The tank can be left in place, but it is better to remove it to avoid damage. The tank is held in place by two bolts at the coolant outlet of the intake manifold. Use a 9/16-in. wrench to remove them. Any coolant left in the tank may gush out, so be ready for it.

Air-Conditioning Compressor—If the vehicle has air conditioning, (A/C), remove the compressor next. The A/C system contains freon under high pressure, so be very careful to protect yourself—especially your eyes. Safety glasses are a must.

On most vehicles, the compressor can be unbolted from the engine and moved out of the way without disconnecting a line and discharging the system. If you can do this, disconnect the compressor-clutch wire and remove the compressor bolts. Be ready for a heavy load—about 15 lb—when the compressor comes loose.

The Thunderbird shown in the photos was a special case. I unbolted the condenser from the radiator support and swung the compressor and condenser over the side of the car as one unit—lines intact. I placed a rug on the fender to protect it and wired the compressor *and* condenser to the hood hinge.

This is not possible on all vehicles because the lines to the condenser pass through the radiator support. This type of installation requires discharging the system and disconnecting at least one line.

Before you disconnect an A/C line, bleed the freon. *Wear safety glasses and gloves when working on a charged A/C system. Freon is under extremely high pressure. When it is released it goes through a huge temperature drop, and will freeze anything it contacts—fingers, hands, eyes....*

Put a wrench on the A/C fitting marked **DISCHARGE** or **D**—it's on the low-pressure side of the compressor or the line coming from the evaporator at the fire wall. Wrap the fitting and the wrench with a rag. Loosen the fitting slowly until you hear hissing, then stop. The sound is escaping freon—stay away from it. When the hissing stops, loosen the line another 1/4 turn. When the hissing stops, you can disconnect the line. Keep the rag wrapped around the fitting just in case.

Remove compressor and store until assembly time. Tape over the ends of the A/C hose and the compressor fitting to keep dirt out.

Power-Steering Pump—The power-steering pump is installed with either a swing or slide bracket. Remove the pump from the bracket and wire it to the side of the engine compartment. If there is room, swing the pump to the side without removing the bracket. On some power-steering pumps, the bracket stays with the pump.

Some T-birds used windshield wipers operated by the power-steering pump. On these models a pair of lines —one from the steering gear and another from the pump—attach to the windshield wiper. These should be re-

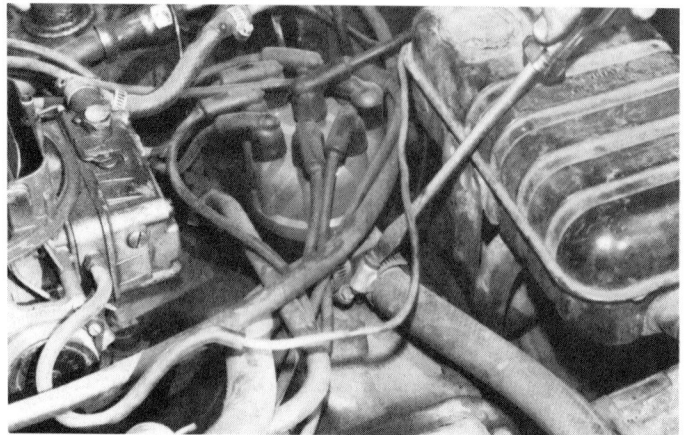

Even if you have good memory, take pictures or notes of hose and wire connections. Better yet, make a "flag" at each hose or wire end. Note on the tape where the hose or wire connects. This is a must on later cars with complicated emission controls.

moved and plugged.

If you have to disconnect any lines to remove the power-steering pump, store it with the pulley facing down to prevent fluid loss. Be sure the cap is tight, otherwise the fluid will seep out. Cap all lines in the engine compartment and on the pump.

Air Pump—If the engine has one of these, disconnect the hoses from the *diverter valve* and remove the bolts from the air pump. The diverter valve is close to the air pump, connected to a 1-in.-OD hose. Unbolt the bracket from the engine and water pump. Store bracket with the air pump.

You can also remove the air-pump lines at the exhaust ports on the cylinder heads. Lines at the exhaust ports will require a good dose of penetrating oil. Use a flare-nut wrench again.

After removing the lines, pull the small funnel-shaped tubes from the cylinder heads. Store the tubes and lines with the air pump to prevent losing them.

Alternator or Generator—Unbolt the alternator or generator from its support brackets. With the brackets off you'll have good access to the wiring. This will make it easier to disconnect the alternator and remove it from the engine compartment. The alternator may have enough wiring to allow you to set it in the battery box. If you leave the alternator in the engine compartment, store it in a plastic bag to protect it.

Hoses, Lines & Wiring—Disconnect the heater hoses from the engine and tuck them in by the right hood hinge. If you wish to remove the hoses completely, be careful not to damage the heater-core fittings. It is better to cut the hoses off with a small knife than to twist off the heater-core connec-

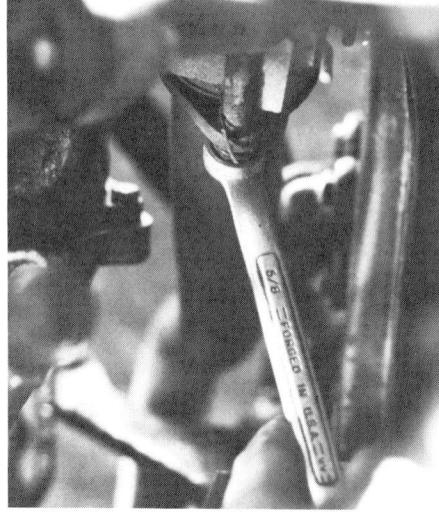

Before raising vehicle, break loose and remove any exhaust-manifold-to-pipe nuts that can be reached from above.

tions. Heater cores are expensive, a pain to remove and time-consuming to repair.

When you disconnect the wiring be sure to disconnect each component. The water-temperature sender is in the front of the intake manifold, behind the distributor. The oil-pressure sender is on the oil-filter adapter. Disconnect the coil wire, throttle-stop solenoid or other carburetor circuits. Fasten the wiring harness to the fire wall out of the way.

Remove the ground wires. One goes from the engine to the battery and the other from the engine to the body. The body ground is normally attached to the rear of the right cylinder head. Remove any vacuum lines for accessories, or the power brakes or wipers. Remove the air-compressor drain line on heavy trucks.

Safety and access are major objectives here—don't take any shortcuts. Block rear wheels, and use high-quality jack stands or ramps. Don't get under vehicle until you are certain it is secure.

Linkages—Take good notes or a picture of the throttle linkage and then remove it. Place all linkage parts together so you know what parts go where. If your vehicle has an automatic transmission, disconnect the kickdown linkage. Remove any bellcrank assemblies from the intake manifold.

Fuel-Pump Lines—Disconnect the fuel-pump-inlet line. Some models also have a return line to the tank that must be disconnected. Plug the hose with a 5/16- or 3/8-in. bolt with an unthreaded shank. Push the bolt in *past* the threads—otherwise fuel will seep around the threads. You can remove the fuel pump now, but it is easier after the engine is out.

Be sure the wire looms are disconnected from the cylinder block and the retaining clips are off the automatic-transmission cooler lines.

Note arrangement of linkages at the carburetor before you disconnect them.

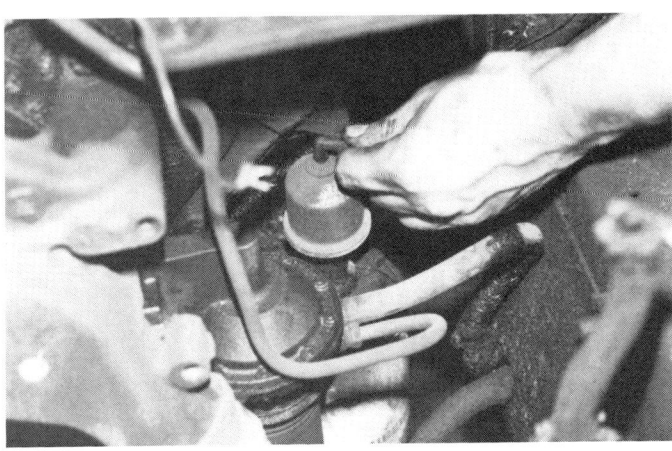
Disconnect oil-sender wire. While on this side, be sure wiring harnesses are free from engine. Remove dipstick.

Floor jack set up to support transmission for engine removal. It also serves as emergency backup should vehicle slip off supports.

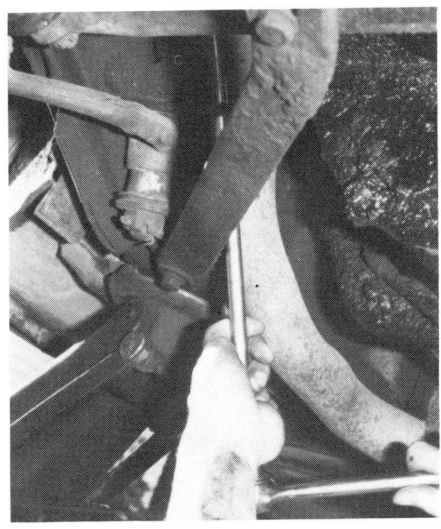
A couple of extensions, with a universal and socket will reach those exhaust-pipe nuts. Use six-point socket to prevent rounding nuts, especially brass ones.

Remove the dipstick and store it where it won't be damaged.

Jack Car Up—I say *car* because most trucks have all the clearance you need. Use a level surface. Block the rear wheels and apply the parking brake. Place the floor jack under the front-suspension crossmember, directly beneath the engine.

Make sure the jack isn't under the steering linkage or oil pan. Jack the car high enough to allow room to work. Place jack stands under the frame—not the suspension. Ramps go under the wheels.

Slowly let the jack down until there is load on the supports. Make sure they are standing up straight and not tipping or buckling.

You will notice I didn't use ramps or stands when I removed the engine. On hot days, anything smaller than 9-in. square sinks into my asphalt driveway. Instead I used a pair of heavy truck rims with added height from some boards placed side by side. I nailed the boards to the pavement to prevent sliding. When installation time came, the weather was cool enough to use ramps.

Use caution when working under a vehicle supported by stands or any other means. If the vehicle should fall, you could be killed or injured. There must be no shortcuts, either in setup time up or in the quality of supports.

As I mentioned, I like to use a backup support. Place a large chunk of wood or some other support under the vehicle. Don't use concrete blocks—they are brittle and can't handle single-point or impact loads. Position the backup support so that the vehicle will still be at least a foot off the ground if a ramp or stand fails.

Be sure the support is forward of the vehicle's center of gravity so it will tilt the vehicle to the rear. The center of gravity is near the end of the transmission for most cars. Place the supports forward of that point. It is safest to place them under a frame rail or unit-body side rail just behind the front wheels.

Exhaust—With the vehicle securely supported, move underneath and disconnect the exhaust system. Wear glasses to protect your eyes. The exhaust-pipe-to-manifold fasteners are usually brass nuts. Use a six-point 5/8- or 11/16-in. socket after giving them a shot of penetrating oil to prevent damaging them.

Unless the nuts were cross-threaded, the exhaust-manifold studs don't usually break like they do on other engines. The studs may come out with the nuts. If they break or the studs pull out, don't worry about it until you have the engine out and the exhaust manifolds off.

Clutch Linkage—Disconnect the clutch-release rod, spring and any stabilizing linkage. Remove the idler, or cross shaft between the engine block and the frame. Save the bushing halves from the male end of the pivot. Even if the cross shaft is not in the way, you still need to watch for the bushing halves. They may stick to the block.

Cooler Lines & Starter Wire—While under the car, disconnect the starter-motor wire and remove it from the clips on the side of the block. Also remove the automatic-transmission cooler lines from their clips. If the cooler lines are in the way, wire them off to the side.

19

Removing bellhousing cover exposes converter-to-flexplate nuts.

Make sure you only remove nuts like this one. Bolt at upper right holds ring gear to flexplate.

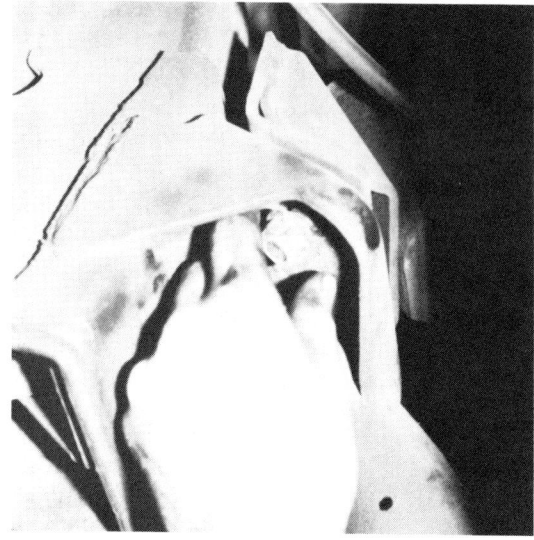
Engine mounts vary in design. Here, only nut and washer from each side need be removed. Some mounts will have through bolts that will have to be removed also. Make sure you disconnect mounts at the frame side of the insulators.

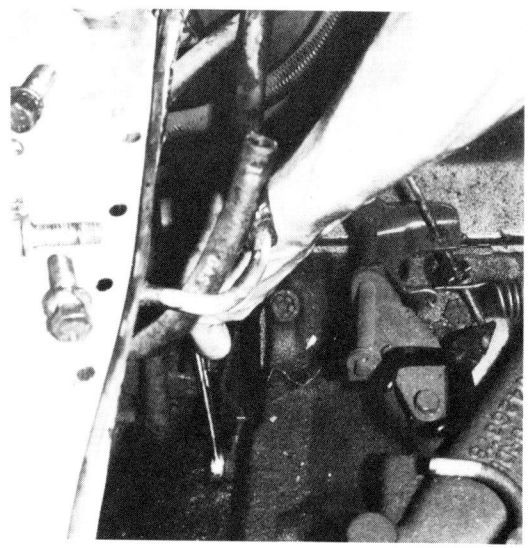
Remove upper converter or bellhousing bolts. Then hook up chain.

Inspection Cover—All automatic and some manual transmissions have inspection covers on the bellhousing. Automatics have a flat plate that bolts to the front of the bellhousing, on the engine side. Manual transmissions have a cover under the lower third of the flywheel and clutch assembly. Both covers are retained by four bolts.

Remove the cover and store it in a safe place. If the cover is missing, replace it when you install the engine. The cover keeps the clutch or converter free from dirt and water.

Converter Bolts—The torque converter is attached to the flexplate by four studs and nuts. Be careful here. On 428s and early 410s the outer starter ring is bolted to the inner ring with the nuts on the same side as the converter nuts. There are also one or two *bolts* on the engine side of the flywheel for draining the converter. Look closely to make sure you're removing the correct fasteners.

To unbolt the converter, place a 9/16-in. wrench on one of the nuts. Use the wrench to turn the crankshaft until the nut is at the top right corner of the inspection opening. Hit the end of the wrench with the palm of your hand to break the nut loose. The nut can then be removed without the crank turning.

After you have removed that nut look to the far side of the inspection opening for the next nut. Move this one around with the wrench and repeat the procedure.

The advantage of this method is that you can do it alone and only need one tool. A similar method is to remove the nuts while holding the flexplate with a flywheel turner or by holding the front of the crank with a 15/16-in. socket and breaker bar. The flywheel turner or breaker bar can also be used to turn the engine over.

Another option is to have someone turn the engine over with the starter motor. Use this method only with the ignition system disabled. *Be sure the person working the ignition switch does so only at your instruction. Otherwise he should keep his hands off the key.*

Before leaving the underside of the engine, place a dab of paint over one

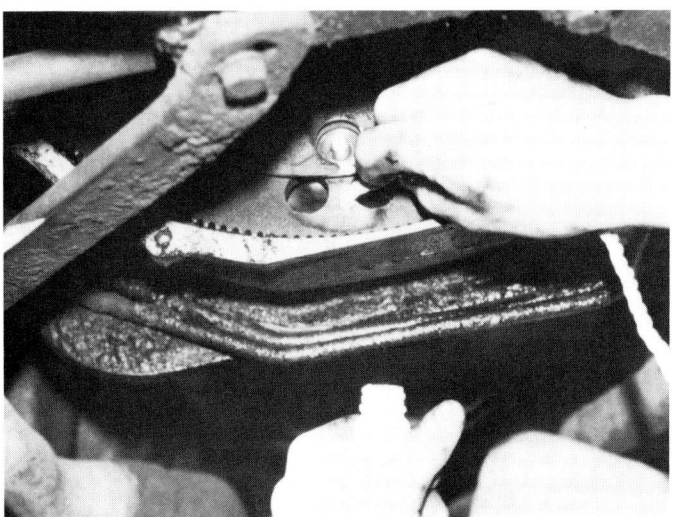
Mark converter stud and flexplate hole for reassembly.

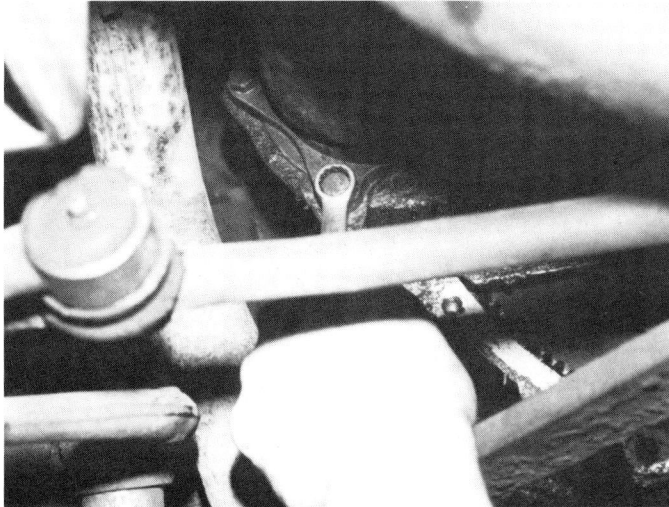
Remove wire from starter. Bottom starter bolt is easiest to reach. Except on trucks, top starter bolt will be hard to reach. Several tool companies make box-end wrenches, with crescent-shaped handles for removing big-block Ford starter motors. Don't bother to purchase one unless you do a lot of starter replacements on big-blocks. With bolts out, either snake starter through steering linkage or let it rest on linkage until engine is removed.

converter stud and next to its hole in the flexplate to ease installation.

Starter Motor—There are three bolts and one wire on the starter motor. Remove the nut holding the wire with a 9/16-in. wrench. Get ready to go through some contortions. The bottom bolt is easy. The other two are harder and hardest. There's no room to move on either of them, and you won't be able to see the top one. The starter motor weighs about 15 lb., so be careful as the last bolt comes out.

There is a crescent-shaped, 1/2-in. starter wrench that can simplify the job. Without it, the bolts can be removed with many 1/8 turns with a box-end wrench. Another option is to use a socket and extension to get past the length of the starter. You can either use a long extension to get between the starter and the engine mounts, or use a *really* long extension to go past the engine mounts.

Removal isn't really that bad because you don't have to start the bolts and hold the starter in place. You will during installation; one more thing to look forward to.

You have to thread the starter motor out through the steering linkage, frame and engine. In most cases you can wiggle the starter through. If the car is on stands, you can turn the steering wheel to gain additional clearance. Don't move the linkage without being sure the starter will stay in place. On the T-bird, I left the starter on a support bar and the steering link until I removed the engine.

Engine Mounts—On passenger cars and pickups, there are two styles of engine mounts. One has a stud pointing down and away from the engine. The other has a through bolt parallel to the engine block.

On the stud type, remove the nut and washer under the support bracket on each side of the engine. On the through-bolt type, remove the bolt and nut from both sides of the engine. The through bolts will have some load on them, so you may have to raise the engine slightly to remove them. If this is the case, wait until you are ready to pull the engine to remove the bolts.

Medium- and heavy-duty trucks have a mount fastened to the timing cover and front of the block. Leave this type of mount alone until you have the engine hoist attached to the engine and the transmission supported by a jack.

Engine Hoist—Before disconnecting the bellhousing bolts, position the hoist over the engine and fasten it to the engine. You can bolt it to the factory brackets at the exhaust manifolds, if they are still on the engine. Otherwise, bolt the chain or cable to the rear of the right cylinder head and the front of the left head.

If you prefer a four-point hookup, remove the exhaust manifolds. Install a bolt in the lower bolt holes at the front and rear of each head.

When using any two-point hookup, be sure the chain runs diagonally across the engine. Otherwise, the engine will tip when you lift it. Use Grade-6 or stronger bolts to attach the chain to the engine.

Check the distance from the ground to the engine oil pan. If you're using a cherry picker, lower the boom until the distance from the boom to the intake manifold is about 2 in. greater than the distance from the oil pan to the ground. This should leave enough boom travel to clear the front of the vehicle *and* allow setting the engine on the ground. If you are putting the engine on a stand or platform, you will not have this problem.

Remove the Engine—Now for the easy part—removing the engine. Roll the floor jack under the transmission. Place a board on top of the jack if you're working with an automatic transmission.

Raise the jack until it *starts* to raise the transmission. Now go to the engine hoist and place slight tension on the chain. Remove the six bellhousing-to-engine bolts. If the bolts come out hard, adjust the tension on the engine or load under the transmission. Readjust the jack or hoist until you can get the bolts loose.

With the bolts removed, make sure no other items are connected to the engine. Be sure the jack and hoist are positioned where you want them. Raise the engine and transmission

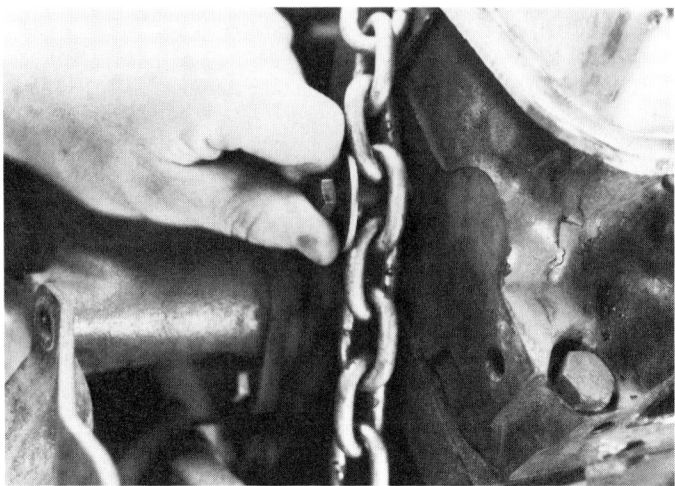 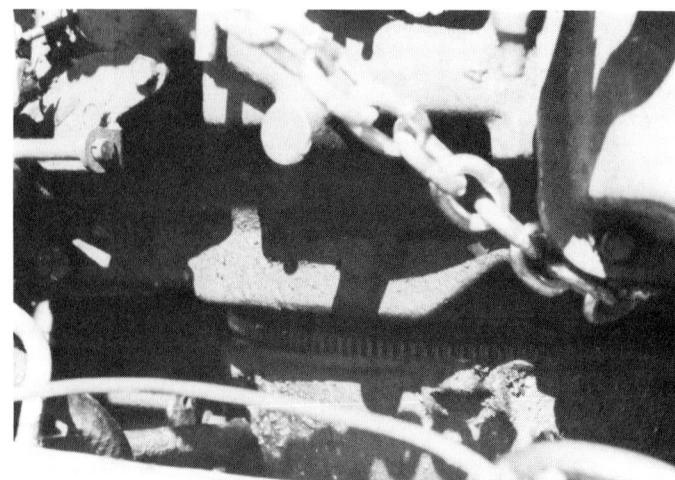

Fasten hoist chain diagonally from left front to right rear. Raise engine hoist so slack is out of chain, but with no tension. Place floor jack under transmission and remove remaining bellhousing bolts. With an automatic, place a 2x4 between transmission pan and jack pad.

Raise hoist until engine mounts are free from their supports or saddles. Readjust jack. Make sure transmission-cooler lines, hoses, wires and linkages are free from engine. Pull engine forward while wiggling until flywheel or flexplate clears the housing and clutch or convertor.

Once free, raise engine so it will clear front of car. Keep checking to make sure nothing is still connected or hanging up while raising hoist.

A well-placed foot and a grunt was needed to move hoist after it sank in asphalt. Use care when moving hoist with engine raised—center of gravity is high.

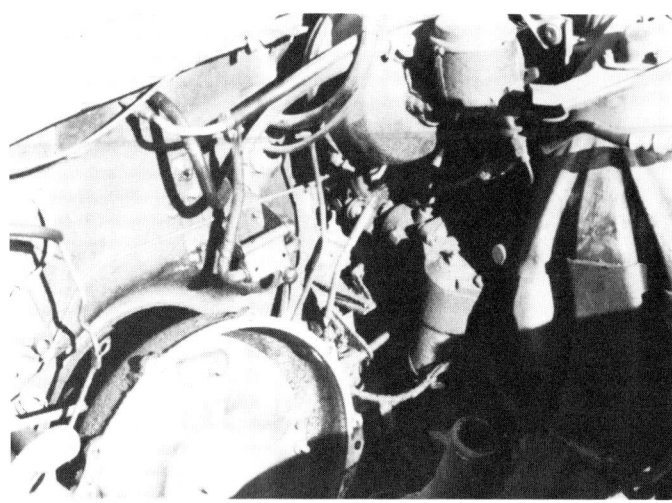 

Don't leave transmission unsupported. If you can't leave floor jack underneath, prop transmission up from below or support with wire strung between housing bolt holes and cowl or pole placed across hood hinges. Before leaving vehicle make sure everything is secure. If car will be left for long period, temporarily reinstall hood.

If you plan to replace front transmission seal, drain converter immediately before you install engine, no sooner.

simultaneously. If you're working alone, alternate between the hoist and the jack, but don't get one too far ahead of the other.

Raise the engine far enough for the engine mounts to clear the supports. Then with the hoist only, raise the engine another inch or so until it begins to separate from the transmission. Finally, pull the engine hoist or engine away from the transmission and they should separate. Take care that the converter doesn't hang up on the flexplate and pull out of the transmission.

When the engine comes loose, be careful not to bang it into the engine compartment. If the engine is a little stubborn and won't separate, you may have to pry it loose. While a friend pulls on the hoist, slip a large screwdriver or pry bar between the engine or engine plate and the bellhousing and pry them apart.

Work your way around and don't use a lot of force. Grab the front of the engine and push up and down and side to side while pulling forward. This may break it loose. If the engine still won't separate, double-check that everything is disconnected and the engine isn't against something.

Once the engine breaks loose make sure the engine plate stays with the engine. Be careful not to bump the aluminum bellhousing with the flywheel.

Lift the engine slowly out of the compartment, making sure it doesn't hang up and there are no wires or lines still attached. Once clear of the vehicle, lower the engine close to the ground. This keeps the center of gravity low and lessens the chance of it tipping over with the hoist. Move the engine out of the way temporarily and return to the vehicle.

The transmission must be supported before you remove the jack. Either wire the bellhousing to the fire wall or place some blocks under it, then lower the jack. The T-bird shown has a bracket at the front of the transmission so it doesn't need any additional support.

Before you turn back to the engine, gather any loose pieces and make the vehicle safe for a short storage. If you plan to leave the vehicle outside, reinstall the hood. Otherwise, stand the hood somewhere safe. Place a couple of wood blocks under the edge and wire it to a wall stud so it can't fall over.

CHAPTER 3
Parts identification and interchange

If you need parts, this chapter tells you which ones fit and how to identify them. Parts identification is no small task, because during the big-block's 20-year life a wide variety of parts were manufactured. The big-block Ford engine was called on to perform many different tasks, everything from powering one of the world's fastest production cars, the AC Cobra, to a heavy-duty truck, the LN800 Linehauler. Most big-blocks are found in family sedans or light trucks. Millions were produced, so parts are readily available at junkyards and parts stores.

This chapter describes differences between parts, tells what they fit and how they perform. Parts interchanging is extensive—there are five different bore sizes and four different strokes to work with.

All big-blocks have the same *bore spacing,* same *deck height* and the same basic external dimensions. *Bore spacing* is the distance between the bore centers of adjacent bores. *Deck height* is the distance from the crankshaft main-bearing-bore center line to the block's *deck surface,* or head-gasket surface.

Because the large Ford-truck big-blocks have major differences from passenger-car and light-truck big-blocks, they are referred to as *FT* engines: 330MD (medium duty), 330HD (heavy duty), 359, 361, 389 and 391. Unless mentioned separately, the 359 and 389 will be considered the same as the 361 and 391, respectively, as they share bore and stroke.

Engines used in passenger cars *and* pickup trucks are referred to as *FE* engines: 332, 352, 360, 361 Edsel, 390, 406, 410, 427 and 428. To make matters more confusing, many FE engines were available in different states of tune, like the 352HP, 390HP, 428CJ and SCJ. Although rebuilding the single-overhead-cam 427 is not cov-

Few engines perform as well as they look—but 1964 427 does! Features: reinforced and cross-bolted block, oil-pressure-relief system in block and oil pump, Clevite 77 bearings, high-performance induction system with Holley carburetion, free-flow exhaust manifolds, and forged pistons. Late models have cap-screw connecting rods and forged-steel crankshafts. Heads have machined combustion chambers and sodium-filled exhaust and hollow-stem intake valves. And, there were special over-the-counter parts from Ford Dealers. Photo courtesy Ford Motor Co.

ered, some identification numbers are listed to help distinguish parts.

In addition to the passenger-car and truck applications, the 352, 390 and 427 were also made for marine use. These engines were available with reverse, or counterclockwise rotation for tandem-engine installations. These engines have unique crankshafts, camshafts and distributors.

IDENTIFICATION
Is it a 352 or a 428?—On numerous trips to locate a 427 in a Shelby Mustang, GTE Cougar or a full-size Ford, I've found a 428, 390 or 352 instead. These engines can easily be made to look like a 427.

Parts such as intake manifolds, cylinder heads, distributors, valve covers and exhaust manifolds are easily fitted to other FE engines. Unless you know what to look for, it's difficult to identify a particular engine. Using casting numbers and physical features are the only positive ways to identify engines or parts.

Vehicle Identification Number—If an engine is still in the vehicle in which it was originally installed, identification can be made from the vehicle identification number—*VIN*. Using the year and engine code, you can identify an engine using the nearby table as reference.

For passenger cars, the first character—a number—represents the year; 8 could be '58 or '68. The fifth character—a letter—is the engine code. For trucks, the first character represents the year. The fourth character is the engine code. The engine code will

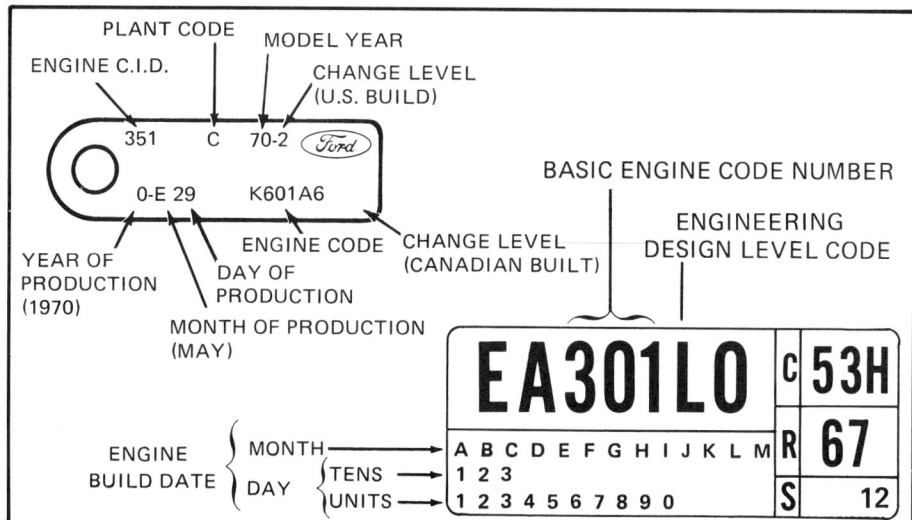

Tags or decals greatly simplify engine identification. Metal tags were used until February 1, 1970, then replaced by decals. Tags are under a coil or intake-manifold hold-down bolt. Decals affixed to valve cover. Both give engine size, change level, assembly date and special options.

Vehicle Identification Numbers are either on plates at the base of the windshield on driver's side or attached to body at driver's front-door pillar. VIN indicates year and code of originally installed engine. Codes are listed above. Drawing courtesy Ford Motor Co.

When Carroll Shelby wedged a 427 into the Cobra, he produced one of the most-potent combinations ever to put tire to pavement. Cobras still rank among the world's quickest production cars.

identify which engine was originally installed in the vehicle.

Serial numbers for '58—'67 models are at either the left front-door hinge pillar or near the door latch. On '68—'71 cars, the number is on the top left corner of the instrument panel, visible through the windshield. A taped-on number remains at the front door-hinge pillar or near the door latch. All truck serial numbers are located near the driver's-door striker, or latch.

Engine Tag—Engines produced from January, 1964 through February, 1973, have a metal identification tag, usually under the coil bracket. This tag gives engine CID, model year, year of production, engine code number and *change level*. The change-level code reflects component changes on an engine, explained on page 27.

On the second line of the tag is the production date. The first digit is a number representing the year, followed by a letter for the month. The letter A represents January and continues on through M for December—I is not used. The one- or two-digit number following the month code is the day of production.

The final code is the engine description: engine size, horsepower, carburetion and various other options, such as transistorized ignition. If the engine tag is missing you've lost an easy way to identify the engine.

Engine Decals—From February, 1973 on, a decal was attached to the front of a valve cover. It provides the same basic information as the metal tag except for engine size. Additional decals on some engines provide engine displacement and basic tuneup specs.

Stamped Numbers or Codes—Most engines have their production dates stamped on machined bosses on the block. There are four bosses, two on each side just below the deck surface. The date code is *alphanumeric*—letters and numbers—usually on the left side of the block on the front or rear boss. The date may also be on the two right bosses or any other machined surface on the block.

For example, the code on the left rear boss of a 1970 428 Cougar reads 9L136. The first four figures are the date. 9 is for the calendar—not model—year, 1969. L indicates November. The next two digits are the day of the month, so the production date of this engine is November 13, 1969. The final digit, 6, is an in-plant control number.

This stamped *production date* is the best one for reference. In fact it's the only one you can use if the tag or decal is missing. It's just more difficult to find and read.

25

Casting numbers and letters on this head give date head was cast. Code 9C12 translates to March 12, 1969.

HP is cast below block casting number of this 406. HP is also stamped on machined surfaces of many early high-performance blocks.

Although most cylinder blocks are cast with 352 on left front, engine can range from 330 to 428 CID.

Most casting numbers are not this legible. Casting number C5AE-F and date code 7A5 tell you this is a 427 Medium-Riser head cast January 5, 1967.

Casting Numbers—Most major parts have cast-in numbers to identify the year, change level and vehicle line for which the part was originally *designed*. Some parts may wind up in other applications.

The casting number represents the original casting or casting change level. *A casting number does not represent machining changes.* It gives you a place to start, but you cannot be sure about identification until you do more checking.

Other numbers, letters or symbols in random locations on the block can provide additional information. HP is cast on the back face and sides of some high-performance blocks: the '61–'62 390HP, the 406 and 427 for example. X is cast into some heavy-duty truck and some race blocks.

Change level may be the only identification cast into a part. For example, the letter J may be the only identification on a cylinder head. Physical identification of a part is a must if you want to be absolutely sure of what you're getting.

Another problem: Different parts may have the same casting number! For example: **C6ME-A** appears both on a 428 FE block with a 4.13-in. bore *and* a 330HD FT block, which has a 3.875-in. bore. The 428 is *cast* with larger bores, so the two blocks are cast differently!

Many cylinder blocks have **352** cast in front of number-5 cylinder. It is found on some FE engines from the 332 through the 428—including the 427! So **352** doesn't mean the engine is a 352. On some FE blocks, **352** is cast on the back, behind number-8 cylinder.

This number does not refer to engine size—it refers to the original FE engine, the 352. So use casting numbers, measurements and physical features to determine what engine you have.

FE CYLINDER BLOCKS

FE big-blocks were installed in passenger cars and light trucks. Major differences between FE blocks are in camshaft retention, thrust-bearing surfaces and engine-mount bosses. Although other changes were made to the blocks, the following are those that can cause problems with interchange.

The first change, affecting the 352, 390 and 406, was in 1963. Original big-blocks in '58 held the camshaft in position with the thrust exerted by the distributor-drive gear on the camshaft. Thrust is caused by the resistance of the oil pump, which is driven by the distributor gear.

With this system oil-pressure surges caused the camshaft to wander back and forth. This caused excessive wear on the lifters and cam lobes.

After the first 90 days of production, a spring and thrust button were added to the end of the camshaft. The

UNDERSTANDING THE PART NUMBERS

If you are like most people, the first time you see Ford part numbers they may seem unnecessarily complicated. This impression changes with an understanding of what the numbers and letters mean.

Engineering & Service Numbers—Each *finished* part has two numbers: an *engineering* or *production* number and a *service* number. The difference is easily explained.

The engineering number is assigned by Ford engineering when a part is approved for production. This number is used by engineering and the assembly plants.

The service number is assigned when the part is ready for the parts-distribution system. A different number is used because the finish or packaging of a part for service is different than it is for production.

The service number is the one the Ford parts people use. They don't care about the engineering number. But you may need to—the engineering number is the one appearing on many parts.

Casting Numbers—Casting numbers are special engineering numbers on a casting to assist in identification at the plant. As explained in the text, the numbers cast into the part apply *only* to the basic *casting*. One casting can be machined several different ways, creating different engineering and service numbers.

Using casting numbers to identify a part is like playing horseshoes. It doesn't count much when you're close, but it does count—and a casting number may be the only number you have.

Dissecting the Numbers—All Ford part numbers—regardless of whether they are engineering or service numbers—consist of three distinct groups: *prefix, basic part number and suffix*.

Prefix—The four-digit *alphanumeric* prefix tells the year the part was released by engineering for production, the car line the part was originally released for and by what Ford engineering group—chassis, body, engine, etc. In the case of a service part, the prefix identifies the division the part is for—Lincoln-Mercury or Ford.

For example, the following part numbers are for a 427 Medium-Riser cylinder head originally released in 1965. The year is indicated by the first two digits. The decade is indicated by the first letter—D for the '70s, C for the '60s, B for the '50s and so on. The following number is the year in that decade—5.

The third digit—usually a letter—indicates the car line for which the part was originally designed. For example, most 427 engine parts have the A designation, because they were originally designed for Ford cars. The accompanying list gives most car-line designations.

The last digit, or letter in the prefix in this part number indicates the part was released for production by the engine group: A is for chassis, B is for body and E is for engine—makes sense because it's an engine part. This applies to engineering parts.

The last figure in a service-number prefix refers to the car division—Z for Ford, Y for Lincoln-Mercury. Other letters refer to special parts, such as X for the extinct Muscle Parts program or M for the current Motorsports program.

Basic Part Number—Regardless of whether it's an engineering or a service part number, the basic part number is the same. For example, 6049 is for all finished cylinder heads, 6303 is for finished crankshafts and 6010 is for finished blocks.

The basic part number for the unfinished casting for these parts is different. Referring to the cylinder head again, the basic part number for the *finished* part is 6049, but the basic part number for the *casting* is 6090.

It's relatively easy to put a casting number on a part while it is being cast, so it's the casting number that appears on the part—great for Ford and terrible for the guy who's trying to identify it. Also, the number that appears on a casting may not include the basic casting number—you don't need a number to tell a block from an intake manifold. The casting number generally consists only of the prefix and the suffix, or C5AE-F for the cylinder head.

Suffix—A part-number suffix generally tells you the *change level* of a part, regardless of whether it is applied to the casting, the finished part or the service part. A signifies a part produced as originally designed; B indicates it was changed once, C twice and right through the alphabet, excluding the letter I. When the alphabet has been gone through once, the suffix grows to two letters—AA, AB, AC and so on.

How does a change at one stage affect the other two numbers? A casting change affects both the finished and the service parts, so the suffixes on all three numbers will change. An engineering change will change the service part—and its suffix—but won't necessarily change the casting.

A service part and its number can change independently of the casting part and its number and the engineering part and its number. Got that? This is simply because the service part is generated after the others in the scheme of things.

A finished, or engineering part can change independently of the casting, but not of the service part. This is why the suffixes of all three numbers rarely match.

---

thrust force in the cam pushes the button against the timing cover to hold the camshaft in position.

These retention methods are interchangeable by switching timing covers and camshaft components from the fuel-pump eccentric out. No changes were made to the cylinder block.

In 1963 the final camshaft-retention change was made. The front of the block was faced, drilled and tapped for screws to retain a thrust plate that in-

Engineering Casting Number	Production Number	Service Number
C5AE-6090-F	C5AE-6049-GC	C5AZ-6049-C

Read "Understanding the Part Numbers" and study these charts to reveal the "secret" of the Ford part numbers.

A—Ford
G—Comet, 66—68
   Montego, 69—70
J—Industrial & marine engines
K—Edsel
M—Mercury
O—Fairlane, 66—68
   Torino, 69—70
S—Thunderbird
T—Truck
W—Cougar, 67—70
Z—Mustang

Wider thrust flanges on late-model engines require wider machined surface on number-3 bearing saddles (arrows).

Early two-bolt engine mounts were replaced with four-bolt mounts in 1965. Two-bolt type may need to be reworked to fit later-model vehicles.

stalled between the camshaft sprocket and the front cam-bearing journal. The front cam bearing was narrowed and a spacer inserted between the camshaft sprocket and the thrust plate.

The next major change coincided with the introduction of the FT engines in 1964. FT engines were introduced with approximately 1/8-in.-wider thrust flanges at the number-3 main bearing. This bearing was also incorporated into FE engines, which required wider machined surfaces on the faces of the number-3 main-bearing web.

1964-and-later engines can use either style of thrust bearing. Pre-'64 engines *must* use the narrow-flange (0.465-in.) thrust bearing. The later-style wider (0.535-in.) thrust flanges may distort if installed in an early block. This can bind the crankshaft or crack the bearing insert.

Unlike the previous changes, the next one was not a simple machining operation. Two engine-mount bosses were added to each side of the block. Pre-'65 blocks use two-bolt engine mounts; '65-and-later blocks use three-bolt.

A later block can be used in place of the earlier one with no problems. Using the early block in a '65-or-later passenger car often requires drilling or reworking the engine-mount supports. Note: 1964—'78 FT blocks use four-hole engine mounts.

One additional change to watch for: In 1963, Ford introduced alternators as a mid-year option. The alternator required an additional tapped hole for mounting to the block. An alternator can be adapted to an earlier engine, but may require drilling and tapping an additional hole in the block to accommodate the mounting. Now for the specific blocks:

332 & 352—These blocks are identical. They use the same casting and 4.00-in. bores. Early-'58 engines were originally equipped with solid lifters. These blocks have no oil galleries for hydraulic lifters; solid lifters must be used. The '58 blocks also have six additional core holes in the block, two at the front and four at the rear.

After 90 days of production the solid-lifter engine was dropped and the hydraulic-lifter engine was introduced. This new block had oil galleries cast and drilled for hydraulic lifters. The addition of low-maintenance hydraulic lifters brought these engines in line with applications for family-type cars.

In 1960 the lifter oil galleries were left undrilled in the 352HP, so hydraulic lifters cannot be installed. Except for the undrilled lifter oil galleries, the block is the same as a standard 352. It did, however, go through closer production inspection.

In 1961 the block received deeper holes for the cylinder-head and main-bearing-cap bolts. Holes were drilled and tapped for 1/4-in.-longer bolts, then counterbored 1/4 in. This means the cylinder-head bolts reach deeper into the block for better retention.

The counterbore prevents the threads at the top of the block from pulling up when the bolts are torqued. The deeper holes also minimize block distortion as the head bolts are tightened. Consequently, the cylinder-head gaskets receive even pressure to minimize head-gasket failure.

1960-and-earlier cylinder heads use 10 4-7/32-in.-long bolts; 1961-and-later use five 2-7/8-in.-long bolts on the exhaust side. On the intake side, '61-and-later engines use 4-19/32-in.-long bolts, 3/8-in. longer than the early style. The only other changes were the camshaft retention and engine-mount holes mentioned earlier.

361—The passenger-car version of the 361 was used only in '58—'59 Edsels and as a '58 Ford Police Power Option. It has the same 4.05-in. bore as the 390. Otherwise it is essentially the same as the 352 block. The 361 used only hydraulic lifters and was machined from the same casting as the 332 and 352. The 361 was the first FE engine to use 4.05-in. bores.

360, 390 & 410—These three blocks all have 4.05-in. bores; displacement varies with crankshaft stroke. The 390 is the most common of the three and

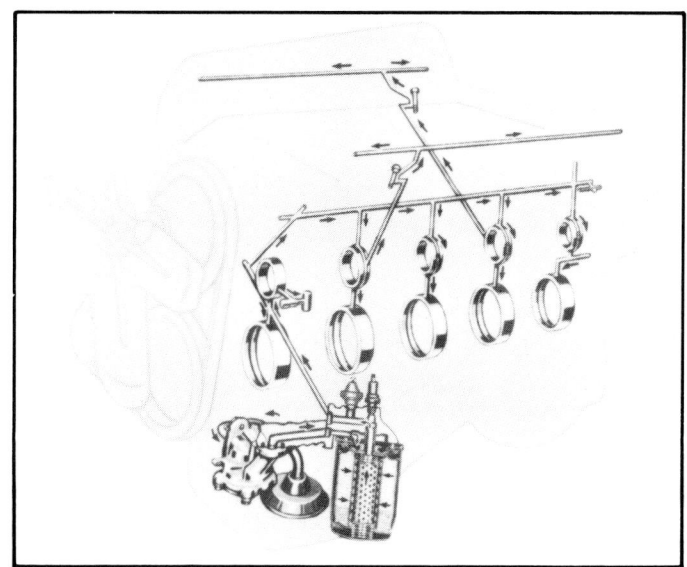
High-performance solid-lifter lubrication system. Note absence of center galleries that were added for hydraulic lifters.

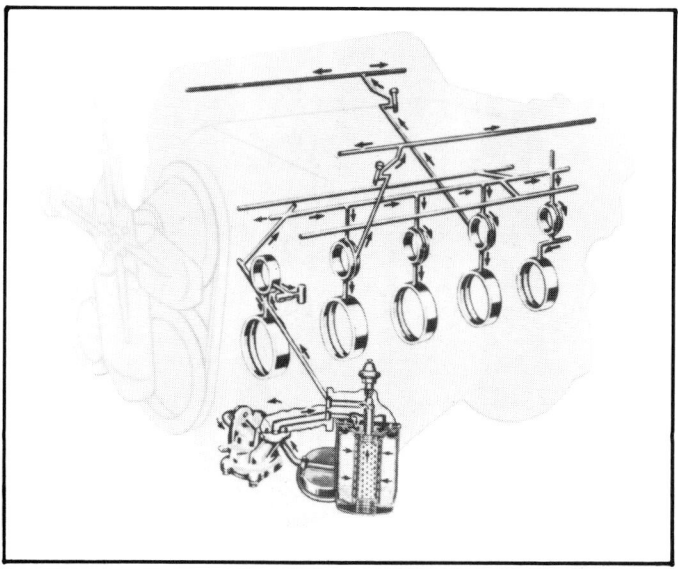
Hydraulic lifters required additional oil galleries parallel to center gallery. Lubrication system is otherwise identical to early solid-lifter system. All FT and most FE engines use this lubrication system.

uses a 3.78-in. stroke. The 360 was only installed in pickup trucks. It uses the 3.50-in.-stroke 352 crankshaft. The 410 is found in '66–'67 Mercurys. It uses the standard 428 crankshaft with a 3.98-in. stroke.

In addition to the 390-block's camshaft and engine-mount changes, ribbing was added around the sides of the block on some late-'66 models. This was to strengthen the area around the water jackets. Otherwise, the standard 390 block was not changed significantly during its 15-year life.

From '61–'65, Police Interceptor engines received thicker main-bearing caps and webs but otherwise were the same as the standard 390. Police engines used solid lifters. The lifter galleries through the block were drilled, but the short passages connecting the lifter galleries with the main oil gallery were not.

390HP—The 390HP engine was available in 1961 and '62 only in 375- and 401-horsepower versions installed in full-size Fords, not T-birds. The T-bird engines had different horsepower ratings but used standard 390 blocks.

The 390HP was the first FE engine with major improvements to the block. The main-bearing caps and webs were cast thicker, plus horizontal ribs were cast between the vertical main-bearing webs. See top center photo on page 30.

Main oil galleries were enlarged and an oil-pressure relief valve added to the rear of the block between the main oil gallery and camshaft plug. This relief valve maintains constant oil pressure at the rear crank and cam bearings. It also prevents extremely high oil pressure during cold starts.

There are no oil galleries for the lifters, as only solid lifters were used. The main-bearing oil-supply holes for number-2 and -4 main bearings are chamfered to align better with the main-bearing inserts.

406—The 406 was built for one purpose—racing. The 406 block casting has thicker cylinder walls and a 4.13-in. bore. This was the first time larger bores required a new casting.

Otherwise, the 406 resembles the 390HP block with its thicker main-bearing webs and horizontal support ribs, larger oil galleries, chamfered main-bearing holes and oil-pressure relief valve at the back of the block.

With the 406's increased displacement and power, a new problem surfaced under racing conditions. On the high-speed, long-distance NASCAR tracks, the main-bearing caps would loosen due to the block deflecting and the caps *walking*—moving in their registers. The solution was the introduction of the famous cross-bolted mains in mid-'62.

The three center main-bearing caps were cast with bosses on each side. They were then drilled and tapped to accept 3/8-16 (3/8-in. diameter, 16 threads per inch) bolts. Cross bolts are installed through drilled bosses in the right and left sides of the block. Selected spacers and washers installed between the cap and the inside skirt of the block complete the assembly.

This fix proved so reliable that the big-block Fords were considered bulletproof in racing circles. Cross-bolted 406s and 427s are very rare and command a high price. Cross-bolted engines are easily recognized by the bolt heads and bosses just above the oil-pan flange at both sides of the block.

427—The 427 has the largest bore of all big blocks—4.23-in. It also produced the most power. The 427 blocks were cast from a high nickle iron and *stress relieved*.

Stress relieving involved sending the blocks from the molds through a long tunnel. The tunnel slowed the cooling of the block, and minimized internal-stress buildup. The result was stronger blocks that maintained straighter cylinder, main- and camshaft-bearing bores.

All 427 blocks were also machined on a special line. First, two feeler probes were inserted through core holes into the water jackets to align with two of the cylinders. This aligned the block to the boring bar more accurately so cylinder-wall thickness was more consistent. The standard method used the outside of the block for alignment.

This special locating method was used because the 4.63-in. bore spacing and 4.23-in. bores left only 0.400-in.

Lifter-gallery connecting passages on some high-performance solid-lifter engines were left undrilled. Lifter oil galleries themselves were undrilled on some other high-performance blocks.

Early high-performance block shows additional support ribbing for main-bearing saddles (arrows). Support ribs are found on heavy-duty applications only.

Early oil-pressure relief valve in block above rear cam-bearing bore. Forward hole (arrow) routes oil back to pan. Smaller hole behind is air relief for plunger. This pressure-relief system is used on '61—'62 390HP, 406 and '63—early-'65 427 blocks.

between cylinder bores. Fitting two cylinder walls and an opening for coolant flow between the bore surfaces required near perfection during casting and machining. Minimum cylinder-wall thickness was limited to 0.110-in., which left 0.180-in. for coolant flow.

Because of the relatively thin cylinder walls, the 427 block can only be bored 0.030-in. oversize. If you find a used block requiring boring to larger than 4.26-in., *it must be sleeved.*

The top—or *deck*—of the 427 block was cast thicker than the standard block to withstand the higher combustion-chamber pressures developed in this high-performance engine. Like the late 406s, all 427s use cross-bolted mains, thicker main-bearing caps and webs.

Early 427s use the same oiling system as the 390HP. This system feeds oil to the front main bearing, camshaft and rocker shafts before the remaining main bearings. This means that an oil-pressure loss anywhere in the lubrication system starves one or all of the four rear main bearings.

To prevent oil-pressure loss to the four rear main bearings, a new and expensive block casting was made. Check the drawing, page 31. The new casting allowed an oil gallery along the left side of the block, from front to rear, hence the name, *side-oiler.*

Diagonal passages drilled through the left side of the block connect this side gallery with the rest of the oiling system. Each diagonal passage runs into a short gallery connecting each

TOP-OILERS & SIDE-OILERS

Early and late 427 blocks. At left, '63—early-'65 *top-oiler,* or *center-oiler,* with casting and machining for oil-pressure relief valve just above rear cam-bearing bore. Note lifter-gallery bosses are not drilled.

The 427 *side-oiler* at right has main oil gallery and relief valve on lower left side. Gallery plugs above rear cam plug show block is drilled for hydraulic lifters—indication this is either a '68 block or a service block. Lifter galleries are undrilled on late-'65—'67 solid-lifter engines.

Note boss between rear-main and cam-bearing supports. This boss provides for an oil gallery between those bearings. Also note boss parallel to number-8 cylinder. Similar boss at number-4 cylinder is just barely visible. This allows block to be drilled for an oil drain for use with single-overhead-cam (SOHC) cylinder heads. Side-oiler 427 was the basis for 427 SOHC engine.

camshaft bearing and main bearing. Each main bearing is lubricated the same way as the front main bearing. This system ensures that the *main-* and *cam-bearing bores* receive oil pressure directly from the oil pump.

Allen plugs seal the rear four connecting passages at the side of the block. Three plugs are immediately above the left main-bearing cross bolts. The relief valve is on the left side of the block in the back, just below the end of the side oil gallery.

Side-oiler 427s require a unique camshaft. As with *top-oilers*—frequently referred to as *center-oilers*—the oil galleries for the rocker-arm shafts tap off the number-2 and -4 cam bearings. Top-oiler engines use grooved cam-bearing *bores* to supply oil to the rocker-arm shafts. But side-oilers use grooves in the number-2 and -4 camshaft-bearing *journals* to allow oil to pass to the rocker shafts.

The original main oil gallery on top and the two lifter galleries are not drilled on the solid-lifter-equipped side-oilers—they are not needed. In 1968 the 427 was cast as a side-oiler but used hydraulic lifters, so the lifter oil galleries were drilled.

Cross-bolted mains used on late 406 and all 427s. If you find one of these blocks, be sure you get its main-bearing caps, bolts *and* spacers (arrows). *Spacers and caps are matched sets,* so if different spacers are used, they must be fitted to the block. Spacers should have 0.000—0.001 clearance with main caps installed and vertical bolts torqued. Non-original main caps will require align boring.

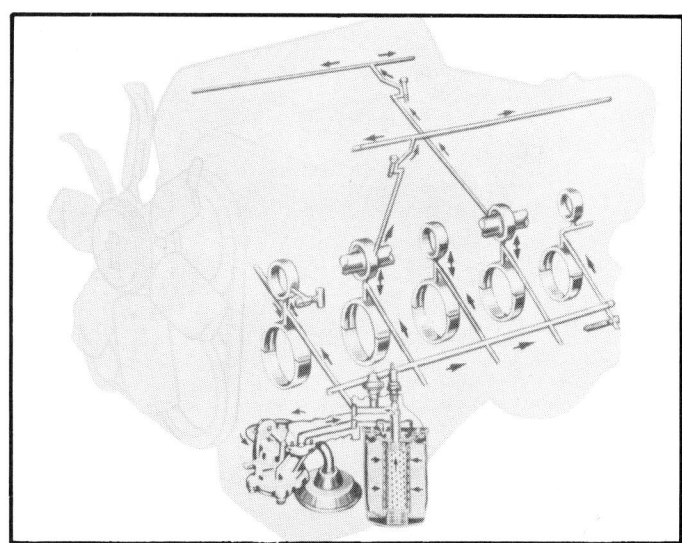

Side-oiler-427 lubrication system feeds the main bearings before the cam bearings. Hydraulic lifters were added in 1968, which required additional galleries drilled parallel to camshaft.

Side-oiler 427 is easily distinguished from top-oiler by screw-in oil-gallery plugs immediately above cross bolts (arrows). Screw-in core plugs are further indications of high-performance nature of this engine.

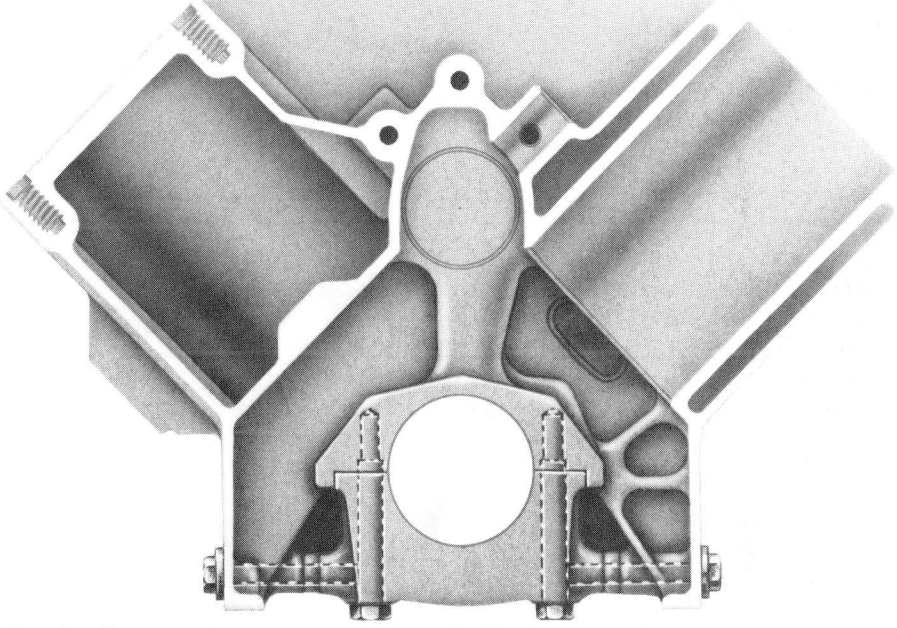

Drawing illustrates improvements made to big-blocks: counterbore for head bolts and for vertical main-cap bolts. Note additional ribs in lower right corner. Cross-bolts were introduced on 406 and used on all 427s.

This block, C8AZ-6010-G, was once available as a Ford service part, but not anymore. The cylinder bores are notched to clear the valves, particularly the intake valve. The notched bores allow using *Tunnel-Port* heads—covered later in this chapter. The block also comes with screw-in core plugs, as did all production-car side-oilers.

This 427 block can be equipped with solid or hydraulic lifters, but the lifter oil galleries *must* be plugged if solid lifters are used. Block has its oil-passages threaded already, so half the job is done. At one time, Ford sold these plugs, C6AZ-6026-A. Your dealer may still have them, but Ford Parts no longer carries them.

If you have these plugs, or can get them, and want to plug the lifter galleries, remove the two plugs angled toward the lifter galleries. Check the drawing, page 32. Screw in the 7/16-in. plugs. If you don't have these plugs you will have to drill the lower hole below the center oil galley to accept a plug. Do this for both sides. Install screw-in plugs in the lower galleries and then replace the upper plugs. Make sure you clean the shavings out before assembling the block.

The first side-oilers were installed in passenger cars on March 1, 1965. If you want the best block for high-performance use, the side-oiler is it.

One further change was made to the side-oiler block in 1966. Under racing conditions there was a problem with cylinder-wall failure. Often mistakenly blamed on a dropped valve, this was actually caused by the piston breaking through the cylinder wall.

As a solution, cylinders were strengthened by increasing the thickness of the cylinder walls at the outside "corners." See the drawing. A number of blocks were also cast with a higher nickel content, but these were sent to Holman and Moody to ensure they would be used for racing.

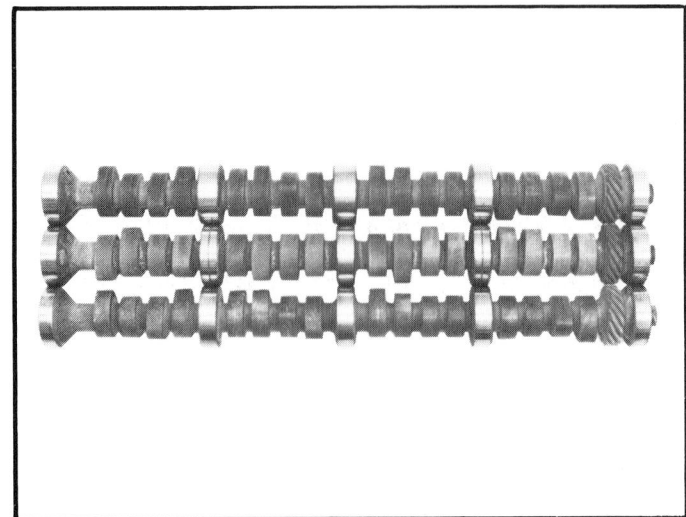
In addition to variety of camshaft *grinds*—lobe designs—three different-type camshafts are used in big-blocks: Top is FT cam, which uses wider cam-bearing journals. Center camshaft is for side-oiler 427; note grooves at number-2 and -4 bearing journals. Bottom is standard FE cam.

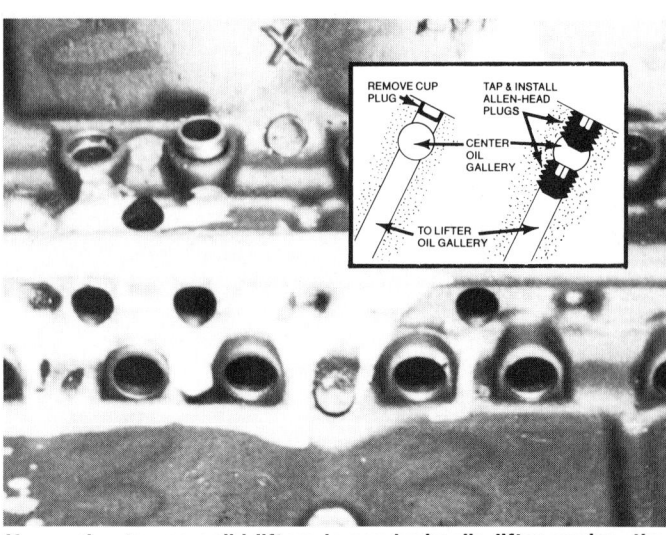
If you plan to use solid lifters in any hydraulic-lifter engine, the lifter galleries must be plugged. Easiest way is to plug short galleries tapping off center gallery. Remove cup plugs. Drill and tap passage to accept two Allen-head plugs, one above center gallery and one below as shown. Use *small* amount of sealer or Teflon tape on plug threads. Repeat on other side.

428—This block uses the same bore as the 406—4.13 in.—but is closer in design to the 390. Oil passages are the same as in the standard 390. It has no oil-pressure relief valve at the rear of the block. The '66—'70 Police Interceptor and '68—'70 Cobra Jet models use thicker main-bearing supports, caps, webs and extra ribs.

Some '66 428 Police Interceptor engines used solid lifters with blocked oil galleries.

FT CYLINDER BLOCKS

FT big-blocks were installed in medium- and heavy-duty trucks. All FT blocks are cast of high-grade-alloy iron with manganese, silicon and other alloys added to improve durability. Most blocks are strengthened in the crank-support areas with much thicker main-bearing caps and webs than those in the FE blocks.

FT blocks use the standard-FE oiling system. Hydraulic lifters are used in all FT engines, so the oil galleries are drilled accordingly. Four-bolt engine mounts are used on all FT engines. FT blocks share the same external dimensions and bellhousing bolt pattern as the FE.

All FT blocks have a hole drilled in them for an air-compressor oil drain. The drain is low on the right side near the center of the block. The compressor is for air brakes and is lubricated by engine oil. This drain returns the engine oil to the oil pan.

Now for specific FT blocks:

330MD & 330HD—This block has the smallest bore of any big-block—3.875 in. The 330MD block uses the FE 352 cast-iron crankshaft, so it has the same bearing-bore diameter, 2.9416 in.

Thicker main-bearing webs and caps are used in the 330HD block. The '72 330HD engine was the last engine to use the HD block, although the other HD components were retained. These include a heavy-duty oil pump with a 5/16-in.-diameter drive shaft.

The larger shaft requires a larger-diameter distributor shaft, a larger bore in the intake manifold for the distributor housing and a larger distributor-housing bore in the block.

In the 330MD, the standard FE oil pump and 1/4-in.-diameter drive shaft are used. 1973—'77 330MD blocks came with a replaceable distributor-shaft bushing. This bushing can be removed to allow using the heavy-duty oil pump with the 5/16-in. drive shaft. The distributor must also be changed for this modification.

361 & 391—These FT blocks have the same 4.05-in. bore as the 390 FE block. The 361/391 FT block has the larger distributor-shaft bore as it uses the heavy-duty, high-capacity oil pump with the 5/16-in. drive shaft.

Some 361/391 FT blocks are cast with thicker main-bearing webs and additional ribs in between. This increases their weight about 20 pounds over the FE 390. Heavier 361/391 FT blocks do not have the 352 as cast on the left front of the standard FE block. Most heavy blocks have a mirror-image 105 on the left front, though this is missing on some blocks.

INDUSTRIAL BLOCKS

There are many different industrial blocks, some are standard FE or FT blocks, others have modifications in machining or casting. I cannot cover them all—or even tell you why they were all made or what parts they used.

Many of these blocks will fit no particular category. You will find blocks cast as 427 side-oilers but machined as top-oilers. You will find blocks cast as standard on the outside but cored for extra-small bores not matching any standard big-block. Some industrial engines were rejected for passenger-car use, but were deemed usable at the lower rpm required of an irrigation pump.

If you find an industrial block, measure all the dimensions and you may be able to figure out what combination of blocks it was cast from or machined to. If you are willing to invest the time and machine work, you may be able to come up with some valuable parts from these industrial engines. For example, many 427 blocks were used for marine or industrial applications.

CRANKSHAFTS

All FE and FT crankshafts have 2.7488-in. main- and 2.4384-in. rod-

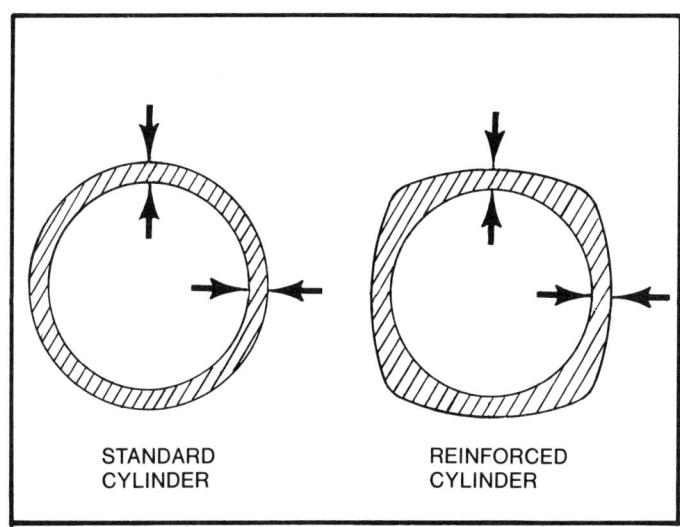

Late-'66 427s had reinforced cylinder walls to prevent failure under racing conditions.

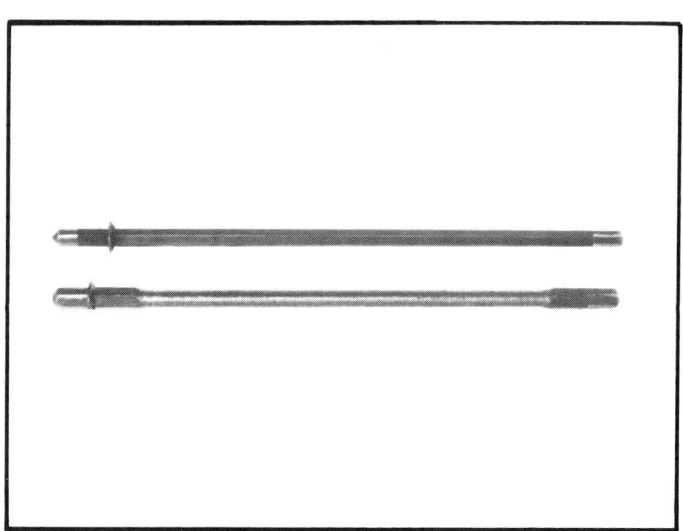

Heavy-duty FT engines use larger oil-pump drive shaft, oil pump and distributor housing. 1973-and-later 330 FT blocks used smaller drive shaft. But 330 came with removable bushing in the block that allowed adapting larger oil-pump drive shaft and larger distributor.

bearing-journal diameters. All have the same bolt pattern at the rear flange. Except for the 410 and 428, FE crankshafts are *zero balanced*—balanced internally by six counterweights.

The FT 330 with the cast-iron crankshaft uses the standard 352 crank, so it is also balanced internally. *Detroit balance* is used on FT engines with forged-steel crankshafts, and the FE 410 and 428.

On the FE 410 and 428 final balance is obtained by counterweighting the flywheel or flexplate. On the FT engines with forged-steel cranks, final balance is obtained by adding weight to the flywheel or flexplate *and* the vibration damper, or harmonic balancer. On the FE 428SCJ, the LeMans rods require an additional counterweight on the vibration-damper spacer.

You cannot mix parts using two different balance techniques. For example, the flywheel from an engine with zero balance should not be used on a crankshaft with Detroit balance unless the crank, vibration damper and flywheel or flexplate are rebalanced. Consequently, when interchanging crankshaft-related parts, use parts with the same balance technique, or have them rebalanced before assembling the engine.

Larger counterweights are used on the FT 330HD, 361 and 391 truck forged-steel cranks, so they weigh about 10 lb more than a standard FE crank. External balancing at the flywheel or vibration damper is required because crankshaft-to-crankcase clearance prohibits larger counterweights.

The heavier flywheel and damper present no problems, because most trucks use manual transmissions. The heavier flywheel supports the heavy-duty-truck clutch and provides additional inertia needed to get a big truck moving. A thicker rear flange and deeper pilot-bushing bore are used.

Crankshaft-nose diameter is larger on the forged-steel FT crankshafts to accommodate a power takeoff. FT noses measure 1.750-in. compared to 1.375-in. for FE cranks.

An FT crank can be used in an FE engine if the crank nose is machined to a smaller diameter. The rear flange must also be shortened 3/16 in. Machining the nose is not necessary if the timing chain, sprockets and front cover from the FT engine are also used. In either case, you have to machine the rear flange and have the bottom-end assembly balanced.

High-Performance Crankshafts—
The first performance-related crankshaft changes were heavier counterweights and grooved main-bearing journals. These modifications were made to the 390HP, 406 and 427, except for the solid-lifter side-oilers. The 352HP had heavier counterweights with non-grooved journals.

Heavier counterweights were added to these cranks to offset the additional weight of the beefier rods. A 1/8-in.-wide groove machined in the main-bearing journals ensures oil flow to the rod journals at high rpm.

In place of the 352 found on most FE blocks, many heavy-duty FT cylinder blocks have mirror image 105 on left front. In some cases, this area is left blank on heavy-duty blocks. HD blocks have additional main-bearing support and thicker main-bearing webs.

Forged-steel FE crankshafts appeared on race tracks shortly after the '64 FT truck engines were introduced with forged-steel cranks. Truck forged-steel cranks used the same oiling system as the standard cast-iron cranks, with a single diagonal hole drilled from each rod journal to the nearest main journal.

Forged racing cranks used *cross-drilled* main and rod journals. These were released for production late in the '65 model year. Cross drilling connects the main- and rod-journal oil galleries the full length of the crankshaft. This creates an uninterrupted gallery from the number-1 through the number-5 main-bearing journal.

Cross drilling was made possible by hollowing the rod journals from both ends. Hollowing the journals ensured that the cross drilling from the main journals connected to the rod journals. See drawing on opposite page. The hollowed rod journals are capped to seal them.

A 1-1/2-in.-long groove cut across one of the oil inlet holes on number-2 and -4 main-bearing journals improves oil flow to the rod journals. Neither the main-bearing journals nor the lower bearing halves are grooved on the cross-drilled crankshafts. This way they retain the full main-bearing-surface area. Top main-bearing halves are grooved.

This crankshaft, known as the *LeMans* crank, proved reliable in all forms of racing and street use. It makes an excellent package when combined with the side-oiler 427.

This forged crank was installed in all side-oiler 427s except for the 1968 model year. The '68 cranks are basically the same as the early top-oiler cranks—high-nodular cast iron with grooved main-bearing journals.

Although no longer available, the ultimate forged-steel crank is the NASCAR unit. This crank and connecting-rod combination was made for sustained running above 7000 rpm.

The NASCAR crank is like the LeMans crank with one exception: Rod journals are 0.080-in. wider to accommodate the wider, extra-heavy-duty, NASCAR connecting rods. Counterweights are heavier on this crank to counterbalance the NASCAR rods, which weigh nearly 1000-grams (more than 2-lb) each.

The 428 crankshafts are unique. From 1966—'68, one crankshaft was used in all 428s, including Cobra Jet Mustangs.

In the 1969 model year, four more crankshafts were added. These crankshafts have weight changes to counterbalance piston-weight changes made December 26, 1968. The '69-and-later Cobra Jet had two cranks. The Super Cobra Jet, with its LeMans rods, accounted for the final two cranks.

Pay close attention to crankshaft numbers in the identification chart. Each crankshaft is matched to a specific piston. If you change to a different-type crankshaft, you must have the bottom-end assembly rebalanced.

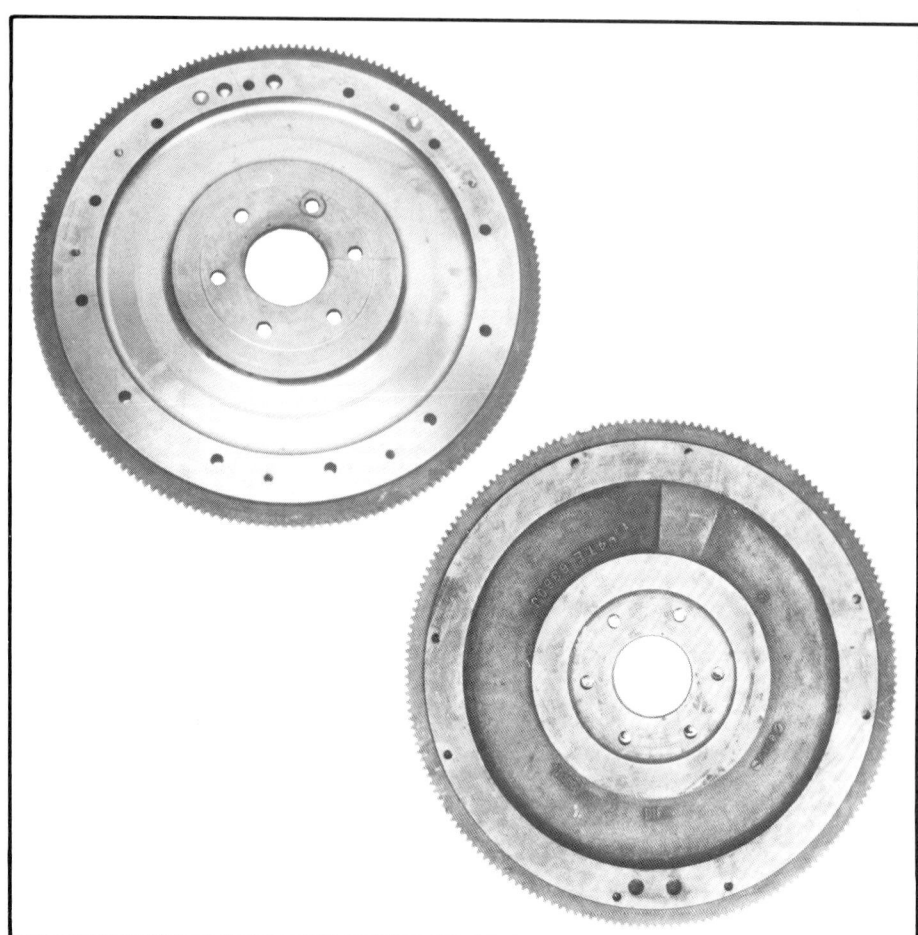

Difference between *Detroit* and *zero* balance is demonstrated by these flywheels. Both are used with forged-steel cranks. Note counterbalance on FT flywheel (below) is missing on FE 427 flywheel. FT flywheel is Detroit balanced, 427 flywheel is zero balanced. You cannot mix parts such as these without having engine-balance trouble.

CONNECTING RODS

Although there are several connecting rods, only two lengths are used. The first has a 6.540-in. *center-to-center length*—distance between the pin-bore and big-end-bore centers. The other rod uses a center-to-center length of 6.488 in. All are forged steel.

Long Rods—The 6.540-in. long rod is used in the FE 332, 352, 360, and 361 Edsel engines, and the '64—'78 330MD and '73—'78 330HD FT engines.

The long rod has remained unchanged except for the 1960 352HP rods. Early 352HP rods use standard 3/8-in. bolts. Later 352HP uses 13/32-in. bolts. Both have wider beams.

Short Rods—The 6.488-in. short rod is used in the FE 390, 406, 410, 427 and 428 engines; and the FT 361, 391 and '64—'72 330HD engines. The *short* rod has a *long* story! Two 390 rods were introduced in 1961. The first short rod is similar to the standard long rod, but is 0.052 in. shorter.

The second short rod, known as the *Police Interceptor rod,* is about 50 grams (1.8 oz) heavier. It has wider and thicker beam flanges and 13/32-in. bolts and nuts. This connecting-rod carries various numbers. It is found in the '61—'62 390HP, late-'62—'65 Police Interceptor 390s, 406s and 427s with the cast-iron crankshafts. It is also used in 428 Police Interceptor and Cobra Jet engines.

In late-'63 a third rod was added. This rod is basically the same as the Police Interceptor rod, except that it uses 3/8-in. bolts.

LeMans Rods—The next step up in performance connecting rods is the LeMans rod, installed in the side-oiler 427. The LeMans rod weighs about 70 grams (2.5 oz) more than the Police

ratio is also affected by these changes. As a piston travels from BDC to TDC, it *sweeps* through, or displaces, a volume called *swept volume* (SV). Swept volume is a cylinder's displacement.

The other determining factor in compression ratio is *clearance volume* (CV). Clearance volume is the area above the piston at TDC. This includes combustion-chamber volume in the head; volume created by the head gasket spacing the head above the block deck; volume created by the block deck clearance and shape of the piston top. Of all these, cylinder-head combustion-chamber volume has the greatest affect on clearance volume.

A cylinder's compression ratio (CR) is *directly proportional* to its *swept volume* (SV) or *displacement*, and *inversely proportional* to its *clearance volume* (CV): $CR = SV/CV + 1$. In simple terms, increasing swept volume increases compression ratio, but increasing clearance volume decreases compression ratio.

The compression-ratio formula says that as an engine's displacement is increased, clearance volume must be increased proportionally to maintain the same compression ratio. That's why oversize pistons are manufactured to increase clearance volume with a smaller-than-original dome or larger-than-original dish. Otherwise compression ratio would be increased by the larger bore.

Compression ratio is an important consideration when interchanging heads, so check the following head-interchange section. Be sure to compare combustion-chamber volume when interchanging heads. It's best to work with a performance engine shop when doing this.

With today's fuel, compression ratios over 9:1 can cause detonation. These engines are not prone to detonation, even with the stock, 10.5:1 compression ratio. But swapping heads without checking combustion-chamber volume can cause severe problems. Severe detonation will damage the engine. At the same time, changing to a head with a too-large combustion-chamber volume will result in a low-compression, poor-performing engine.

The final consideration in bottom-end interchange is *balance.* Changing *any* lower-end parts—from a long to a short rod or a light-duty to a heavy-duty rod—changes engine balance. As a general rule, any interchange of lower-end parts should begin by balancing at a machine shop.

Don't forget to take into account valve lift and size when selecting parts: Changing the camshaft or rocker-arm ratio or block and head milling can eliminate valve-to-piston clearance.

FE CYLINDER HEADS

Major specifications for the cylinder heads—casting numbers, valve sizes and combustion-chamber volume—are listed in the charts on pages 55 and 56. What follows is simply a short history on the evolution of FE heads.

332, 352 & 361 Edsel—The first heads used on the 332 and 352 engines had *machined* combustion chambers with a 69—72cc combustion-chamber volume. Valves were 2.02-in. intake and 1.55-in. exhaust.

After only 90 days of production these heads were replaced with heads having *as-cast* 70.4—73.4cc combustion-chambers. This eliminated the expensive machining process on the combustion chambers. The 361 Edsel also used this later head.

In later years of production combustion-chamber volume changed slightly, but no other significant changes were made.

352HP & 390HP—The first performance head appeared on the 360-horsepower 352HP. This head used *taller* intake ports and forged intake valves. Intake-valve-seat angles were ground to 30° on all HP engines, instead of the normal 45°. Valve sizes remained the same, 2.02-in. intake and 1.55-in. exhaust.

This same head was used on the '61—'62 390HP or *Special* engines in full-size Fords.

360, 390 & 410—These heads went through a number of changes in combustion-chamber volume, so check the charts, pages 55 and 56. These heads used 2.02-in. intake and 1.55-in. exhaust valves.

The only major change in the heads came in 1966. The 390 was installed in '66—'67 Comets and Fairlanes and '67—'69 Cougars and Mustangs. Because of the small engine compartment, new exhaust manifolds were used. New manifolds required a new bolt pattern on the heads, photo page 45.

Earlier cylinder heads are not easily converted to this application, as additional metal was cast in to accommodate the new bolt pattern.

406—The first 406 heads, casting C2SE-6090-B, had a combustion-chamber volume of 62.6—64.6cc.

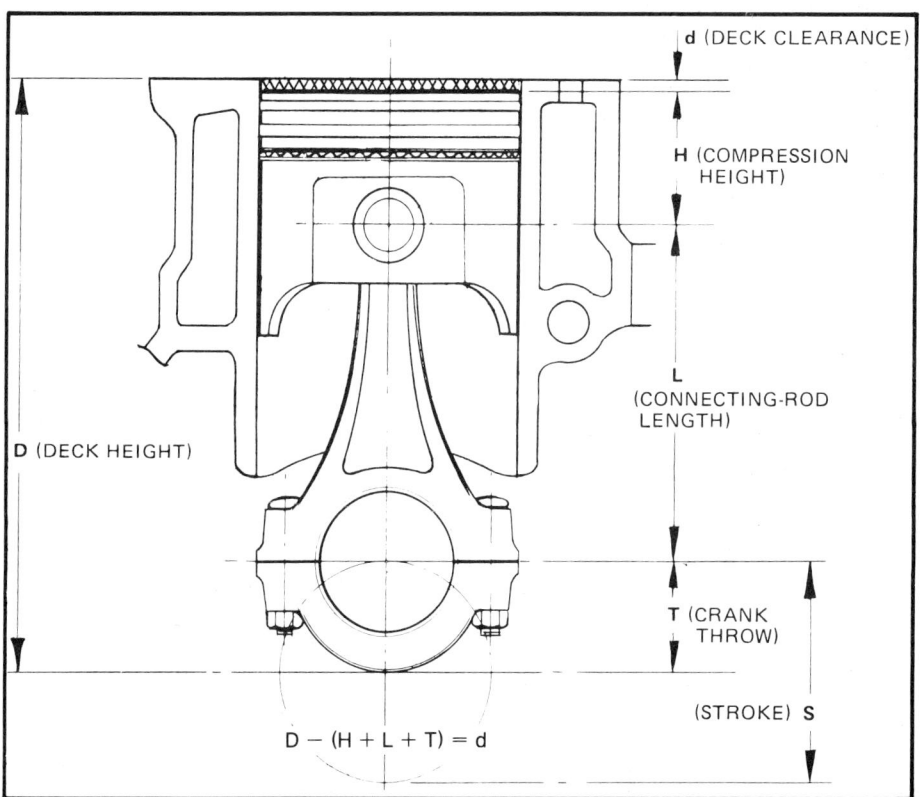

When interchanging parts, use chart at left to ensure engine will have sufficient deck clearance.

Port sizes and combustion chambers changed radically during the life of the big-block Ford. Be sure to check combustion-chamber volume and provisions for intake-manifold heat before interchanging heads. See charts, pages 55 and 56.

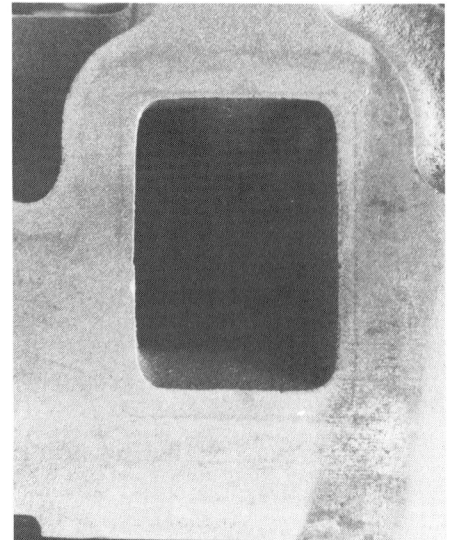
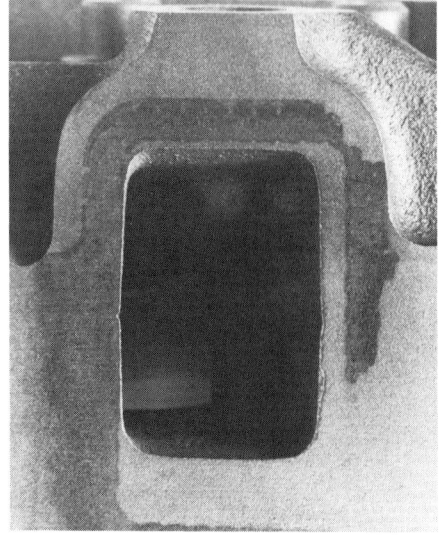
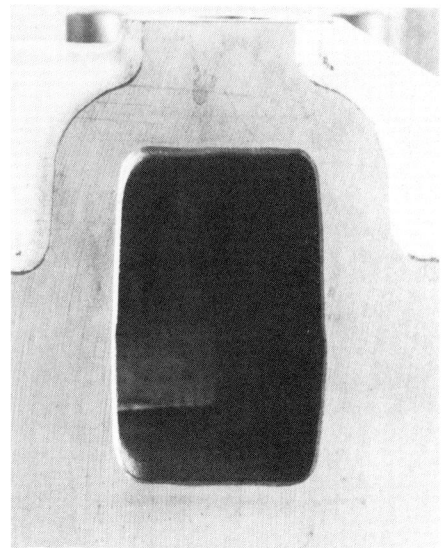

Standard-FT combustion-chamber. Note as-cast chamber, small valves and valve-seat inserts.

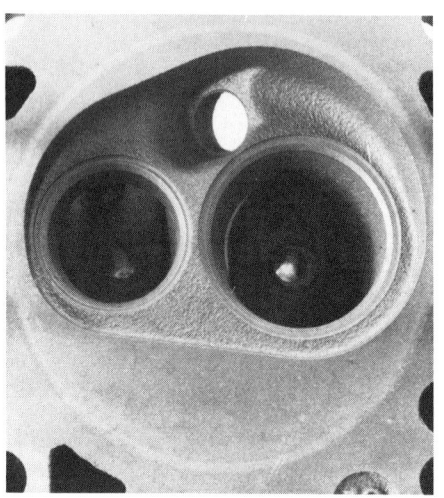
Most-common FE cylinder head. Standard 390 with as-cast combustion chamber.

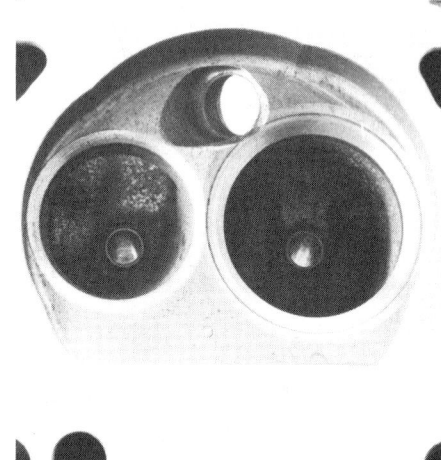
Medium-Riser 427. Machined combustion chamber, larger-than-standard ports and valves.

Valve diameters of 2.02-in. for intake and 1.55-in. for exhaust were used.

Valve seats on these heads were cut to the outer limit of the valve diameter, so Ford literature often lists these valves as 2.04-in. intake and 1.57-in. exhaust.

The late-'62—early-'63 head, C2SE-6090-C, has 1.66-in. exhaust valves. This head has larger combustion chambers at 64.0—67.0cc. This lowers the compression ratio slightly.

The late-'63 406 with three 2-barrel carburetors used another head. This head has smaller combustion chambers—56.4—61.0cc—which raise the compression. This head is the first big-block head machined for separate spring seats for intake and exhaust valves. The inserts reduce spring chafing on the head and also minimize side-loading the valve stems.

427—With the high-performance and racing applications of the 427, there were many changes to the heads during production. Some heads used cast combustion chambers, others were machined, some had sodium-filled exhaust valves and hollow-stem intakes, certain heads had separate valve-spring seats, some did not, and so on. The heads were even changed after the production of the engine had stopped!

The first 427 head, casting C3AE-6090-D, resembles the late-406 head—it has the same valves and spring-seat inserts. But it is thicker—stronger—at the head-gasket surface. The intake ports are larger, although the openings at the manifold-mating face are the same size as the 406.

The next change was to add larger intake valves to increase horsepower. Valve diameter increased from 2.04-in. to 2.09-in. This change was made on March 15, 1963 to castings, C3AE-6090-G and C3AE-6090-H.

The next change also improved breathing. The floor of the intake port was raised and a raised ring cast above the intake-valve seat. Flow testing showed that these improvements let the head flow more air—thus producing more horsepower. The ring and hump were used to direct the air and

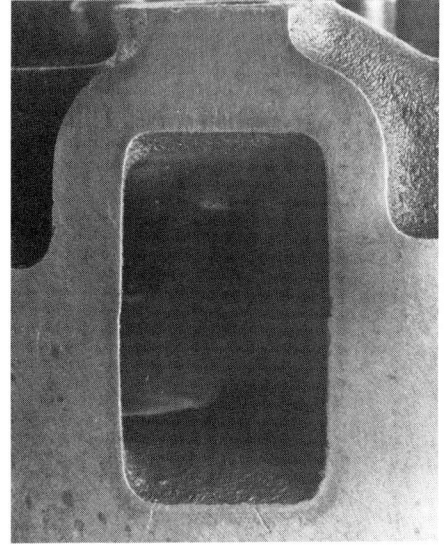
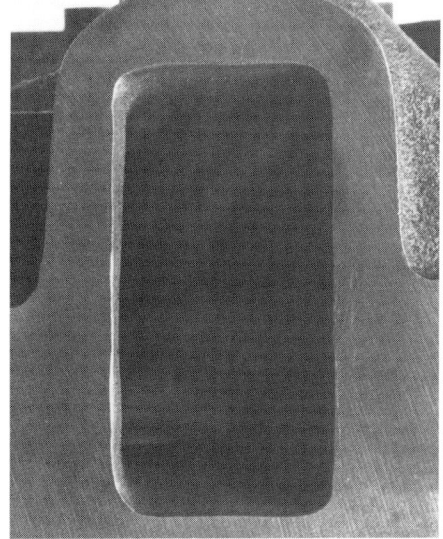
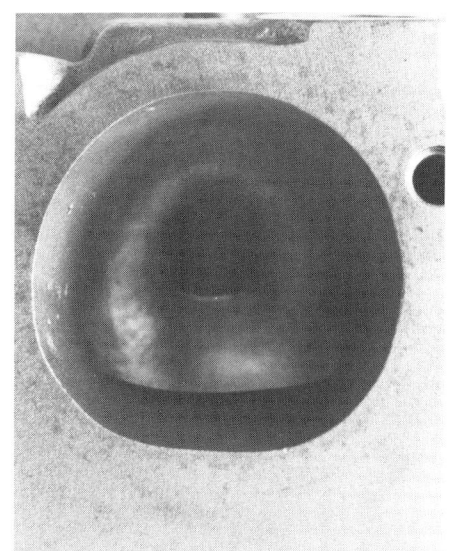
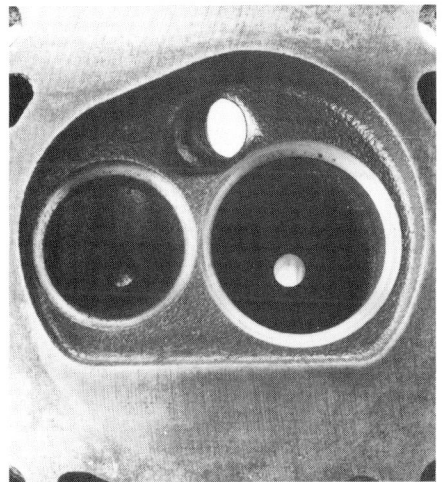

Early 427 Low-Riser uses non-machined combustion chambers, larger intake ports, but standard-size FE valves.

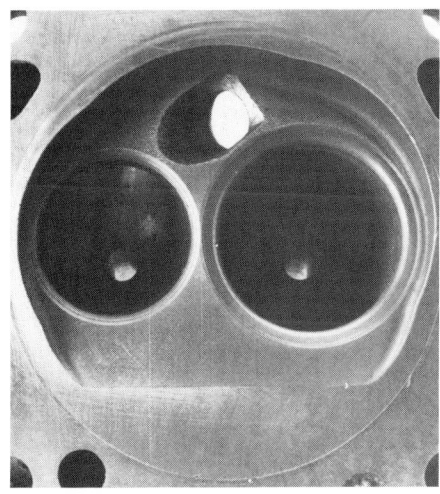

High-Riser 427 features machined combustion chambers, tall intake ports.

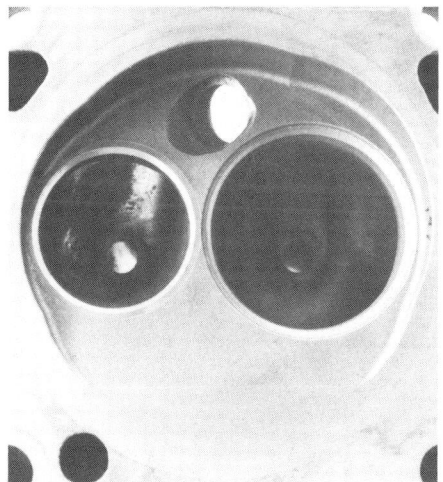

Tunnel-Port 427 has round intake ports, machined combustion chambers and exhaust-valve-seat inserts. Tunnel-Ports also have largest intake valves of any FE head at 2.25 in.

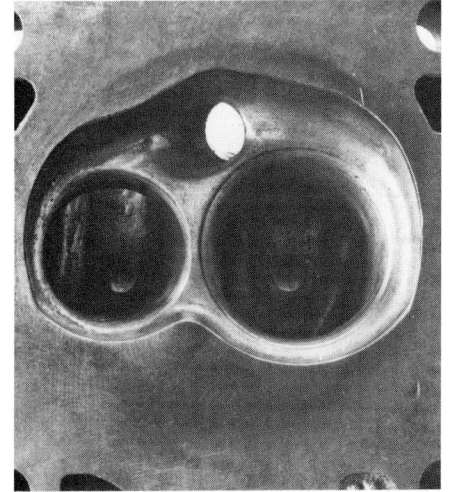

352HP and 390HP use same port and valve sizes as small-chamber Low-Riser. Note: Chamber has been modified.

fuel mixture toward the center of the port while the ring increased air/fuel-mixture velocity. The casting, C3AE-6090-J, came with '64 and early '65 *Low-Riser* engines.

Before proceeding with the 427 heads, I should explain some terminology used in reference to them. Although 427 cylinder heads are often referred to as Low-, Medium- or High-Risers, the terminology stems from intake-manifold design. Check the photo, page 43.

In general terms, a riser is simply the vertical distance from one level to the next. On an intake manifold, the riser is the distance between the port floor at the cylinder head and the plenum-chamber floor. It's easiest to visualize with a single-plane intake manifold.

The Medium-Riser has a 2—3-in. higher rise than the Low-Riser. The High-Riser has an even steeper rise, 2—3 in. more than the Medium-Riser.

The reason the cylinder heads are also termed Low-, Medium- or High-Riser is that the intake-port angle in the head matches that of the intake manifold. As the height increases, so does the angle of the intake port. The Low-Riser is angled about 20°, the High-Riser is close to 45°.

Some confusion comes about because cylinder heads that are referred to as *Low-Risers* are actually used in both Low- and Medium-Riser applications. Another problem is that the

 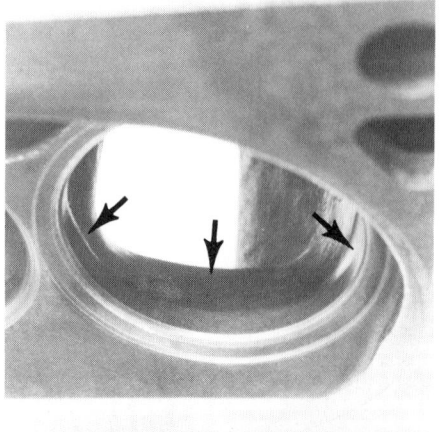 

Early High-Riser intake port (left) and late High-Riser with modification described in text (right). Raised ring (arrows) that appears to restrict port actually improves mixture velocity and flow, thus produces more power. Ring was later incorporated into all FE heads.

High-Riser and Tunnel-Port heads require notching at top of cylinder bores for valve clearance. If block deck or cylinder heads are milled notches must be restored.

Limited-production '64 Fairlane *Thunderbolt* used High-Riser 427. Induction system required tear-drop hood scoop for carburetor clearance.

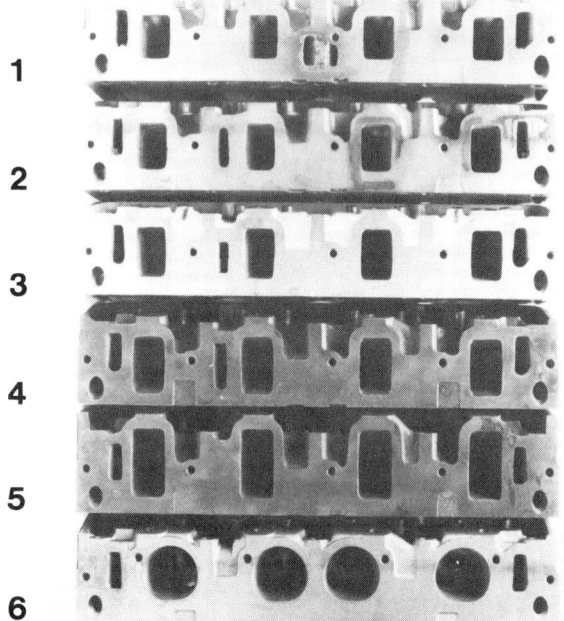

Intake-manifold heat is one consideration in cylinder-head interchange. Exhaust heat is routed through intake manifold by several methods (arrows): FT heads (1) use exhaust heat routed through exhaust manifolds to center of each head. Most FE heads (2—4) have exhaust routed through ports between end cylinders. Heat riser runs diagonally across intake manifold. High-Riser (5) and Tunnel-Port (6) heads have no provision for intake manifold heat, in keeping with their high-performance/racing design.

cylinder heads referred to as *Medium-Riser* have shorter ports—measured at the head—than most Low-Riser heads.

Low-Riser heads are actually used on some Medium-Riser applications. The 1968 427 is actually a Medium-Riser as are the 428CJ, SCJ and 428 Police engines.

The reason for the height differences came in the horsepower race. The High-Riser design increases the ram effect and gives the fuel mixture a straighter path into the combustion chambers. Problem is, the High-Riser manifold is tall—required a bubble or teardrop-shape hood scoop to clear the carburetor/s and air-cleaner assembly.

The Medium-Riser was introduced so the standard hood could be used on production cars. The Medium-Riser was actually a compromise design. Tuned for more mid-range power, the Medium-Riser performs exceptionally well on the street.

Getting back to production history, High-Riser heads, casting C3AE-6090-K, appeared on the race circuit late in '63. This head is used on early-production '64 engines. The modified head casting, C4AE-6090-F, is used on late-'64 and early-'65 engines.

The High-Riser head was discontinued as a production option in '65, but remained an over-the-counter

Note runner angle on High-Riser intake manifold. Manifold gives air/fuel mixture straighter path into cylinder head and combustion chamber for increased flow.

Height difference between High- and Low-Riser intake manifolds. Although 427 cylinder heads are often described as Low-, Medium- or High-Riser, terminology stems from intake manifolds.

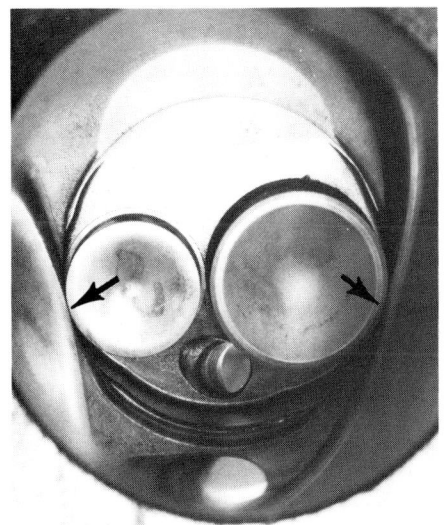
Keep valve-to-bore clearance in mind when interchanging heads. Tunnel-Port heads installed on 428 bored 0.030-in. oversize shows insufficient intake and exhaust valve-to-bore clearance (arrows). Tunnel-Port heads use 2.25-in. intake valves, which only fit 4.23-in. bores of 427. Still, notches are required so valves will clear cylinder walls.

option for about 10 years. "Production" cars that came with this head were the '64 and '65 full-size cars and the '64 Thunderbolts, Falcon and Comet FX drag cars—if you could call a Thunderbolt or FX production cars.

The High-Riser heads have the tallest intake ports of all the big-blocks, 2.72 in. High-Riser heads also used machined combustion chambers for accurate volume control and to provide a smooth surface to eliminate hot spots in the combustion chambers. A *cast* combustion chamber has a rougher surface than a carefully machined one. The resulting hot spots cause preignition in a high-compression engine.

The late cylinder head—C4AE-6090-F—has the same modification as those mentioned for the '64—early-'65 Low-Riser head: raised intake-port floor and cast-in ring above the intake-valve seat. Racing High-Riser heads were actually the first with these improvements, but Low-Risers were the first street engines to use them.

Medium-Riser heads, as mentioned earlier, were introduced on production cars on March 1, 1965. They eliminated the need for the hood scoop and provided nearly the same performance as the High-Riser setup.

Medium-Riser heads also featured machined combustion chambers. These heads came on production 427s through 1967. In 1968, the 427 was again equipped with the Low-Riser heads with cast combustion chambers.

In 1968, 427 Cougars and Mustangs used heads with the same exhaust-flange bolt pattern as the 390 Fairlane. Remaining '68 427s had the standard *vertical* bolt pattern. These heads were similar to the Cobra Jet heads for the 428. They were the only 427 heads without separate spring seats. The seats required additional machining on the tops of the cylinder heads.

The last head style to be used on the 427 was the *Tunnel-Port*. It was a radical departure from all earlier big-block heads. The intake ports on other big-block heads are rectangular. The rectangular ports are offset to one side, because the intake-manifold ports bend around and are necked down to clear the pushrods.

The Tunnel-Port uses nearly round ports from the intake-manifold *plenum*—the chamber below the carburetor/s—all the way to the valve. To make this possible, the pushrod runs in a sleeve through the center of the intake port.

A major advantage of this design is that round ports have less flow loss for a given area as compared to rectangular ports. Rounded ports also tend to *wire-draw* the mixture. This wire-draw action resembles the whirlpool action you see when draining a sink or tub.

Round passages also reduce the bends that tend to separate the fuel from the air. Overall, the Tunnel-Port heads increase top-end rpm and horsepower over the High-Riser heads.

Tunnel-Port intake runners are actually elliptical- or oval-shaped, measuring 2.34 X 2.17-in. These heads are particularly effective at high rpm, so much so that if you operate the engine at less than 6000 rpm, you are wasting your money on a Tunnel-Port manifold and heads.

Early Tunnel-Port heads used 2.19-in. intake valves. Later Tunnel-Port heads have the largest intake valves offered on the 427— 2.25-in.—to match the large ports. These were the ultimate wedge heads and were used for many stock-car victories in the '68—'70 period. A 1.73-in. exhaust valve is used.

Tunnel-Port heads were first cast in 1967. Casting continued at least as late as 1970. And they were never offered on production cars. These heads—in fact, all of the 427 heads—are no longer available from Ford.

FE intake-manifold heat passage (arrows). FT heads route heat straight across manifold. Some high-performance intake manifolds have no provision for heat. Street-driven vehicles should use heated manifolds, so be sure heat passages in manifold and heads match.

Big-blocks are equipped with a number of different manifolds, many with multiple carburetors. See caption at right for an explanation of this manifold and those at right.

Specialty Heads—In addition to cast iron, the Medium-Riser, High-Riser and Tunnel-Port heads were also cast in aluminum. A set of these is a golden investment—they command high prices from collectors.

You will also find many of the heads with reworked ports—some by the factory, others by specialty shops like Holman and Moody, and still others by individuals. Some heads were only partially machined by the factory, so they could be custom machined for specific applications. For this reason you may not be able to tell what engine a particular cylinder head originally came on, or what the factory dimensions were.

FT CYLINDER HEADS

As with the FE heads, FT heads went through a number of changes during their production history. I give a brief description of each type here. Check the chart for details, page 56.

330, 361 & 391—Compared to the standard FE cylinder head, the FT cylinder head has a larger combustion chamber and smaller valves and ports. Combustion-chamber volume is 78.5—81.5cc. Intake valves are 1.75-in.; exhaust valves 1.5-in.—the smallest valves of all the big-block heads. There's a reason: Smaller ports limit high rpm but increase low-rpm torque. Trucks need low-rpm torque and the improved durability that results from low rpm.

FT ports appear *slightly* smaller than those on the standard FE head. Actually, they get *substantially* smaller as they taper toward the intake valve. Port dimensions are 1.81 X 1.32 in. at the intake manifold, and 1.81 X 1.22 in. at the exhaust manifold.

All 361 and 391 heads have intake- and exhaust-valve-seat inserts. Heads for the 330HD have seat inserts on the exhaust side. Except for those on the 330MD, all FT exhaust valves are sodium-filled, like those in the 427. The heads are also stress relieved.

Intake valves are nickel-chrome-alloy steel for the heavy-duty engines and aluminized steel for the 330MD engine. With the exception of the the valve-seat inserts, the 330 head is identical to the 361 and 391 heads.

FT cylinder heads route the heat passage to the intake manifold between the two center cylinders rather than between the two end ones on the FE heads.

CYLINDER-HEAD SWAPPING

FE and FT cylinder heads are one of the most-common and easiest

FE & FT INTAKE MANIFOLDS		
Cyl. Head	Intake Manifold	
1.81 x 1.34	1.69 x 1.16	FT truck
1.93 x 1.34	1.82 or 1.75 x 1.16	FE standard Pass car and pickup truck, '66-'76
2.06 x 1.34	1.94 x 1.24	427 Medium-Riser
2.26 x 1.34	2.14 x 1.16	FE standard Pass car and pickup truck, all '58-'65 and some '66-'76
2.34 x 1.28	2.20 x 1.14	352HP, 390HP, 406, 427 Low-Riser
2.34 x 1.34	2.00 x 1.24	428 CJ, SCJ
2.72 x 1.34	2.60 x 1.24	427 High-Riser
2.17 x 2.34	2.05 x 2.20	427 Tunnel-Port (Oval ports)
1. Measured at cylinder-head-to-manifold mating face. 2. Medium-Riser intake manifold adapts to FE cylinder heads measuring 1.93 in. or 2.26 in. vertically.		

components to interchange. Any FE or FT cylinder head will bolt to any FE or FT block. In most cases the combination will work—but there are problems.

Problems have nothing to do with bolting the heads to the block. They concern valve-to-bore clearance, exhaust-manifold bolt pattern and intake-manifold matching. Let's look at each problem separately:

VALVE-TO-BORE CLEARANCE

3.87-in. Bore—The small bores of the 330 only accept FT cylinder heads. The only change to consider here is using the 361, 391 or 330HD heads, all of which have hardened exhaust- or intake-seat inserts for durability.

4.00-in. Bore—Any cylinder head with 2.04-in. intake and 1.56-in. exhaust valves—found in the majority of FE engines—can be used. Heads with larger intake ports—found on the 352HP and 390HPs of the early '60s—also used this valve size.

Low-Riser 427 and 428 Cobra Jet heads with 2.09-in. intake and 1.66-in. exhaust valves will also work,

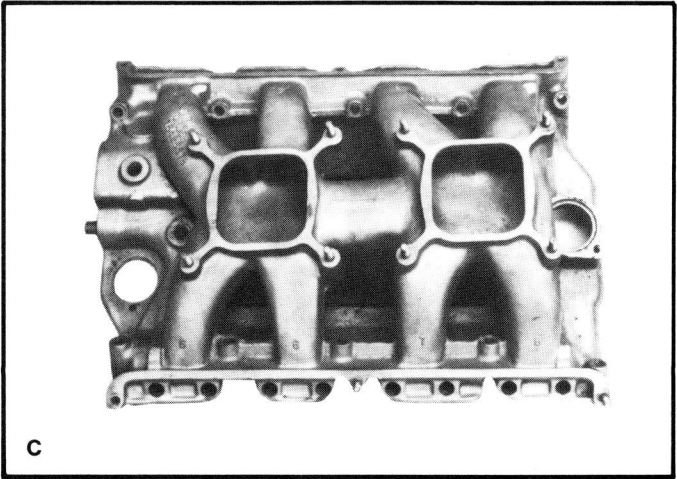

Left to Right: (A) Medium-Riser with single 4-barrel; (B) Tunnel-Port with single 4-barrel (note pushrod tubes through intake ports); (C) two 4-barrel Tunnel-Port-style intake, which fits Medium- and standard FE engines. Be sure to match intake-port sizes if you change either heads or manifold.

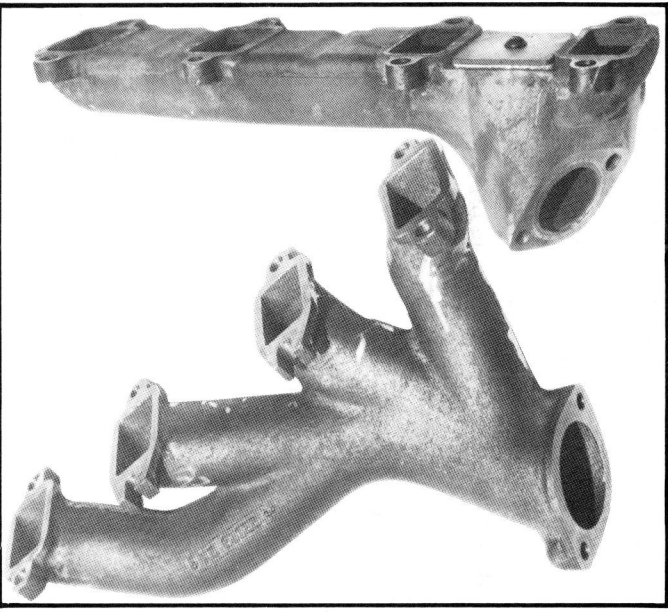

In head interchange, also watch exhaust-manifold bolt pattern. Top to bottom, 16-bolt, 14-bolt, 8-bolt and FT. Unique FT exhaust manifolds can't be interchanged with FE manifolds. Diagonal bolt patterns were required to accommodate manifolds for tight engine-compartments. See text.

First exhaust-manifold improvement was made on 352HP engine in 1960. Improvement over standard manifold (top) is obvious. A number of FE engines were factory equipped with cast-iron headers. Later engines were equipped with longer headers, page 24.

but notches, or local chamfers must be ground into the top of each bore for valve clearance. The process involves cutting crescent-shape notches into the side of the bore, concentric with the valve heads.

Notches are needed only at the top of the bores. They taper down to meet the cylinder wall, about 1/4 in. from the deck surface at the intake valves and about 1/8 in. at the exhausts.

4.05-in. Bore—Obviously, 4.05-in.-bore blocks accept any cylinder head originally installed with 4.00-in. bores. They will also accept Low-Riser 427 and 428CJ heads without notches.

Although notches are not *required* for valve-to-bore clearance with the 427 and 428CJ heads, breathing will be improved if it is done. Otherwise, the valves are shrouded by the cylinder walls. Notching is particularly important if you plan to operate the engine at high rpm.

4.13-in. Bore—All but the Tunnel-Port 427 heads can be used. Medium- and High-Riser heads, with their 2.19-in. intake and 1.73-in. exhaust valves, require bore notching for valve clearance.

4.23-in. Bore—Any big-block cylinder head will fit the 427 block. Keep in mind that all side-oiler blocks were notched to improve breathing. Many engines originally equipped with High-Riser heads were notched for the same reason.

Milling the cylinder heads or decking the block means the valves enter the cylinder closer to the bore edge. If the block deck or heads are milled the notches must be restored or enlarged.

INTAKE-MANIFOLD INTERCHANGE

With all of the cylinder-head interchanging possible, don't make up your mind yet. You still have to consider intake manifolds.

One consideration with the

Four different-style rocker stands are available. Left to right: aluminum stand used on standard FE and FT engines; cast-iron stand for 406 and Low-Riser 427; cast-iron stand for 427 Medium-Riser and Tunnel-Port; and short cast-iron stand used only with 427 High-Riser heads. Aluminum stands should be replaced with stronger cast-iron stands if high-performance camshaft or valve springs are used.

Undersides of rocker stands show oiling provisions for rocker shafts.

Several types of lifters have been used in big-blocks. At left is shell-type solid lifter used in some High-Riser 427s. Center is standard solid lifter. Hydraulic lifter at right is used in most big-blocks.

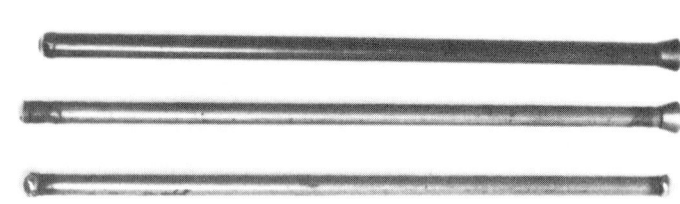

Be sure to match length and style when interchanging pushrods. Adjustable rocker arms require cupped ends, upper right.

head/intake-manifold interchange is the carburetor-heat passage. The heat passage is between the two center cylinders on the FT head. The heat passage on the FE engines angles across the intake manifold between the two front cylinders on the right and the two rear cylinders on the left. If you use FT heads, you must use an intake manifold from an FT engine. Otherwise, the heat passage will be blocked. The same goes for FE heads and manifolds.

The next positive no-interchange items are the Tunnel-Port and High-Riser intake manifolds. High-Riser heads must have High-Riser intakes and Tunnel-Port heads must have Tunnel-Port intakes.

There is a Tunnel-Port-*style* intake manifold that will match the *standard* FE heads with the 1.93-in.-tall intake ports. It was originally designed for the 427 Medium-Riser with the 2.06-in.-tall ports. It can be used with the Low-Riser head if the ports in the manifold are matched to the 2.34-in. head.

So there are basically eight different intake ports: seven rectangular and one round. They will interchange, except for the aforementioned exceptions. See listing of intake-port-heights and applications, page 44.

EXHAUST MANIFOLDS

The major concern is the bolt pattern. Most cylinder heads mount the exhaust manifolds with eight bolts—two for each exhaust port. Holes for these bolts are in a vertical pattern, or one above and one below each exhaust port. All pickup trucks and full-size cars use this design.

FT heads use this basic bolt pattern, with the addition of two holes between the two center exhaust ports. This provides additional exhaust-manifold mounting at the unique FT carburetor-heat passage, which is incorporated into the exhaust manifolds, heads and intake manifold.

During the muscle-car era in the '60s, Ford installed the 390 in lightweight Fairlanes and Comets. The engine fit, but there was insufficient head-to-manifold clearance to install the lower bolts. New manifolds gave sufficient clearance by using a diagonal bolt pattern.

The diagonal pattern required a different head casting to accommodate additional bolt holes. The *Fairlane head* has 14 exhaust-manifold bolt holes with *both* vertical and diagonal patterns. Holes below the center two ports are deleted and the top end holes are lower than on other FE heads.

These cylinder heads were used on 390s installed in Comets, Fairlanes, Cougars and Mustangs. The 427 installed in '66–'68 Comets and Fairlanes used the standard eight-bolt vertical pattern. Holes were provided in the spring towers on the '66 and '67 models for bolt access. The '68 427 Cougars and Mustangs use the 14-hole pattern.

The final exhaust-manifold bolt pattern appeared with the 428 Cobra Jets. It uses the combination vertical-/diagonal pattern found on the 14-hole *Fairlane head,* but has drilled and tapped holes below the two center ports. This makes the 428CJ head a *16-holer,* enabling it to use any FE ex-

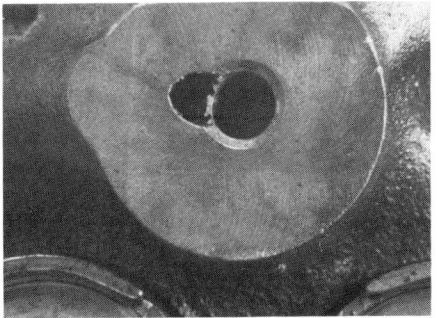
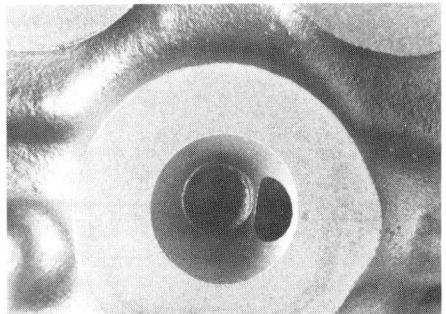

If you change rocker stands, be sure these holes align with the oiling holes in the heads.

Two rocker-arm types used on big-blocks: adjustable rocker with 1.76:1 ratio and non-adjustable with 1.73:1 ratio. Rockers require different pushrod styles.

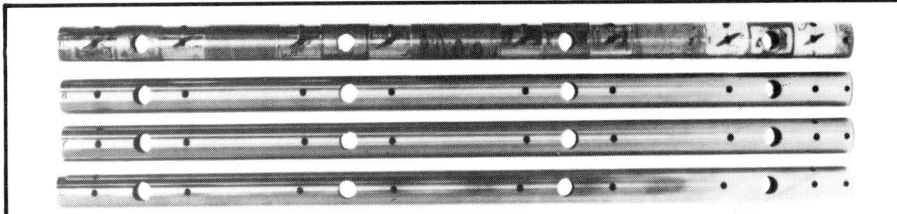

Rocker shaft at bottom is for Medium-Riser and Tunnel-Port heads. Shaft is slightly longer due to wider valve spacing on these heads. Rocker-arm oil holes are also spaced farther from rocker-stand bolt holes, so they'll line up with center of rockers. Second shaft from bottom fits Low-Riser heads with adjustable rocker arms. Third one from bottom is for early-'58 engines with adjustable rocker arms. Shaft at top is for hydraulic-lifter engines.

haust manifold.

FE cylinder heads cast after 1966 had the additional bosses cast into the heads for the combination bolt patterns—but the holes were not drilled and tapped. Exceptions are the Medium-Riser and Tunnel-Port 427s, which retained the eight-bolt pattern.

There are two of other exhaust-manifold considerations. First, the High-Riser and Tunnel-Port heads don't come with a passage cast for intake-manifold heat. Also, if you use FT heads, you must use the FT exhaust and intake manifolds if intake-manifold heat is wanted.

AIR-PUMP AND PCV APPLICATIONS

There is one last item to consider on cylinder-head and intake-manifold swapping. Some '66-and-later heads have drilled and tapped holes above each exhaust port. These holes allow plumbing for air to be injected into each port behind the exhaust valve. The additional air reduces hydrocarbon and carbon-monoxide emissions by allowing more complete burning of the fuel charge.

On engines not requiring air injection, the holes can be plugged using eight C6AZ-6052-A plugs. Engines equipped with the *Thermactor emission system*—air pump—require these passages. Thermactor heads are available with all three exhaust-manifold bolt patterns. Some high-performance heads such as the Medium-Riser, High-Riser and Tunnel-Port were not Thermactor equipped.

Intake manifolds are also made to accept PCV valves and plumbing. Be sure the intake manifold you choose can be adapted if needed. Many high performance intakes, especially over-the-counter 427 manifolds, are not cast to accept PCV fittings.

VALVE-TRAIN COMPONENTS

Rocker Arms—Two types of rocker arms are used: adjustable with a 1.76 ratio and non-adjustable with a 1.73 ratio. The adjustable arm is part B8A-6564-B; the non-adjustable is B8AZ-6564-C.

Adjustable rocker arms are often used in hydraulic-lifter engines because their higher ratio gives a slight increase in valve lift. Adjustability also makes it easier to obtain correct valve "clearance," or adjustment.

Adjustable rocker arms use an adjusting screw with a ball-type end to mate with the pushrod socket. A 0.0023-in.-oversize adjusting screw, C2AZ-6549-A, is available to replace one that becomes loose in the arm.

Rocker Stands—There are four different rocker stands, or supports. The first, C2AZ-6531-B, is used as stand-

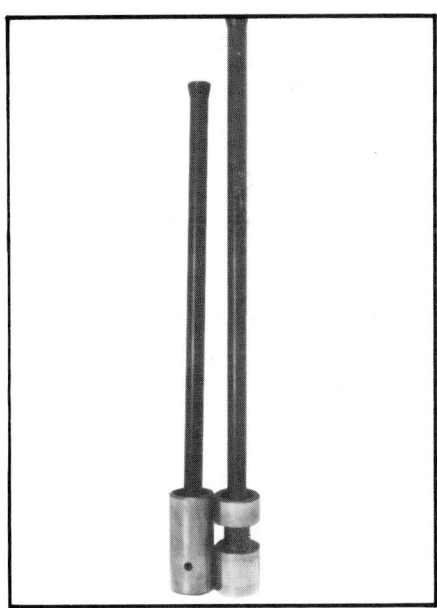

Shell-type lifter and standard solid lifter with same-length pushrod. Pay strict attention to pushrod and lifter length when interchanging or installing new parts.

ard equipment on all engines except the 406 and 427. This aluminum stand should be replaced with a cast-iron one if heavier valve springs and a higher-lift cam are used.

The last three stands came on factory race big-blocks—406 and 427. All are cast iron, with a split on one side and a notch on the other. This makes the stand easier to install and allows the stand to clamp onto the shaft.

The second stand, C3AZ-6531-A, is used on the 406 and the Low-Riser 427. It has the same 0.98—1.02-in.

47

width as the aluminum stand: This stand replaces the aluminum stand with no problems. In fact, the split and notched sides make it easier to install.

The third stand, C3AE-6531-A, is used on High-Riser heads only. It is 1.10-in. wide. Wider stands move the rocker arms slightly to match them to the wider-spaced valves. These stands are shorter to compensate for the taller head—due to taller intake ports. These stands are 1.72-in. tall; all others measure 2.2 in. They are no longer available from Ford.

The last cast-iron stand, C5AZ-6531-A, is 1.10-in. wide, like the High-Riser stand, and 2.2-in. tall. It is used on Medium-Riser and Tunnel-Port heads.

The only 427 not to use cast-iron stands was the '68 hydraulic-lifter 427. It uses aluminum stands.

Rocker-Stand Bolts—All rocker-stand bolts are 3/8-16. Only the length varies. Standard-height stands use three 2-31/32-in.-long bolts and one 3-19/64-in.-long bolt. The long bolt fits the stand with the oil hole, see page 47. The only exception is High-Riser heads—all four bolts are 2-1/2-in. long.

Rocker Shafts—Four rocker shafts are use in the big-block Ford. The first, B8AZ-6563-D, is used on hydraulic-lifter engines. It has 16 oiling holes, eight at the very bottom and eight slightly offset toward the pushrod side. The eight at the bottom have either a spiral groove cut around them or a counterbore. The eight facing the pushrod side are offset about 1/4 in.

The second shaft, B8AZ-6563-A, is for solid-lifter engines and adjustable rocker arms. This shaft was used on the early-'58 engines with solid lifters and later Police engines. It has eight oiling holes on the bottom.

The last two are for high-performance engines. Rocker-shaft C3AZ-6563-A is the same as that used with adjustable rocker arms, but with an additional hole drilled at each end of the shaft as with the hydraulic-lifter shafts. This gives a total of ten holes on the bottom of the shaft. Low- and High-Riser heads used this shaft. But on High-Riser heads, it should be replaced with the following shaft, used on Medium-Riser and Tunnel-Port heads.

The last shaft, C5AZ-6563-A, was first used on the Medium-Riser. It was later installed on the Tunnel-Port heads and should be used on all 427 heads except the Low-Riser. This shaft has the same oil-hole arrangement as C3AZ-6563-A, but rocker-arm oil holes are spaced 1/10-in. farther apart. Wider rocker stands should be used. This shaft is 18.1-in. long; the standard-length shaft measures 18.0 in.

All rocker shafts have the same 0.840—0.841-in. OD, and will fit any of the stands. Bolt holes for the stands have the same spacing, so any shaft can be used on any head. The longer shaft is best choice for for the Medium-, High-Riser and Tunnel-Port heads with their wider-spaced valves.

VEHICLE IDENTIFICATION CODES Engine Code (fifth digit)*			
Year	Engine	Code	Comments
TRUCK *65—78	engine code 4th digit		
64	330 2V	A	medium-duty
	330 2V	M	heavy-duty
	361 2V	K	
	391 4V	9	
65	352 2V	D	
	330 2V	A	medium-duty
	330 2V	M	heavy-duty
	361 2V	K	
	391 4V	9	
66—67	352 2V	Y	
	330 2V	C	medium-duty
	330 2V	D	heavy-duty
	361 2V	E	
	391 4V	F	
68—72	360 2V	Y	
	390 2V	H	
	330 2V	C	medium-duty
	330 2V	D	heavy-duty
	330 2V	U	heavy-duty propane gas
	360 2V	Y	
	361 2V	E	
	361 2V	W	propane gas
	391 4V	X	propane gas
	391 4V	F	
	390 2V	H	
73	330 2V	C	medium-duty
	330 2V	D	heavy-duty
	361 2V	E	
	391 4V	F	
	330 2V	U	heavy-duty propane gas

(VEHICLE IDENTIFICATION CODES CONT'D)			
Year	Engine	Code	Comments
	361 2V	W	propane gas
	391 4V	X	propane gas
	330 2V	3	low-compression
	330 2V	4	low-compression
	361 4V	5	low-compression
	360 2V	Y	
	390 2V	H	
	360 2V	8	low-compression
74	330 2V	C	medium-duty
	330 2V	D	heavy-duty
	361 2V	E	
	391 4V	F	
	330 2V	U	propane gas
	361 2V	W	propane gas
	391 4V	X	propane gas
	330 2V	4	low-compression
	361 2V	5	low-compression
	360 2V	5	low-compression
	360 2V	Y	
	390 2V	H	
	390 4V	M	
75	330 2V	C	medium-duty
	330 2V	D	heavy-duty
	361 2V	E	heavy-duty
	391 4V	F	heavy-duty
	330 2V	U	propane gas
	361 2V	W	propane gas
	391 4V	X	propane gas
	330 2V	4	low-compression
	361 2V	5	low-compression
	360 2V	Y	
	390 2V	H	
	390 4V	M	
76	330 2V	C	medium-duty
	330 2V	D	heavy-duty
	361 2V	E	heavy-duty

(VEHICLE IDENTIFICATION CODES CONT'D)

Year	Engine	Code	Comments
	361 4V	E	heavy-duty
	391 4V	F	heavy-duty
	361 2V	M	propane gas
	361 2V	P	heavy-duty
	330 2V	U	propane gas
	361 2V	W	propane gas
	361 4V	W	propane gas
	391 4V	X	propane gas
	330 2V	4	heavy-duty low-compression Export
	361 4V	5	heavy-duty
	361 2V	7	heavy-duty
	360 2V	8	
	360 2V	Y	
	390 2V	H	
	390 4V	M	
77	330 2V	C	medium-duty
	330 2V	D	heavy-duty
	361 4V	E	heavy-duty
	391 4V	F	heavy-duty
	361 2V	M	propane gas
	361 2V	P	heavy-duty
	330 2V	U	propane gas
	361 2V	W	propane gas
	361 4V	W	propane gas
	391 4V	X	propane gas
	330 2V	4	heavy-duty low-compression Export
	361 2V	5	heavy-duty low-compression Export
78	330 2V	D	heavy-duty
	361 4V	E	heavy-duty
	391 4V	F	heavy-duty
	361 2V	M	propane gas
	330 2V	U	propane gas
	361 4V	W	propane gas
	391 4V	X	propane gas
	330 2V	4	heavy-duty low-compression Export
	361 2V	5	heavy-duty low-compression Export
EDSEL			
58	361 4V	W	Ranger & Pacer only
59	332 2V	B	
	361 4V	W	
60	352 4V	Y	
MERC			
61	352 2V	X	
	390 4V	R	low-compression
	390 4V	Z	
62	352 2V	X	
	390 4V	R	low-compression
	390 4V	Z	
	390 4V	P	Police Interceptor
	406 4V	B	
	406 6V	G	
63	390 2V	Y	
	390 4V	Z	
	390 4V	P	Police Interceptor
	390 4V	9	low-compression
	406 4V	B	
	406 6V	G	
	427 4V	Q	
	427 8V	R	
64	390 2V	Y	Reg
	390 2V	H	Special
	390 4V	Z	
	390 4V	P	Police Interceptor
	390 4V	9	low compression
	390 4V	M	
	427 4V	Q	
	427 8V	R	
65	390 2V	H	C6 automatic transmission
	390 4V	P	Police
	390 2V	Y	manual transmission
	390 4V	Z	
	427 8V	R	
66	390 2V	H	C6 automatic transmission
	390 2V	Y	manual transmission
	390 2V	X	Premium fuel
	390 4V	Z	
	390 4V	S	GT
	410 4V	M	
	427 4V	W	
	427 8V	R	
	428 8V	Q	
	428 4V	P	Police Interceptor
67	390 2V	H	C6 automatic transmission
	390 2V	Y	Manual Transmission
	390 2V	X	Premium Fuel
	410 4V	M	
	427 4V	W	
	427 8V	R	
68	390 2V	Y	Regular Fuel
	390 2V	X	Premium Fuel
	390 4V	Z	
	390 4V	S	
	427 4V	W	
	428 4V	Q	
	428 4V	P	Police Interceptor
69	390 2V	Y	Reg fuel
	390 2V	X	Premium fuel
	390 4V	S	GT
	428 4V	J	CJ w/mech lifters
	428 4V	Q	CJ w/hyd lifters
	428 4V	R	CJ w/hyd lifters and ram air
	428 4V	P	Police Interceptor
70	390 2V	Y	
	428 4V	Q	CJ
	428 4V	R or Q	CJ w/ram air
	428 4V	P	Police
FORD			
58	332 2V	B	
	332 4V	G	
	352 4V	H	
	361 4V	W	Police Power Pack
59	332 2V	B	
	352 4V	H	
60	352 2V	X	
	352 4V	Y	
	352 4V	G	Export
61	352 2V	X	
	390 4V	Z	Includes T-bird & early 390 4V & 6V Specials
	390 4V	R	Export
	390 4V	Q	Special
	390 6V	Q	Special
62	352 2V	X	
	390 4V	Z	
	390 4V	P	Police Interceptor
	390 4V	R	Export
	390 4V	Q	Special
	390 6V	M	Special
	406 4V	B	
	406 6V	G	
63	352 2V	X	
	390 4V	Z	
	390 4V	P	Police Interceptor
	390 6V	M	Thunderbird only
	406 4V	B	
	406 6V	G	
	427 4V	Q	
	427 8V	R	
64	352 4V	X	
	390 4V	Z	

(VEHICLE IDENTIFICATION CODES CONT'D)

Year	Engine	Code	Comments
65	390 4V	P	Police Interceptor
	427 4V	Q	
	427 8V	R	
	352 4V	X	
	390 4V	Z	
66	390 4V	P	Police Interceptor
	427 4V	Q	
	427 8V	R	
	352 4V	X	
	390 2V	Y	3-spd manual transmission Fairlane
	390 2V	Y	automatic transmission Ford
	390 2V	H	automatic transmission Fairlane
	390 4V	Z	
	390 4V	S	GT
	427 SOHC 8V	D	
	427 SOHC 4V	L	
67	427 4V	W	
	427 8V	R	
	428 4V	Q	
	428 4V	P	Police Interceptor auto
	390 2V	H	
	390 2V	Y	3-speed MT Fairlane
	390 2V	X	Premium Fuel
	390 4V	Z	
	390 4V	S	GT

(VEHICLE IDENTIFICATION CODES CONT'D)

Year	Engine	Code	Comments
	427 SOHC 8V	D	
	427 SOHC 4V	L	
	427 4V	W	
	427 8V	R	
	428 4V	Q	
	428 4V	P	Police Interceptor
68	390 2V	H	Automatic transmission
	390 2V	Y	3 spd MT Fairlane
	390 2V	X	Premium Fuel
	390 4V	Z	
	390 4V	S	GT
	427 4V	W	
	428 4V	R	Cobra Jet (Ram Air)
	428 4V	Q	
	428 4V	P	Police Interceptor
69	390 2V	Y	
	390 2V	X	Premium Fuel
	390 4V	S	GT
	428 4V	J	CJ w/solid lifters
	428 4V	Q	CJ w/o Ram Air
	428 4V	R	CJ w/Ram Air
	428 4V	P	Police Interceptor
70	390 2V	Y	
	428 4V	Q	CJ
	428 4V	R	CJ w/Ram Air
	428 4V	P	Police Interceptor
71	390 2V	Y	

CYLINDER BLOCK

PASS CAR & LIGHT-DUTY TRUCK

Engine/Bore

332/4.00
EDC	58 solid lifters two core plugs at front of block, four in rear, three on each side
575063	58 hydraulic lifters
5751091	58—59
B9AE-B	59

352/4.00
EDC	58 solid lifters two core plugs at front four in rear, three on each side
5750603	58 w/hyd lifters
5751091	58—59
B9AE-B	59—60
EDC-B	60 HP
EDC-C	60 HP
CIAE-G	61—62
C3AE-A	63
C3AE-F	63
C3AE-G	64
C4AE-A	64
C5AE-C	65
C6TE-C	66
C6TE-L	66
C7AE	67
C6ME-A	66—67

360/4.05
C6ME	68—76
C6ME-A	68—76
C8AE-A	68—76
C8AE-C	68—76
C8AE-E	68—76
D3TE	73—76 reinforcement webs,

Cylinder Block (cont'd)

Engine/Bore

360/4.05 (cont'd)
D3TE-1	some have SPECIAL cast below number
D3TE-AC	73—76 reinforcement webs
D3TE-HA	
D4TE-AC	74—76 reinforcement webs
D7TE-BA	service block

361/4.05
EDC	58
EDC	59

390/4.05
	61—62 Police solid lifters
C3AE-KY	63 Police solid lifters
C3ME-B	63 Police solid lifters
C4AE-F	64 Police solid lifters
C5AE-B	65 Police solid lifters

390/4.05
CIAE-V	61 HP solid lifters—oil-pressure relief valve at rear of block
C2AE-BC	
C2AE-BE	62 HP solid lifters—oil pressure relief valve at rear of block
C2AE-BR	
C2AE-BS	

390/4.05
C1AE-C	61—62
C1AE-G	61—62
C2SE	62
C3SE-A	63

Cylinder Block (cont'd)

Engine/Bore		
390/4.05 (cont'd)		
C4AE-D	64	
C3AE-AY	65	
C5AE-A	65	
C6ME	66—76	
C6ME-A		
C8AE-A	68—76	
C8AE-C	68—76	
C8AE-E	68—76	
D3TE	73—76 has reinforcement webs, some have	
D3TE-1	SPECIAL cast below number	
D3TE-AC	73—76 has reinforcement webs	
D3TE-HA	73—76 has reinforcement webs	
D4TE-AC	74—76 has reinforcement webs	
D7TE-BA	Service Block	
406/4.13		
C2AE-J	62 solid lifters—oil-pressure relief valve at rear of block	
C2AE-K		
C2AE-V		
C2AE-BD	62—63 cross bolts, solid lifters, oil-pressure relief valve at rear of block	
C3AE-D	63 w/bosses but no cross bolts	
C3AE-V	solid lifters—oil-pressure relief valve at rear of block	
410/4.05		
C6ME	66—67	
C6ME-A	66—67	
427/4.23		
C3AE-M	63	top-oiler
C3AE-AB	63	top-oiler
C4AE	64	top-oiler
C4AE-A	64	top-oiler
C3AE-Z	65	top-oiler
C5AE-A	65	top-oiler
C5AE-E	65	top-oiler
5AE-D	65	side-oiler
C5AE-D	65	side-oiler
C5AE-H	65—66 to 12-9-65	side-oiler
C6AE-B	65—66 to 12-9-65	side-oiler
C6AE-B	66	
C5JE-D	66 irrigation	
C7JE-E	68 irrigation	
C6JE-B	66 marine	
C7JE-A	68 marine	
6AE-C	65—66	side-oiler
C6AE-C	65—66	side-oiler
C6AE-D	66-67	side-oiler
C7AE-A	67	side-oiler
C8AE-B	68 hydraulic lifters	side-oiler
C8AE-H	68 hydraulic lifters	side-oiler
C8AE-A	68 hydraulic lifters	side-oiler
428/4.13		
C6AE-A	66—67	
C6AE-B	66 Police solid lifters	
C6AE-F	66 Police solid lifters	

Cylinder Block (cont'd)

Engine/Bore	
428/4.13 (cont'd)	
C6ME	66—70
C6ME-A	66—70
C7ME	67—70
C7ME-A	67—70
C8ME	68—70
MEDIUM- & HEAVY-DUTY TRUCK	
330MD/3.875	
C4TE-C	64—77
C4TE-F	64—77
C6ME	
C6ME-A	
D3TE	reinforcement webs
D3TE-1	reinforcement webs
D3TE-BA	reinforcement webs
D3TE-BC	reinforcement webs
D4TE	reinforcement webs
D4TE-1	reinforcement webs
C7ME-A	
330HD/3.875	
C4AE-A	64—78 has bosses for cross-bolted mains partially cast from 427 molds, reinforcement webs
C4TE-A,G	reinforcement webs
C5AE-A	
C6ME	
C6ME-A	
D3TE	reinforcement webs
D3TE-1	reinforcement webs
D3TE-BA	reinforcement webs
D3TE-BC	reinforcement webs
D4TE	reinforcement webs
D4TE-1	reinforcement webs
D4TE-BC	reinforcement webs
D7TE-GA	
C7ME	
C8ME	
359, 361/4.05	
389, 391/4.05	
C4AE-A	64—78 has bosses for cross-bolted mains partially cast from 427 molds, reinforcement webs
C4AE	
C4TE-D	reinforcement webs
C4TE-B	reinforcement webs
C6ME	
C6ME-A	
D1TE-AA	
D2TE-C	
D3TE	reinforcement webs
D3TE-1	reinforcement webs
D3TE-EA	reinforcement webs
D3TE-EB	reinforcement webs
D3TE-EC	reinforcement webs
D4TE	reinforcement webs
D4TE-1	reinforcement webs
D4TE-BC	reinforcement webs
D4TE-CC	reinforcement webs
D7TE	
D7TE-CA	
D7TE-DA	
C7ME	
C8ME	

FE CRANKSHAFTS

Engine 332
- Stroke: 3.30
- Journal Diameters — Main: 2.7488, Rod: 2.4384

Casting or Forging Number	Year/Application
EDC	58
5752421	59
COAE-C	replacement

Engine 352
- Stroke: 3.50
- Journal Diameters — Main: 2.7488, Rod: 2.4384

Casting or Forging Number	Year/Application
EDD	58
5752420	59
COAE-A	60 352HP
COAE-D	60 352HP
COAE-B	60–62
C3AE-A	63
C4AE-A	64
C4AE-E	64
C5AE-A	65
C5AE-B	65
C6AE-B	66-67
C6AE-D	66–67
C6AE-D	marine standard rotation
C6JE-H	marine reverse rotation

Engine 360
- Stroke: 3.50
- Journal Diameters — Main: 2.7488, Rod: 2.4384

Casting or Forging Number	Year/Application
2T	68–76
2TA	68–76

Engine 361
- Stroke: 3.50
- Journal Diameters — Main: 2.7488, Rod: 2.4384

Casting or Forging Number	Year/Application
EDD	58 Edsel
5752420	59 Edsel

Engine 390
- Stroke: 3.784
- Journal Diameters — Main: 2.7488, Rod: 2.4384

Casting or Forging Number	Year/Application
CIAE-A	61–62 Ford, 61–early-62 Police to 1-15-62. Mercury & T-bird 61–62
* CIAE-D	61 390HP
* CIAE-H	61 390HP
C2AE-B	62 390HP, Ford late-62–63 Police from 1-15-62 Mercury & T-bird 61–62
C3AE-B	early-63 Ford to 11-1-62,
C3AE-C	early-63 Merc. to 10-15-62, early-63 T-bird to 11-10-62
C3AE-E	late-63 Ford from 11-1-62, late-63 Merc. from 10-15-62, late-63 T-bird from 11-10-62
C4AE-C	64–65, Police
C4AE-D	Police 64–65
C5AE-C	65
C6AE-A	66

(Crankshaft cont'd)

(390)

Casting or Forging Number	Year/Application
C6AE-C	66
C6JE-J	marine reverse rotation
2U	66–73 to ser. Q80,001 truck
3U	73–76 from ser. Q80,001 truck
2UA	73–76 from ser. Q80,001 truck

Engine 406
- Stroke: 3.784
- Journal Diameters — Main: 2.7488, Rod: 2.4384

Casting or Forging Number	Year/Application
* C2AE-D	62–63
* C3AE-D	63

Engine 410
- Stroke: 3.948
- Journal Diameters — Main: 2.7488, Rod: 2.4384

Casting or Forging Number	Year/Application
IU	66–67

Engine 427
- Stroke: 3.784
- Journal Diameters — Main: 2.7488, Rod: 2.4384

Casting or Forging Number	Year/Application
* C3AE-G	63 cast iron
C3AE-U	63 cast iron
* C3AE-V	64 cast iron
* C4AE-B	64 cast iron
C4AE-H	64–65 forged steel, cross drilled
C4AE-AJ	64-65 forged steel
C5AE-C	65–67 forged steel, cross drilled
* C8AE-A	68 cast iron
* C8AE-B	68 cast iron
C9AE-6303-A	NASCAR forged steel, cross drilled has longer rod journals at 1.835 compared to standard 1.755
C5JE-A	marine standard rotation, cast iron
C6JE-B	marine standard rotation, cast iron
C6JE-C	marine standard rotation, cast iron
C5JE-B	marine reverse rotation, cast iron
C6JE-D	marine reverse rotation, cast iron

Engine 428
- Stroke: 3.984
- Journal Diameters — Main: 2.7488, Rod: 2.4384

Casting or Forging Number	Year/Application
C6ME	66–68 428, including 68–early-69 CJ to 11-13-68 number found on #7 counterweight or cheek
IU	66–68 428, including 68–early-69 CJ to 11-13-68 number found on #7 counterweight or cheek
IUB	69–70 police, mid-69 CJ from 11-13-68 to 12-26-68 number found on #7 counterweight or cheek
A	late-69–70 CJ from 12-26-68 #1 counterweight may also have IUB on #7 counterweight
IUA	early-69 SCJ to 12-26-68 #7 counterweight or cheek
B	late-69–70 SCJ from 12-26-68 #1 counterweight may also have IUA on #7 counterweight

FT CRANKSHAFTS

Engine 330	Stroke 3.50	JOURNAL DIAMETERS	
		Main 2.7488	Rod 2.4384

Casting or Forging Number	Year/Application
C4AE-E	64 MD, cast iron
EDD	64 MD, cast iron
2T	65—73 MD, 73—78 HD, cast iron
C6AE-B	65—73 MD, 73—78 HD, cast iron
C6AE-D	65—73 MD, 73—78 HD, cast iron
2TA	74—78 MD, 73—78 HD, cast iron
C4TE-A	64—72 HD, forged steel
C4TE-AF	64—72 HD, forged steel
C6TE-B	66—72 HD, 1-3/8-in.-dia. nose forged crank
C6TE-D	66—72 HD, forged steel
C7TE-A	67—72 HD, forged steel
D2TE-A	67—72 HD, forged steel
D2TE-AA	72 only

Engine 361	Stroke 3.50	JOURNAL DIAMETERS	
		Main 2.7479	Rod 2.4381

Casting or Forging Number	Year/Application
C4TE-A,AF	64—78 forged steel
C6TE-D	66—78 forged steel
C7TE-A	67—78 forged steel
D2TE-A	67—78 forged steel
D2TE-AA	72—78 forged steel

Engine 391	Stroke 3.79	JOURNAL DIAMETERS	
		Main 2.749	Rod 2.4381

Casting or Forging Number	Year/Application
C4TE-B	64—78 forged steel
C4TE-G	64—78 forged steel
C6TE-C	66—78 forged steel
C6TE-E	66—78 forged steel
C7TE-B	67—78 forged steel
D2TE-EA	72—78 forged steel

Engine 427	Stroke	JOURNAL DIAMETERS	
		Main	Rod

Casting or Forging Number	Year/Application
C6JE-A	cast iron industrial
C6JE-F	cast iron industrial

*Crankshaft main journals have 1/8-in.-wide groove for additional lubrication.

CONNECTING-RODS		
LONG ROD		
Wt. (grams)	Number	Application
716—728	EDC	58 332, 352, 361
716—728	EDC-A	58—59 332, 352, 361 60 352
727—740	C1AE-A	60—67 352, 68—76 360, 64—77 330MD, 73—78 330HD
762—774	C0AE-A	60 352HP early to 5-15-60 3/8-in. cap bolts
762—774	C1AE-B	60 352HP late from 5-15-60 13/32-in. cap bolts
716—728	C7TE-A	67—77 330, 73—78 330HD
SHORT ROD		
Wt. (grams)	Number	Application
716—728	C1AE-C	61—early-62 390 Police to 1-15-62, 61—early-63 390 Ford to 11-1-62, 61—early-63 Merc to 10-14-62 (includes T-bird Special)
762—774	C3AE-A	late-63—65 Ford 390 from 11-1-62, late-63—65 390 Merc from 10-14-62, (includes T-bird Special) 64—65 330 HD, 361, 391
762—774	C6AE-A	66—67 390, 410, 428
762—774	C6AE-C	68—70 390, 410, 428 (includes GTs) 66—70 330HD, 361, 391
762—774	D1TE-A	71—76 390, 71—72 330HD, 71—78 361, 391
762—774	C1AE-E	61 390HP
762—774	C1AE-F	61 390HP
762—774	C1AE-G	61—62 390HP
762—774	C2AE-B	62 390HP, 62—63 406, late-62—65 390 Police from 1-15-62
762—774	C2AE-D	62—63 406, late-62—65 390 Police from 1-15-62
762—774	C3AE-C	63—early-65 427 to 3-1-65
762—774	C3AE-F	63—early-65 427 to 3-1-65
762—774	C6AE-D	66—70 428 Police, 68—70 428CJ, 68 427
762—774	C7AE-B	66—70 428 Police, 68—70 428CJ, 68 427
833—845	C5AE-B	65 427 LeMans—original
833—845	C5AE-C	65—66 427 LeMans
833—845	C5AE-D	66—67 427 LeMans—heavier cap added
833—845	C6AE-E	66—70 427, 428SCJ (428SCJ uses shorter cap screws)
986	C7OE A	67—70 427 NASCAR*

*Requires 0.080-in.-wider rod journals.
Requires use of NASCAR crank with 0.080-in.-wider rod journals or standard crank with rod journals widened 0.080 in.
NOTE: 428s must use rod-bolts w/shorter cap screws when using LeMans rods.

STANDARD CYLINDER HEADS

Casting Number	Year/Application	Combustion-Chamber Volume
EDC, EDC-E	58 332, 352*	69.0—72.0
5752142	58—59 332, 352, 361 to code 96A	70.4—73.4
5752143	59 332, 361 from code 96A	70.4—73.4
COAE 6090 C	60 352	72.8—75.8
C1AE A	61—63 352, 390	71.2—74.2
C1SE A	61	71.2—74.2
C4AE	64—65 352, 390	71.2—74.2
C4AE-G	64—65 352, 390	71.2—74.2
C6TE-B,G	66 std. 352, 390, 410, 428 w/o AIR	71.2—74.2
C6AE K	66 352, 390, 410, 428 w/o AIR	71.2—74.2
C6AE AA	66—67 352, 390, 410, 428 w/o AIR	71.2—74.2
C7AE A	67 352, 390, 410, 428 w/o AIR	71.2—74.2
C8AE B	68 std. 360, 390, 428 w/o AIR	67.1—70.1
D2TE AA	72—76 std. 360, 390, 428 w/o AIR	68.1—71.1
D3TE,B,C,E,F		
C6AE A		71.2—74.2
C6AE D	66 std. 352, 390, 410, 428 w/AIR	71.2—74.2
C6AE R	66 352 only	71.2—74.2
R	66	71.2—74.2
C6AE J	66	71.2—74.2
C6AE RVL	66	71.2—74.2
C6AE AB	66—67	71.2—74.2
C8AE A	68 std. 360, 390, 428 w/AIR	67.1—70.1
C8AE H		68.1—71.1
D3TE D		
D5AE		
C6AE L	68 Mercury 390 4V, high-compression 2V	67.1—70.1
C6AE U	68 Mercury 390 4V, high-compression 2V	67.1—70.1
C6OE 6090 R		
C6OE 6090 H[1]	66 390 w/AIR holes Comet, Fairlane	71.2—74.2
C6OE AB[1]	66-67 390 w/AIR holes Comet, Cougar, Fairlane, Mustang	
C8OE A[2]	68 390 w/AIR holes Comet, Cougar, Fairlane, Mustang	67.1—70.1
C6OE Y	GT 4V Fairlane heads	
C6OE AC		
C6OE AA[1]	66—67 w/AIR bosses only Comet, Fairlane	71.2—74.2
C7AE A		
C8OE B[2]	68 390 w/AIR bosses only Comet, Cougar, Fairlane, Mustang	67.1—70.1
C8OE F[2]	69 390 w/AIR bosses	68.1—71.1
XX	69 390	
C6AE AA	66 428 P/C w/o AIR holes	71.2—74.2
	67 428 P/C w/AIR holes	68.1—71.1
	67 428 P/C w/o AIR holes	68.1—71.1
C8AE F	68 428 P/C w/AIR holes	68.1—71.1
	69—70 428 P/C w/AIR holes	68.1—71.1

[1] Uses 14-hole exhaust-flange bolt pattern.
[2] Cast with bosses for 16-hole exhaust-flange bolt pattern.
*machined combustion chambers, first 90 days of production.

Note: Exhaust-valve-stem diameter is 0.3709 for most pickup trucks. Other engines use 0.3695—0.3705-in.-diameter exhaust-valve stems. Intake valves for all engines are 0.3711—0.3718-in. diameter.

All standard cylinder heads use 2.02-in. intake valves and 1.55-in. exhaust valves. Intake port dimensions at manifold-mounting face measure: 2.26 X 1.34 in. for 58—65 & 2.26 X 1.34 in. or 1.93 X 1.34 in. for 66—76. Exhaust ports measure 1.84 X 1.28 in. at manifold-mounting face.

Numbers listed in cylinder-head charts are the casting numbers as they appear on the head. Casting numbers may contain the full number, the prefix and suffix only, or the suffix alone. Intake-valve stems are 0.3715-in. diameter; exhaust-valve stems are 0.3705-in. diameter.

HIGH-PERFORMANCE CYLINDER HEADS

Casting Number	Year/Application	Combustion-Chamber Volume (cc)	Valve Sizes Intake	Valve Sizes Exhaust	Port Dimensions† Intake	Port Dimensions† Exhaust
COAE 6090 D	60 352HP, 61—62 390HP					
C2SE 6090 A	early-62 390HP w/dished piston 62—63 390 6V T-birds	59.7—62.1	2.022/2.037	1.551/1.566	2.34 X 1.34	1.84 X 1.28
	late-62 390HP w/flat-top pistons	65.5—68.5	2.022/2.037	1.551/1.566	2.34 X 1.34	1.84 X 1.28
C4SE 6090 A		65.5—68.5	2.022/2.037	1.551/1.566	2.34 X 1.34	1.84 X 1.28
C2SE 6090 B	early-62 406 4V & 6V to 1-29-62	62.6—64.6	2.022/2.037	1.551/1.566	2.34 X 1.34	1.84 X 1.28
C2SE 6090 C	late-62—63 406 4V					
	62—early-63 6V to 12-17-63	64.0—67.0	2.022/2.037	1.551/1.566	2.34 X 1.34	1.84 X 1.28
C3AE 6090 C	late-63 406 6V has spring-seat cups	56.4—61.0	2.022/2.037	1.645/1.660	2.34 X 1.34	1.84 X 1.28
C3AE 6090 D	early-63 427 Low-Riser to 3-15-63	64.0—67.0	2.022/2.037	1.645/1.660	2.34 X 1.34	1.84 X 1.28
C3AE 6090 G	late-63 427 Low-Riser from 3-15-63	72.8—75.8	2.082/2.097	1.645/1.660	2.34 X 1.34	1.84 X 1.28
C3AE 6090 H	late-63 427 Low-Riser from 3-15-63	72.8—75.8	2.082/2.097	1.645/1.660	2.34 X 1.34	1.84 X 1.28
C3AE 6090 J	64—65 427 Low-Riser, has improved intake port, see text	72.8—75.8	2.082/2.097	1.645/1.660	2.34 X 1.34	1.84 X 1.28
C3AE 6090 K[1]	63 427 High-Riser	73.2—76.3	2.185/2.195	1.723/1.733	2.78 X 1.38	1.86 X 1.30
C4AE 6090 F[1]*	64 427 High-Riser (used early 65) has improved intake port, see text	73.2—76.2	2.185/2.195	1.723/1.733	2.78 X 1.38	1.86 X 1.30
C5AE 6090 F[1]* R	65—67 427 Medium-Riser	88.0—91.0	2.185/2.195	1.723/1.733	2.06 X 1.38	1.78 X 1.30
C7OE 6090 K[1]*	427 Tunnel-Port	88.0—91.0	2.25[3]	1.723/1.733	2.17 X 2.34	1.78 X 1.30
C8AX[1]*	427 Tunnel-Port	88.0—90.0	2.25	1.723/1.733	2.17 X 2.34	1.78 X 1.30
C8AE 6090 J, C8AE-N	427 Low-Riser 68 Fairlane & Montego to 2-1-68	72.7—75.7	2.082/2.097	1.652/1.660	2.34 X 1.34	1.84 X 1.28
C8WE 6090 A[2], C8AE-N	427 Low-Riser 68 Cougar & Mustang to 2-1-68	72.7—75.7	2.082/2.097	1.652/1.660	2.34 X 1.34	1.84 X 1.28
C8OE 6090 H[4]	68 428CJ	72.8—75.8	2.082/2.097	1.652/1.660	2.34 X 1.34	1.84 X 1.28
C8OE 6090 N[4]	68—70 428CJ & SCJ	72.8—75.8	2.082/2.097	1.652/1.660	2.34 X 1.34	1.84 X 1.28

*machined combustion chambers

†Unlike those on FT and standard-FE heads, intake- and exhaust-port dimensions vary widely on high-performance FE heads. See text.

[1] Valve length is 5.45 in. All other heads use 5.36-in.-long valves.

[2] Uses 14-hole bolt pattern at exhaust flanges. Same as 390GT.

[3] Early heads use 2.185/2.195-in. intake valves.

[4] 16-hole bolt pattern

FT CYLINDER HEADS

Casting Number	Year/Application	Volume (cc)
C4TE-C	64 330MD w/squared combustion chamber	78.5-81.5
C4TE-E	330MD w/squared combustion chamber	78.5-81.5
C7TE-D	330MD w/rounded combustion chamber and bosses for AIR	78.5-81.5
C4TE-A[1]	64 330HD, 359, 361, 389, 391	78.5-81.5
C4TE-D[1]		78.5-81.5
C7TE-C[1]		78.5-81.5

[1] Uses 1.500—1.510-in.-diameter exhaust valves with 0.4338—0.4348-in.-diameter stems. Others use 1.500—1.515-in.-diameter exhaust valves with 0.3701—0.3708-in.-diameter stems.
All FT engines use 1.745—1.760-in.-diameter intake valves.
All 359, 361, 389, 391 have intake-valve-seat inserts
All 330HD, 359, 361, 389, 391 have exhaust-valve-seat inserts

STANDARD PUSHRODS (6565)

Part Number	Length (in.)	Pushrod Ends*	Application
B8A-A	10.70	B & C	58 Ford with solid lifters
B8A-C	9.35	B & C	58 Ford with hydraulic lifters 60 Ford 352HP 61—62 390HP (full-size Fords) 61—65 390 Police 62—63 406
B8AZ-C	9.35	B & C	63—67 427 66 428 Police
COAE-J	9.59	B & B	59 332 & 352 Ford 60—62 352 & 390 (including Thunderbird) 63 Ford to 12-23-63 (except Thunderbird)
C8AZ-A	9.59	B & B	late-68—78 FT engines from change L11 late-68 390 & 427 from 1-15-68 change L10 late-68 428 change L5 68—70 428 CJ & SCJ late-68 360 & 390 trucks from change L2 69—76 360 & 390 including 69—71 car
C4TZ-B	9.62	B & B	64—early-68 FT engines to change L11 late-64 352 & 390 from 12-23-63 65—67 352 & 390 66—67 410 & 428 early-68 390 & 427 to 1-15-68 change L10 early-68 428 (except CJ) to change L5 65—67 352 pickup early-68 360 & 390 pickup to change L2 late-58 361 Edsel 59 332 & 361 Edsel 60 352 Edsel
C4TZ-D	9.56	B & B	early-58 361 Edsel
C3SZ-A	9.77	B & B	63 Thunderbird 390 early-64 Thunderbird 390 to 12-23-63

*B—ball end C—cup end
All pushrods used with hydraulic lifters and non-adjustable rockers are available in 0.060-in. under- or oversize lengths.

CHAPTER 4
Teardown

The first engine-rebuilding step is to disassemble the engine. Though it seems easy, don't rush through teardown. Remove each part *carefully*. Later you'll inspect the parts to determine those that need replacing or reconditioning.

Keep track of the parts that obviously need machining. This way you can keep a rough tab on upcoming costs and remember to take all the parts to the machine shop on one trip.

When disassembling the engine be sure not to damage any machined surfaces, such as camshaft lobes or cylinder bores. Keep them covered with grease or oil to prevent rusting. A machined surface will rust in minutes once it has been cleaned.

ENGINE STILL ON HOIST
Drain Oil and Coolant—Use care while the engine is still hanging on the hoist. With a drain pan or bucket underneath, remove the drain plug and drain the oil. While it's draining, gather some materials to drain the coolant.

Get a large sheet of plastic and four boards. The plastic sheet can be a garbage bag split lengthwise and opened up. Replace the oil-drain plug and lay the large sheet of plastic under the engine. Lay the four boards under the edges of the plastic to form a square. This makes a wide basin to catch coolant as it drains from both sides of the engine.

Lower the engine to reduce splash. To drain coolant from the block, open the petcocks or remove the two block drains behind the engine mounts. Early big-blocks were equipped with petcocks. They were replaced with plugs on most FE engines.

If the coolant doesn't flow when you open the drains, poke a wire or small screwdriver through the hole. Rust or dirt will often block the opening. After draining the coolant, lift the plastic sheet by its four corners and pour the coolant into the oil drain pan.

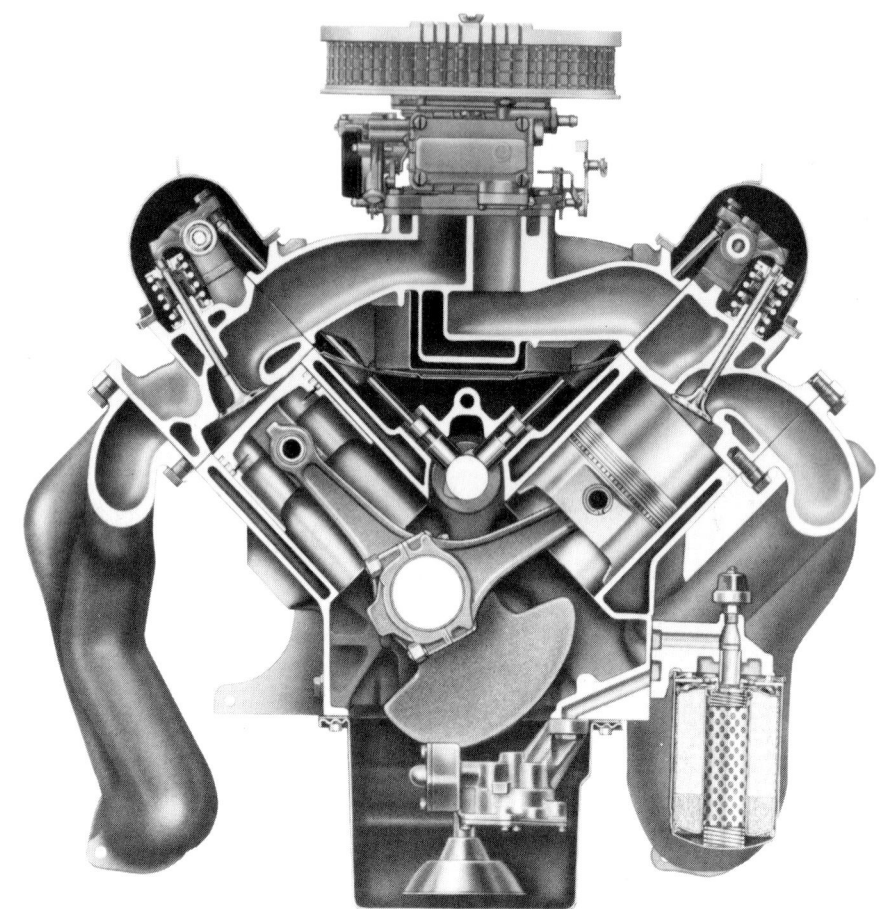

Experienced eye will see difference between this illustration and 63-1/2 427 8V it depicts. Vacuum-diaphragm location on left indicates carburetors face "forward" in normal position. Production-427 8V's had carburetors "backward," with secondaries at front. See photo, page 70. Photo courtesy of Ford Motor Co.

EXTERNAL HARDWARE
If you have an impact wrench you can remove many of the parts with the engine hanging from the hoist, but I don't recommend it. If you don't have impact tools, forget it. You'll waste a lot of energy wrestling with the engine while breaking bolts loose. Dancing with a 700-lb engine is a dangerous proposition, so wait until it's firmly supported.

Engine Stand or Workbench?—There are two common ways to support an engine during a rebuild: with an engine stand or with a workbench.

An engine stand can support a complete engine, and provide access to nearly all components during teardown and assembly. A stand has three or four wheels so you can move the engine easily.

Its best feature is that the engine can be rotated 360° around its longitudinal axis and locked in several positions. This makes it easy to work on any part of the engine.

The one problem is that engine stands are expensive. Equipment-rental outlets will rent them for a weekly or monthly fee. Consider the rental an excellent investment.

Another way is to use a sturdy 30-in.-high bench. Be sure the bench can support the weight of a complete

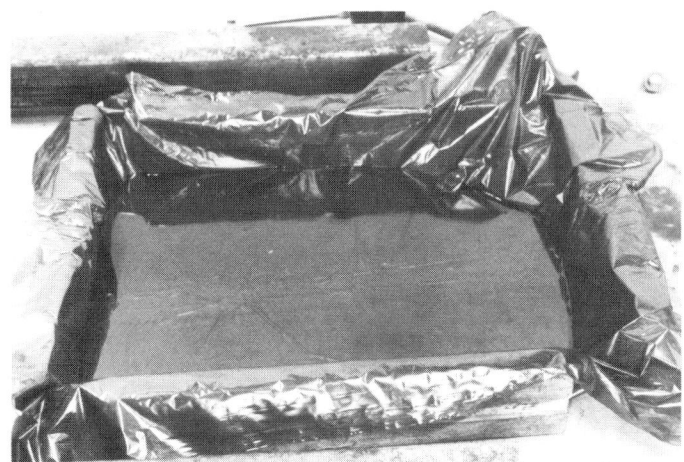
Four boards and a plastic bag are all that's needed to drain coolant.

Remember to allow good access when supporting engine for teardown. Also, the engine must be stable. V-notch cut into rear wooden block gives added stability. Engine hoist is also left attached as long as feasible for added safety.

Jack handle inserted under engine plate keeps flexplate from turning while breaking bolts loose. If engine uses flywheel, use flywheel turner, or a screwdriver wedged against ring-gear teeth and engine dowel. You can also insert a Phillips screwdriver or punch through pressure-plate bolt holes and under bottom edge of block.

Flywheel turner works on flexplates too. Engine was partially disassembled before I removed flexplate.

Seepage from lifter oil-gallery plugs is evident on this engine plate. A set of high-quality expansion or screw-in plugs correctly installed will prevent this.

engine and will allow you to turn the engine on its side. Don't feel that it's necessary to use an engine stand. Many professional engine rebuilders prefer benches.

You can tear down the engine without a bench or an engine stand. You'll need some way to support the engine while removing the external hardware. The engine should be level and high enough to keep it from resting on the flywheel or flexplate. I use two large wood blocks, one with a V-notch cut into it.

The V-notch fits in the shallow rounded part of the pan near the flywheel. The other block fits squarely under the oil-pan rail at the front. The block with the V-notch stablizes the engine while you work on it.

Lower the engine until the oil pan rests on the supports. Leave the hoist connected to prevent the engine from rolling over while you work.

Pressure Plate and Clutch—On manual-transmission engines, mark the pressure plate and flywheel so they can be reinstalled in the same position relative to one another. Use either a center punch or paint. Mark the flywheel and pressure plate at their outer edges. If the flywheel is resurfaced you won't lose the mark.

Remove the six pressure-plate-to-flywheel bolts. Alternate between the bolts, loosening each bolt one turn at a time. This will prevent the pressure-plate cover from bending from uneven loads. Once the disc is loose, remove the bolts.

Place your hand under the assembly or through the center to support it when removing the final bolts. Be prepared to catch at least 30 lb.

Keep your fingers off the clutch-disc surface. Grease on any friction surface of the clutch will make it slip, grab or chatter. If you do get grease or dirt on the surface, use lacquer thinner or brake-cleaning fluid to clean it. Watch the fumes from this cleaner—they are highly toxic.

Flywheel or Flexplate—If the vehicle has a manual transmission, it will

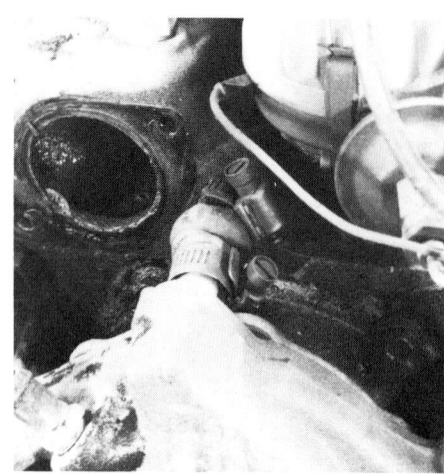
Crimped bypass hose is common on big-blocks. This usually occurs while installing intake manifold or water pump. Correct hose ID and length, plus some soapy water to lubricate, will prevent this.

Break harmonic-balancer bolt loose with socket and long breaker bar. You'll need some way to keep crank from turning as you loosen bolt—here I used magic third hand. You can also use mallet to smack bar sharply to break bolt loose. Pull on bar while striking it.

To remove harmonic balancer you need this special puller—sometimes called a *steering-wheel puller*. Reinstall harmonic-balancer bolt without its washer to protect threads in crank nose. If harmonic balancer is stubborn, tighten puller against bolt. Rap end of center puller bolt sharply with brass mallet. Shock should break balancer loose.

Early-style canister-type fuel pump uses flare nut at pump-to-carburetor line. Use flare-nut wrench and support fitting at pump with open-end wrench.

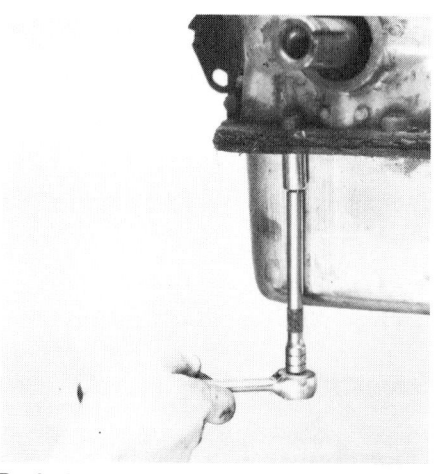
Don't forget four oil-pan-to-timing-cover bolts. Take them out *before* removing timing cover.

Remove remaining timing-cover bolts, damper key, spacer and cover.

have a flywheel. An automatic transmission uses a flexplate. Some 410 and 428 flexplates have an extra-large ring gear, but it's still a flexplate.

Use a breaker bar to remove the six bolts from the flexplate or flywheel. To break the bolts loose, hit the breaker-bar handle with a large soft mallet. Don't let the crankshaft turn. You can hold the crank by inserting a screwdriver through the engine plate and flywheel or flexplate.

Another method is to hold the crank with a 15/16-in. wrench at the vibration-damper bolt. You'll be turning the bolt in the "loosen" direction. But if the vibration-damper bolt doesn't come loose, this will work.

Don't let the flywheel or flexplate fall when you remove the last bolts. Most flexplates have a load-spreader ring under the bolts. Remove the ring, the flexplate and the engine plate. The engine plate serves as the bellhousing-to-engine mating surface. It is usually located by two dowels in the block's rear face.

Water Pump—Loosen the hose clamps on the thermostat-bypass hose between the pump and the intake manifold. Next remove the four water-pump bolts or nuts. In some cases, the bolts have studs attached —the "nut" flats are in the center, with a threaded shank at each end.

Note the location of the studs and bolts. They must go back in the same place. This will prevent having to swap studs around during assembly after half of the accessories are in place.

The water pump weighs about 15 lb, so be careful when removing the last bolt. Step back when you pull the pump off—there may be some coolant left in the block.

Pulley or Vibration Damper—Remove the three bolts holding the pulley to the vibration damper. The damper is a three-piece assembly—the hub, an outer ring and a rubber cushion bonded between them. It's very easy to damage the damper, so don't pry on it.

With cover off, remove oil slinger from crank and store with spacer and key. Note direction slinger faces before removing.

Scribe or punch some matchmarks on base of distributor and intake manifold to simplify reinstallation. On this particular engine I removed distributor after water pump and front cover. Order makes little difference.

Remove emission plumbing and vacuum hoses before removing carburetor.

The damper may have an additional pulley groove cut into it. Don't try to remove it—it's part of the damper. On some engines this groove is covered by the outer pulley.

Remove the damper bolt with a breaker bar and a 15/16-in. socket on the bolt. A sharp blow to the bar with a soft mallet or large block of wood should break the bolt loose.

You may have to install two flywheel bolts in the flywheel or flexplate flange and bridge them with a pry bar to hold the crank. If the bolt is really stubborn, wait until you remove the oil pan. Then you can wedge a block of wood between the block and a crankshaft counterweight.

Once the bolt is removed you will need a *vibration-damper* or *harmonic-balancer puller*, often called a *steering-wheel puller*. If you don't have one, buy one. The price is about the same as renting one. And the cost of a puller is considerably less than that of a damper ruined by removing it the wrong way.

To use the puller, reinstall the damper bolt *without* the washer. Mount the puller to the balancer. Make sure the puller is parallel to the face of the damper and all three bolts are threaded in the same amount. Turn the center puller bolt against the damper bolt until the damper pulls free—don't drop it.

Pull the Woodruff key out of the crank snout. Slide the damper spacer off the crank. The spacer goes through the timing cover.

Fuel Pump—Remove the fuel line between the pump and carburetor/s. Use a tubing wrench on the flare nut and support the pump or carburetor fitting with an open-end wrench. If the line begins to twist, spray the fitting with penetrating oil. Work the nut back and forth to break it loose.

Next remove the two fuel-pump bolts and remove the pump. Seal the fuel-pump openings with tape or rubber plugs. This prevents the remaining fuel from escaping and the diaphragm from drying out while the pump is stored.

Timing Cover—Remove the bolts holding the timing cover. Some models have bolts with different lengths and diameters, so note which bolts go where. Be sure to remove the four oil-pan bolts that thread into the bottom of the cover.

If the cover is stuck, check that *all bolts* are out. A buildup of oil and dirt could be hiding a bolt head. When all the bolts are out, tap the cover with a soft mallet or *gently* pry it loose at the top. The cast-aluminum cover breaks easily. With the cover off, remove the oil slinger from the crank snout.

Distributor—Matchmark the distributor housing and intake manifold with a scribe or center punch so the distributor can be aligned easily during installation. Make sure the coil-to-distributor wire has been removed.

Remove the bolt and hold-down clamp at the base of the distributor. Rotate the distributor slightly when lifting it to break the seal between the housing and intake manifold, then pull it straight up.

If you hear something fall inside the engine when you remove the distributor, don't worry. It was the oil-pump drive shaft. It will be in the oil pan.

Emission Plumbing and Carburetor—Make a sketch or take some pictures of the emission plumbing *before* you remove it. Keep as much of it connected as you can. This will ease the load on your memory during installation.

If you haven't removed the carburetor linkage and springs, take special note here too. There are dozens of ways to hook up the linkage, but only one is right. Remove the carburetor nuts, the carburetor/s and any remaining linkage.

Valve Covers and Rocker-Arm Assemblies—With a 7/16-in. wrench, remove the five valve-cover bolts. If you can't lift the covers with your fingertips, they may be glued on. If you have to pry them off, pry only from the intake-manifold side. The intake side is less prone to leaks if you distort the cover's gasket flange.

A sharp rap with your trusty rubber mallet may also loosen the cover. If you don't smack it too hard you won't have to worry about distorting it.

With the covers off you'll find four bolts holding each rocker-arm assembly in place. Break all four loose. Removing any one, two or three bolts before the load of the valve springs is released can bend the rocker shaft. Starting at one end, loosen each bolt two turns until the shaft is unloaded.

Repeat this until all the spring force

61

On rare occasions, valve covers will lift off. If not, tap them with a mallet rather than prying them off. This keeps gasket surface from being bent.

Loosen rocker-shaft bolts a few turns at time to prevent bending rocker shaft. If a valve or two are open at one end, the shaft will be under considerable load. To keep load to minimum, turn crankshaft over until number-1 cylinder intake valve has just closed, then go 45° past TDC. You can also align timing marks on crank and cam sprockets to find TDC.

Remove rocker shaft, baffle and pushrods for each bank. Store parts together. If you plan to save lifters and rocker arms, keep pushrods marked as to their position.

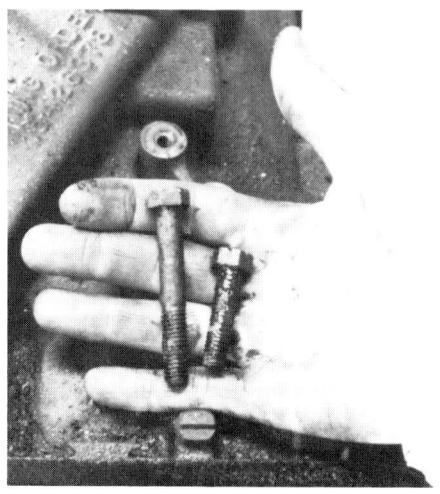
Remove intake-manifold bolts. Two shorter bolts go at rear. All big-block intake manifolds use 10 bolts, except Tunnel-Ports. They use 12.

is released. Remove the bolts and lift the rocker assembly off as a unit. Keep the bolts with each rocker assembly.

Next remove the oil baffle and the pushrods. The sheet-metal baffle routes the oil to drain holes at the ends of the cylinder head or to holes near the pushrods. Oil then drains to the oil pan preventing it from puddling in the heads and draining down the valve guides.

Intake Manifold—With a 9/16-in. wrench, remove the intake-manifold bolts. There are two short and eight long bolts; the short ones are at the rear. Tunnel-Port manifolds use 12 bolts; four short and eight long.

To break the intake manifold loose, insert a pry bar in the space between the intake manifold and cylinder head. If the manifold is stubborn, drive a thin wedge—a screwdriver blade or dull chisel—between the head and intake manifold. Wedge it behind number-4 or -8 cylinder above the water passage in the head.

Next lift the intake manifold off—it's heavy. Cast-iron manifolds each weigh 75 lb, aluminum manifolds 30 lb. Ford manuals show lifting the intake with a hoist—a good indication of its weight. Under the intake manifold is a sheet-metal baffle over the lifter valley. Pry the baffle loose at one side and lift it out. Store it where it won't be damaged.

Lifters—If you plan to reuse the cam and lifters, *be sure to keep the lifters in order.* The best way is with two egg cartons, with each space marked INTAKE or EXHAUST, with the cylinder number. Place a lifter in the appropriate space as you remove it.

If the engine interior is fairly clean you may be able to pull the lifters out with your fingers. Stubborn ones can be pulled out with the tip of a screwdriver hooked under the inner edge of the lifter. Or you could use blunt-tipped needle-nose pliers. A special lifter-removing tool wedges itself inside the lifter body for a better grip. If you can't remove the lifters by any of these methods, wait until you've removed the camshaft and push them out.

Exhaust Manifolds—I explained an easy way to loosen the exhaust-manifold bolts during engine removal, page 14. If you followed those instructions, you should have little trouble removing the exhaust manifolds.

If the bolt-locking tabs are still in the way, bend them flat, away from the bolt heads. To remove the bolts, soak them with penetrating oil for a few minutes. Use a six-point socket or wrench to prevent rounding the bolt head. If you have an impact wrench or driver, use it to remove these bolts.

If a bolt breaks off, leave it for now. The problem can be corrected when you recondition the heads.

Be sure to remove the dipstick tube from the left side of the engine. Pull it straight up once you remove the exhaust-manifold bolt holding it.

Engine Mounts—Use a 5/8-in.

Previous rebuilder was too enthusiastic with gasket sealer—this manifold refused to come loose and had to be wedged off. Use an old gasket scraper or screwdriver to remove the intake manifold. Don't do this with an aluminum intake manifold. Do the following instead.

Pry bar used in open areas between manifold and head will usually break loose stubborn intake manifold and without risk of damage to mating surfaces. Be sure to use head as fulcrum, not intake manifold—especially aluminum intakes.

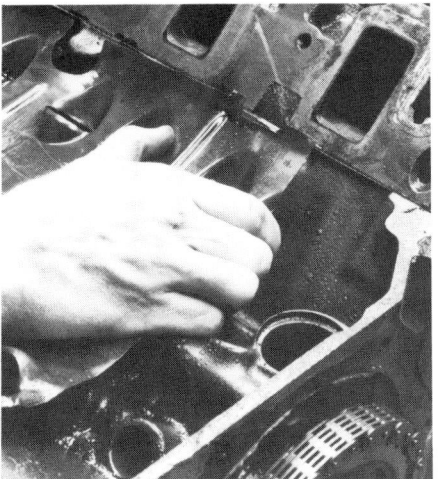
Pry lifter baffle loose at one corner and remove.

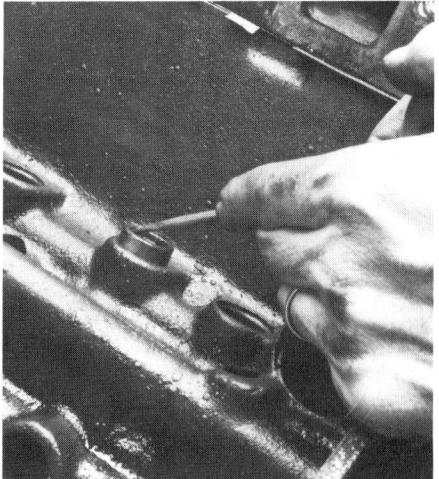
If you plan to save lifters, be careful removing them. Don't score outer surface; use a magnet or catch the inside edge of the lifter with a screwdriver. Keep them in order.

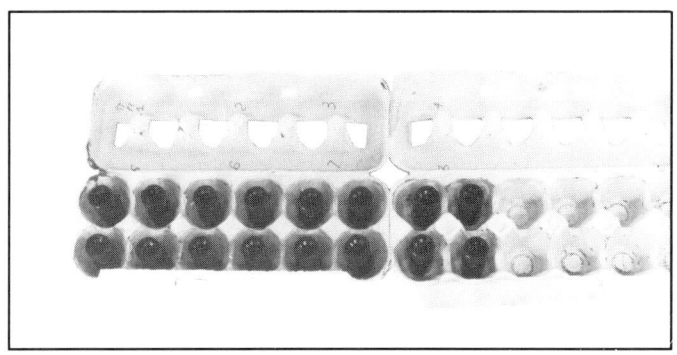
Simplest way to keep lifters in order is with an egg carton. Label compartments with cylinder number and mark sides LEFT or RIGHT.

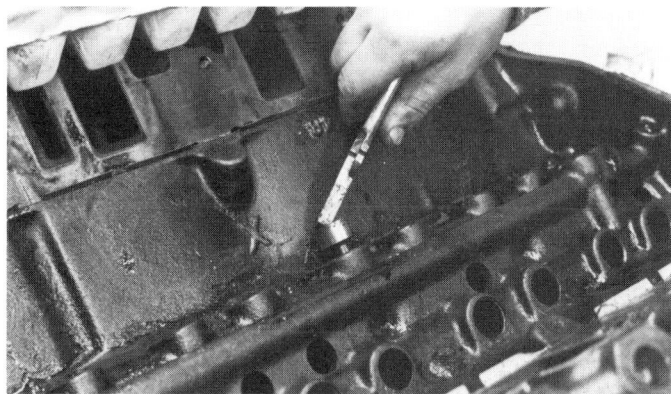
If lifters are stubborn, grip edge with pliers or Vise-Grips and pull them out. This usually scores outside surface, so don't plan to reuse lifters. If they still won't come out, wait until camshaft is removed. You can then push them out the bottom of their bores.

Sharp-eyed observers will note solid lifters used in hydraulic-lifter block. Hydraulic-lifter oil galleries were left open, so "high-performance" modification caused low oil pressure. Engine lasted less than 2000 miles.

If you plan to use solid lifters in a hydraulic lifter block, the lifter galleries must be blocked. See modification, pages 31 and 32.

wrench or socket to remove the two engine mounts. Take special note on the position of the mounts. Mark them **DRIVER** and **PASSENGER** or **LEFT** and **RIGHT** and draw a sketch to show how they mount.

The left engine mount usually has a heat shield to protect it from exhaust heat. The shield is especially important if the exhaust-pipe outlet is in the center of the manifold rather than at the rear.

Oil Filter and Adapter—Remove the oil filter and the four bolts holding the oil-filter adapter to the block. If you plan to replace the oil-pressure sending unit—it's a good idea—remove it before removing the adapter. That way, you won't have any trouble holding the adapter.

Cylinder Heads—Each cylinder head is retained by 10 bolts torqued to about 100 ft-lb. Use a breaker bar with a 3/4-in. socket to break them loose. Remove the bolts and lift the head straight up from the block.

If the cylinder heads are stuck they may have to be pried loose. Run two bolts into the block, through the bolt holes above the end exhaust ports. These will prevent the head from popping loose and landing on the floor, breaking the head or your foot.

Next insert a piece of wood into an intake port and lever the head up. If that doesn't work, give it a couple of sharp blows on each side with a large plastic or rubber mallet.

If both methods fail, you will have to wedge the head off. Use a *wide* chisel between the head and the block on the intake side. Drive it in until the head breaks loose. Remove the safety bolts and lift the head off.

Be sure the engine is well supported before you lift the head, otherwise the engine will roll in the direction of the other head. Check whether the locating dowels stayed with the heads. If they did, remove the dowels and put them away.

Remove the Ridge—As the piston rings travel up and down the cylinder bore, some of the bore surface is worn away. Because the rings stop short of the top of the bore, the topmost part of the bore receives no wear.

Remove exhaust manifolds. I rebuilt two engines during the course of writing this book. This one wouldn't run, so I couldn't use the hot-engine trick on manifold bolts, page 14. Two broken bolts resulted. If this happens to your engine, have bolts removed during head reconditioning.

Remove oil-filter adapter and sender. Store with other aluminum parts that must be cleaned and can't be hot tanked.

Break all cylinder-head bolts loose. Remove all but the ones above each end exhaust port.

Break head loose carefully. Ratchet handle placed in intake port works well. Use block of wood with aluminum heads.

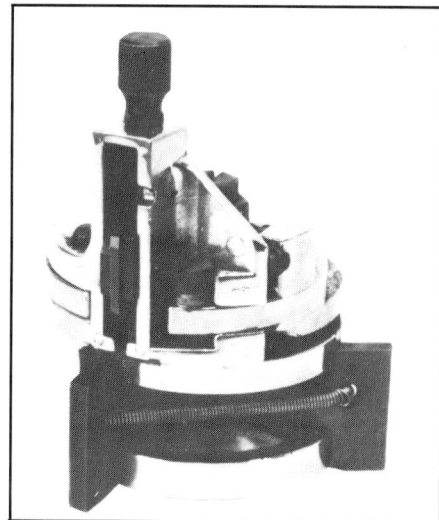
Use ridge reamer to remove ridge at top of bore. This is must if you plan to reuse pistons. It prevents piston-ring-land damage during piston removal.

Use care when removing ridge. Do not cut below the wore bore surface.

This unworn section of bore is called a *ridge*. You can see the ridges or feel them with your fingernail. If you plan to reuse the old pistons, remove the ridges. Otherwise, the piston-ring lands may be damaged when the pistons are removed.

To remove the ridges you need a *ridge reamer*. A ridge reamer has two major parts, a cutting tool to remove the ridge, and a body that pilots the tool in the cylinder bore. There are several types of ridge reamers.

If you buy or rent one, look for a ridge reamer with a strong base and an adjustable cutting head. The cutting head should be sharpened only on the angled part of the blade, not the flat part. The flat part pilots in the cylinder bore. Also, get a reamer that is self-feeding, so it moves up the bore each time it makes a revolution.

To fit the reamer in each bore, you have to turn the crankshaft to move some of the pistons down. Get a slide-handle—T-handle—to turn the reamer. This keeps the cutter straight in the bore.

Read the directions with the reamer. Cut the ridge only to the worn part of the bore—no farther. If the tool has a spring-loaded, adjustable cutting head, set the depth of the cut with the spring force released. The cutter should be extended as far as the spring will push it.

If you set the cutter and then release the spring, you will cut too far into the bore. At best you will then have to bore the block. At worst you may ruin the block.

Start the cut at the *bottom of the*

64

Best thing to do with timing chain and aluminum/nylon cam sprockets—throw them away! If you don't want to spend money on replacements, measure timing-chain deflection. Take all slack out of left side of chain. Lay straightedge across right side of cam and crank sprockets on right side. Push chain away from straightedge. There should be less than 1/2-in. deflection midway between sprockets. Even without straightedge, it's easy to see this chain is shot.

To remove timing-chain assembly, remove cam bolt and fuel-pump eccentric. Use two screwdrivers to work crank sprocket forward until cam sprocket can be pulled free with chain. Slip off crank sprocket.

ridge, just below the ridge's lowest point. This prevents angling the cutting head and cutting a step in the bore that extends below the ridge.

TURN THE ENGINE OVER

If the engine is on a stand, you won't have much problem turning it upside down. If it is on a workbench, you will have to muscle it over. When working on a bench, remove anything that that can scratch the block-deck surface before rolling the engine over.

Oil Pan—The oil pan has 20 bolts. The four long ones that threaded into the timing-chain cover were removed. Remove the remaining bolts and the pan. If the pan is stubborn, a few sharp raps against its side with a soft mallet should break it loose.

If you have to pry the pan off, pry at the sides, not at the front or rear. The front and rear must be kept distortion-free to seal the rear-main-cap and timing-cover joints. The aluminum timing cover may break if unevenly loaded by a distorted pan.

Oil Pump and Pickup—Remove the two bolts from the oil-pump pickup and remove the pickup. Thread the bolts back into the pump. Some engines, particularly four-wheel drives and large trucks, have a long pickup tube. The tube bolts to the center main-bearing cap for support.

Remove the two oil-pump bolts and lift the pump out. If the pump drive shaft didn't come out with the pump, pull it out. Store all the oil-pump parts together.

Timing Chain and Sprockets—If the engine has more than 50,000 miles on it, replace the timing chain and sprockets. On low-mileage engines, check for slack in the chain. Place a straightedge against the side of the chain and sprockets as pictured. Push the chain away from the straightedge, midway between sprockets. If the chain deflects 1/2 in. or more, replace the chain *and* sprockets.

The best way to double-check the sprockets is to compare them with new ones. A worn crank sprocket will have distinct grooves in the sprocket-teeth faces. If the cam sprocket is worn, the teeth will be rounded or chipped, especially if the sprocket has nylon-coated teeth. When in doubt, replace the entire assembly—chain and both sprockets.

To remove the assembly, remove the bolt from the cam-sprocket center, then the fuel-pump eccentric and washers. Slide oil slinger off of the crankshaft nose. Store all the parts together.

With two screwdrivers, pry on the back of the crank sprocket with even pressure. Work the sprocket forward until the chain tightens. Grab the camshaft sprocket and chain and pull forward; the sprocket should come free of the cam.

The chain can now be slipped off of the crank sprocket. Pull the crank sprocket off the crank. With the crank sprocket off, remove the Woodruff key. A tap on one end of the key with a punch and hammer will lift the opposite end. Pry it up from the crankshaft and remove the key with your fingers or a pair of pliers.

PISTON-AND-ROD ASSEMBLIES

With the ridges removed, you still have a bit more preparation to do before removing the piston-and-rod assemblies. First, make sure each rod and cap is marked with its cylinder number. Numbers should be stamped on the right side of the rod and cap on the right cylinder bank and stamped on the left for the left cylinder bank.

If the rods and caps are not marked you can stamp them or mark them with an electric engraver. Electric engravers work fine; just be sure the marks are legible.

Connecting-Rod Side Clearance—Before you remove the connecting rods, check them for side clearance. Connecting-rod side clearance is rarely a problem, but it's easy to check them now. If the crankshaft journals are machined—or the rods or crankshaft replaced—the clearance must be rechecked.

The rod faces must be resurfaced if there is too-little clearance. If the clearance is too large you'll have to replace the rods or crankshaft. If you don't check it now you'll have to

Removing oil pickup with pump attached to block is easier than separating them later.

Late 428CJs are equipped with a windage tray. Remove tray and put it out of harm's way.

Remove oil pump and set it aside. Pump will be inspected later.

Before removing piston-and-rod assemblies, check connecting-rod side clearance. Clearance should be 0.010—0.020 in., with maximum of 0.023 in. If clearance is out of spec, rods or crank must be reconditioned or replaced.

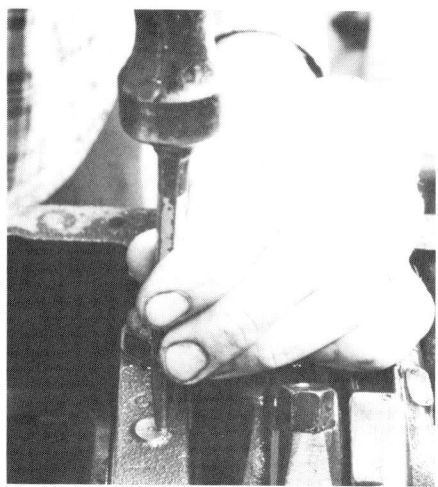
Rods, rod caps and main-bearing caps must all be marked for position and direction. Rods will "fit" any rod journal, and number-1, -2 and -4 main-bearing caps can be accidentally switched. Use center punch and mark main caps on raised bosses with appropriate number of dots—one for number-1, two for number-2, and so on.

Main-bearing-cap direction is usually marked from factory with number or letters stamped in same direction, or with cast-in triangles pointing toward front of engine. A main-bearing cap installed backwards almost guarantees bearing and crankshaft failure.

check it during assembly. Then you risk needing new rods, crankshaft or additional machine work after everything else is done.

Check the clearance between the rods with feeler gages. Standard FE engines and 330 FTs should have 0.010—0.020-in. clearance, with a maximum allowance of 0.023 in. High-performance FE engines should be 0.018—0.028 in., with 0.025 in. recommended. FT engines, except the 330, should be 0.010—0.030 in.

Pistons And Rods—Remove the rod nuts or bolts and cap. The next thing to do is to protect the cylinder walls. Except for the LeMans and NASCAR connecting rods, the rod-bolt threads are exposed when the rod cap is removed. These bolts can damage a cylinder bore or crankshaft journal as the piston-and-rod assembly is removed.

To prevent damage, cover the rod-bolt threads. Plastic sleeves made for this purpose are available, but pieces of 3/8-in. rubber fuel line work as well.

You should be able to push each piston-and-rod assembly out. If you need to tap one out, use a rubber-covered hammer handle or length of wood—a broomstick is fine. Turn the crankshaft to bottom dead center—BDC—and then *tap* on the bearing surface of the rod.

Be ready to catch the piston as it leaves the cylinder bore. Once the rings leave the bore, the piston and rod will fall out.

Save the rod-bearing inserts. Mark them as to their cylinder or keep them in their respective rod for safekeeping. The inserts will help you tell whether a crankshaft journal needs attention, or if a rod is bent or twisted. Reinstall the cap and its nuts or bolts on the connecting rod to prevent damage or mismatching of parts.

Remove the Crank—With the timing chain off and the piston-and-rod assemblies removed, you are ready to remove the crank. Before removing the main-bearing caps, read this

One engine I tore down was a complete nightmare. Renumbered rod, (left) and its journal (right). Crankshaft journal was already 0.020-in. undersize; spun bearing reduced it additional 0.050 in. Crank was junked, though damaged journal could have been welded and remachined.

To remove piston assembly, rotate crank until piston is near BDC. Remove nuts or capscrews.

Tap rod down enough to remove cap.

entire section. Check that the caps are marked for number and direction. They must go back in *exactly* the same position.

Number-1, -2 and -4 main-bearing caps will all fit the same *registers*—the machined locating bosses in the block. Caps number-1, -2, -3 and -4 can be reversed in their registers. Mispositioned main-bearing caps can ruin a crankshaft. Unless the caps can be returned to their original positions, the block must be *align bored*—the main-bearing bores machined.

Most main-caps have numbers on them and a method to show direction. If the caps have triangles on them, the point of the triangle should be toward the front of the engine and the base toward the rear.

If the caps have circles cast in them, there will usually be an arrow or the letter A. Either should have the point toward the front of the engine. All the caps should be numbered with the top of the number toward the front of the block.

If the caps are in any other order or are out of sequence, mark them so you can return them to that sequence or position. If you find the bearings are worn evenly and the crank is OK, the block may have been align bored with the caps in that position. In this case they can be reinstalled in the same order in which you found them—they'll be OK.

If it were my engine, I would rather put the caps in their correct positions and have the block align bored or honed to true up the main-bearing bores. If the caps are not marked at all, mark their position—bearing number *and* direction.

On the 406 and 427 cross-bolted mains, remove the cross bolts before loosening the vertical bolts. With the caps marked, break the bolts loose with a socket and breaker bar.

The caps should pull out without too much trouble, but you may have to persuade some with a soft mallet. Tap from one side and then the other while lifting straight up on the cap.

The number-5, or rear main-bearing cap can be particularly stubborn. It is wider and the side seals wedge it between long side surfaces of the block.

If you do any prying, do it with a wooden handle or pry only on the side of the cap. Place the bolts in the cap holes and gently rock the cap back and forth to work it out.

Once you have all the main caps off, place the bearings back into their caps for future reference. Lift the crankshaft out of the block. Be sure you have a firm grip on it—it weighs about 65 lb. Stand the crankshaft on its flywheel flange, somewhere where it won't be knocked over. Remove the bearings and place them with their caps for future reference.

Now that you've finished with the bottom end, roll the engine back over. If you're working on a bench, lay the block on its side.

CAMSHAFT REMOVAL

If you weren't able to remove the lifters before, do it now. Remember, if you think you might be able to reuse the cam and lifters, keep the lifters in order. Some lifters may pull right out of their bores, others need some coaxing and some are downright rebellious. If they pull out, fine. If not you may want to try the lifter-removing tool I mentioned earlier.

Place protectors over bolts to prevent them from damaging the crank-journal and bore surfaces. Use hammer handle or wooden dowel to push piston-and-rod assembly out of the bore.

Catch piston as it pops out. Once piston is out, loosely install cap on rod.

With main-bearing-cap bolts removed, extra dose of patience will help remove number-3 and -5 caps. They usually give some extra resistance. Best way to remove these caps is to work them back and forth while pulling up.

Cast-iron crank weighs about 60 lb; forged-steel one near 70 lb. Lift crank straight up and be careful not to bump journals when removing it. Store crankshaft where it will not be damaged.
Crankshaft should be stored standing on its flywheel flange or hung from a wire or rack. If placed on its side, crank may bend. Be sure to coat crankshaft journals with WD-40 or CRC 5-56 to prevent rust.

To remove core plugs, drive them into the water jacket. I use an old 1/2-in.-drive extension. Grab the edge of the core plug with a pair of Channellock pliers and lever it out.

If you still can't get some lifters out, lift each one up as far as you can. Grab it with a pair of Vise-Grips and pull it out. Don't clamp too tightly on the lifter; the metal can shatter. Don't worry about damaging the lifters if they are this hard to get out. You will need new ones.

If you can't get the lifters all the way out, put the timing sprocket back on the cam. Turn the camshaft one revolution to get all the lifters to their highest position, then push the lifters out as far as you can with a screwdriver. Just be careful not to damage the camshaft lobes or lifter bores.

Once you have the lifters this far out of their bores, go ahead and remove the camshaft. With the cam out, you'll have enough room to drive the lifters out. Use a wooden dowel or the female end of a 1/2-in.-drive extension to avoid damaging the lifter bores.

Thrust-Plate Removal—If the camshaft is an early thrust-button type, you can now remove the camshaft. Later types use a thrust plate retained by a pair of Phillips-head screws.

You will need a #4 Phillips-head screwdriver or tip to remove the screws. They can be removed with Vise-Grips, but you will need the screwdriver or tip for installation, so you may as well get one. Once the thrust plate is off, reinstall the camshaft sprocket to help steady the camshaft as you remove it. You are now ready to remove the camshaft.

If the engine has a lot of varnish or carbon buildup, cam removal is easier said than done. The camshaft-bearing journals are wider than the cam bearings. Consequently, varnish builds up on the journals on each side of the

Remove two #4 Phillips-head screws and camshaft thrust plate. If you don't have correct bit, break screws loose with Vise-Grips and remove with flat-blade screwdriver. You will need a #4 Phillips-head screwdriver for installation.

With cam out, remaining lifters can be driven down out of their bores. If you are still trying to save lifters, use female end of 1/2-in.-drive extension as driving tool. One-piece construction means solid lifters might still be saved. Hydraulic lifters will be gummed up internally—replace them.

To prepare block for cleaning, remove core plugs. Allen-head oil-gallery plugs may be stubborn. Light rap with mallet should loosen them. If you plan to hot tank block, remove cam bearings.

bearing, just as it does on the lifters. Buildup in effect makes the journal OD larger than the cam-bearing ID.

If the buildup is heavy you may have to turn the camshaft while working it slowly back and forth—in and out—to get it out of the bearings. This is the reason for installing the cam sprocket.

Once the cam pulls out of the bearings, be careful not to let it drop. The cam lobes will score the bearing surfaces. Once the cam clears the first two sets of journals, you won't have any problems. But watch your fingers. The cam lobes are *very* sharp and will cut like a knife!

CORE-PLUG REMOVAL

If the engine has cup-type core plugs, place a blunt tool in the center of each plug and knock it into the block with a hammer. Remove the plug by grabbing its edge with a pair of Channellock pliers. Rest the heel of the pliers against the side of the block and lever the plug out.

If the engine has the screw-in core plugs, I *strongly* recommend leaving them in. They are about as difficult to remove as cup-type plugs are easy. If you decide to remove them, use a 1-in. hex-drive adapter or Allen wrench.

If you can't remove these plugs easily, *leave them in*. Forcing them can overstress the side of the block and possibly crack it. If the engine is fairly clean inside, don't worry about leaving the plugs in. If the engine is full of rust or slime, have the block hot tanked twice.

If you still want to remove these screw-in plugs, you will find them easier to remove after they have been warmed up in the hot tank or the area around the plug heated with a torch.

Oil-Gallery Plugs and Camshaft Plug—Remove the rear cam plug by driving it out from the inside. If you plan to save the cam bearings, be careful not to hit the rear bearing.

With pressed-in oil-gallery plugs, punch a small hole in them with a prick punch or small center punch and pry them out. Or you can drill a small hole in each plug, insert a sheet-metal screw and pull the plug out with pliers.

The galleries in the top of the block are drilled completely through and plugged at both ends. On these, you can remove the plugs at one end and knock the plugs out the other end with a 5/16-in.-or-smaller-diameter rod. A 5/16-in.-or-smaller rod is required because the lifter galleries are drilled from both ends during manufacturing. The two galleries do not always line up perfectly.

If the engine has screw-in oil-gallery plugs with an Allen or square head, leave them in if the block is fairly clean. If you have to remove them, warm the area around each plug with a propane or oxyacetylene torch to free it. Use a *tight-fitting*, 5/16-in. Allen wrench.

With all core plugs out, check for leftover sand in the water jackets. Remove any sand or slag. This will increase the efficiency of engine's cooling system. Most slag can be removed with a triangular file. While you are removing rough-casting edges check the water-passage holes in the block deck. Remove any rough edges in the holes.

CAM-BEARING REMOVAL

If you plan to remove the cam bearings—you'll have to if the block will be hot tanked—you will need some special equipment. There are two types of cam-bearing removers. The first is a threaded rod with a collet—it pulls the bearings out. The other is simply a rod and mandrel—it is used to drive the bearings out.

Most engine rebuilders use the rod and mandrel—driver type—to remove the bearings, so I'll discuss this type. The first thing to do is find the correct mandrels. The mandrels are stepped so that the smaller diameter fits the bearing ID while the step butts against the bearing. The larger diameter must fit through the bearing bore in the block.

Drive out the front bearing first, then simply continue through the block. Be careful not to damage the bearing bores and don't hit the bearings hard—especially when one is almost out. You may drive it into the next web or bulkhead of the engine block.

69

CHAPTER 5
Short-block reconditioning

This chapter covers inspection, cleaning and some reassembly. Take your time, particularly with the inspection and reconditioning. If you are in doubt about any part or assembly, retrace your steps. The time to correct any problem is *now,* before the engine is reassembled.

Cleanliness and accuracy are the major priorities during the remainder of the rebuild. By building the short block right the first time, you avoid costly, time-consuming mistakes.

Some tools mentioned in this chapter are not found in most home toolboxes. Unless you have a machine shop do some of the work, purchase or rent these tools. If you know you are only going to rebuild one engine, rent them. If you think you may rebuild more engines later, consider purchasing some of the less-expensive tools, such as a dial indicator or vernier caliper.

CYLINDER-BLOCK CLEANING AND INSPECTION

Pre-Inspection—Before you go through the trouble of cleaning the block, you should make sure it's worth reworking. Some blocks may require sleeves for one or more cylinders. You may find a cracked block or gouged cylinder wall. Repair of these items could cost more than a replacement block.

Let the value or rarity of the block dictates your course of action. Some high-performance blocks—427s, for example—are very hard to obtain. So it makes sense to spend more on one of these than on a 390 or 428. If the block is cracked, check with your machine shop for an approximate price to repair it.

Checking the Block-Deck Surface—Another area to check before cleaning is the block deck. Before scraping the

Although overshadowed by the 427, 428 Cobra Jet was one of the best high-performance engines Ford ever produced. Photo courtesy of Ford Motor Co.

head gaskets off the block, check them for signs of leakage or burned spots. This can give you a clue as to where the deck surface may be warped or *etched.*

Etching, or *notching,* is where the block is actually burned away by leaking combustion gasses. It usually occurs when an engine is driven a long time with a blown head gasket. Etching can be repaired, but adds to the block-reconditioning cost.

Clean the head-gasket or deck surface and use a straightedge to check it for warping. Lay the straightedge lengthwise across the cylinder bores and check the gap between it and the block with feeler gages. The largest gap should be no greater than 0.007 in.

Recheck the gap with a 6-in. straightedge. The gap should measure no more than 0.003 in. Finally, recheck the deck, holding the straightedge diagonally across the deck surface.

If you find a gap, double-check to be sure the block is at fault. Hold the feeler gage in place and turn the straightedge over; the gap should remain the same. If the gap varies, the problem is with the straightedge—it's not straight. Junk it and get another.

70

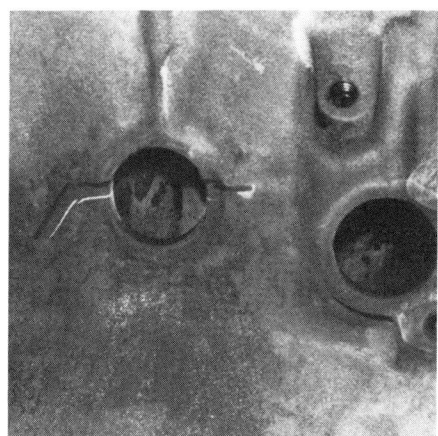
Before you get too far into the rebuild, check block for cracks or other flaws. Crack near core plug could be repaired, but adds to reconditioning cost. Keep cost in mind when reconditioning parts.

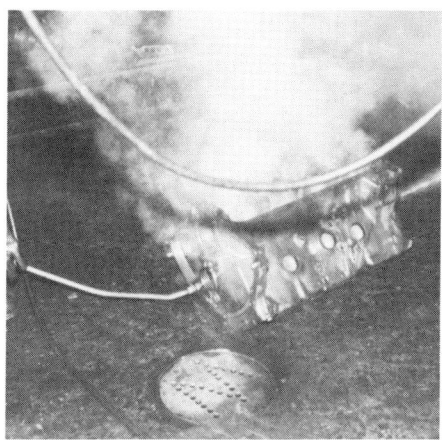
Hot tanking is easiest way to get block thoroughly clean. Take cylinder heads, crankshaft, connecting rods and other pieces of steel or cast-iron along for cleaning, too.

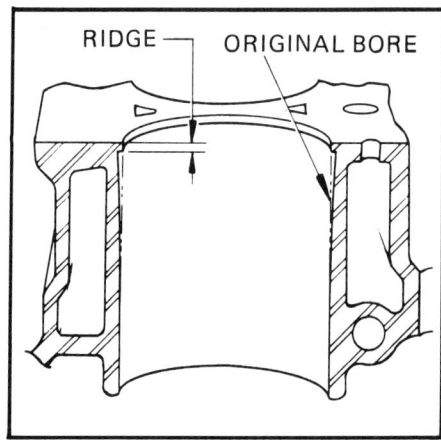
Exaggerated section of worn cylinder bore. Bores wear in a taper: more at top than at bottom. Short unworn section at top of bore is ridge, directly above top-compression-ring-travel limit. Drawing by Tom Monroe.

If you find problems with the block, check the cost of reconditioning it before having it machined. The cost may be less to buy a used block. Depending on the application, you may even want to buy a new one.

CLEANING THE BLOCK

Now you are going to find what's under all the grease and dirt. Do your cleaning where making a mess won't be a problem. If the engine is on a stand, place some newspaper under it. If you have it on the floor, slip some cardboard under the block to keep from staining the floor.

A cup-type wire brush in a drill works well on cylinder blocks. Be sure to wear goggles to protect your eyes from the flying wire strands and engine gunk.

Scraping the Gaskets—Start cleaning by scraping off all gasket material, including the oil-pan surface, water-pump outlets, timing-cover and intake-manifold surfaces. A wide *gasket scraper,* designed for scraping gaskets, works best. Don't use screwdrivers or putty knives. Use care not to gouge any of the machined surfaces, especially the cylinder walls.

As long as you have a mess, scrape out any accumulated sludge in the block. Most of the sludge will be in the lifter valley. Heavy sludge is caused by low operating temperatures or infrequent oil changes—both easily avoided.

Thread Chasing—Chase the threads on the main-bearing- and head-bolt holes. Use a 1/2-13 *bottoming tap*—1/2-in. diameter with 13 threads per inch. A bottoming tap has full threads cut square to the end, while a *taper tap*—as the name implies—tapers to the end.

You will be surprised how much gunk has accumulated in the bolt holes—especially the head-bolt holes. This can cause inaccurate torque readings when you reassemble the engine. Oil the threads afterwards to prevent rust, even if you plan to hot tank the block.

Last of all, clean the lifter bores. To clean the lifter bores use a 10-gage-shotgun brush and one length of cleaning rod.

Insert the rod in a low-speed drill and clean the passages with solvent or penetrating oil. Be sure to wear goggles when doing this. Clean only long enough to remove the buildup—you don't want to gouge the bores or increase lifter-to-bore clearance.

This won't be the last time you clean the block. Clean it after any machine work or any time you notice dirt on it. To keep it clean, keep the block covered while it's not being worked on.

This cleaning is done to allow you to inspect the block and to keep the machining equipment and cutting oil from being contaminated. It also makes final cleaning before assembly much easier.

You can clean the block by a number of methods. It can be hot tanked in a caustic or non-caustic solution, or pressure washed with water or steam. You can also use detergent, scrub brushes, miscellaneous rifle or toothbrushes and elbow grease.

Hot Tanking—The best way to have the block and other cast-iron or steel

Sleeve is simply replaceable cylinder bore. If one bore can't be reconditioned by boring, consider having it sleeved. If more than one cylinder requires sleeving, consider a new or used block. What you should pay for reconditioning depends on value—or rarity—of block. Photo by Tom Monroe.

parts cleaned is in a hot tank. The problem is that solutions used in most hot tanks are caustic. They will eat aluminum, plastic, paint or basically anything that is not cast iron or steel.

The caustic solution will also destroy bearing-insert material, such as camshaft bearings. If you are hot tanking the block, you must remove and reinstall new cam bearings.

The major expense with new cam bearings is in installation, not material. Although cam bearings receive little wear, you're better off replacing them with new ones if you have the block hot tanked.

Do yourself a favor and have all the parts hot tanked at the same time—intake and exhaust manifolds, crankshaft, rods, pushrods, oil pan, rocker cover, bolts and so on.

Disassemble the cylinder heads if you want them hot tanked too.

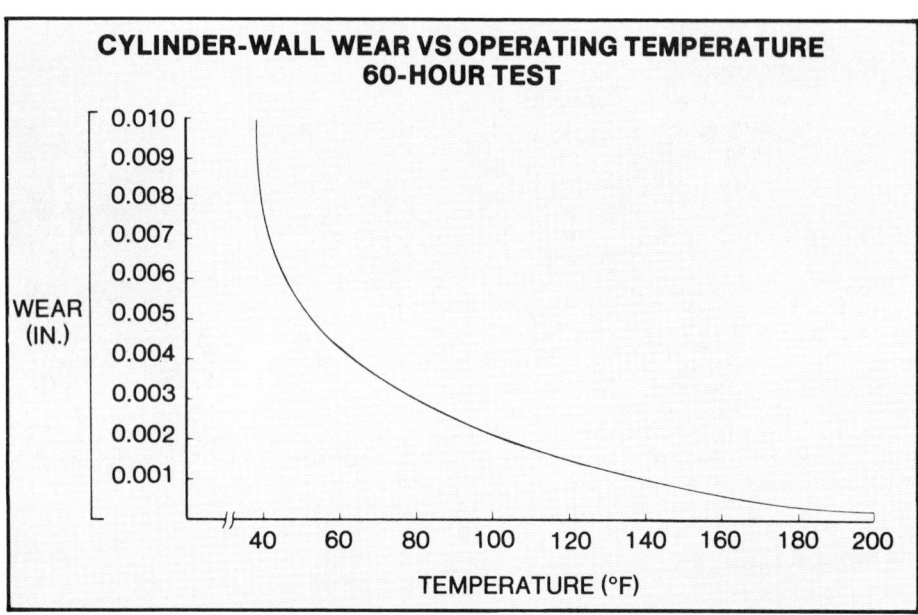

Thermostat does more than provide quick warmup. Bore wear increases dramatically at lower engine operating temperature. Data courtesy Continental Motors.

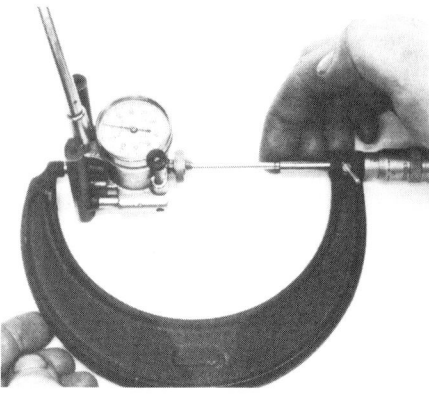

Dial bore gage is easiest and most-accurate way to measure bore wear. Outside micrometer being used to set dial bore gage.

Remember, don't put aluminum parts—timing cover, aluminum intake manifolds, cylinder heads or rocker-arm stands or other soft metals—into the hot tank. Hot tanking will also destroy connecting-rod bushings, but they should be replaced anyway.

The machine shop may charge you one price for everything, for one load or by individual pieces. Weigh the cost against scrubbing the parts at home. Some parts, such as the air-cleaner housing or valve covers, are probably worth doing yourself.

When you take parts to the machine shop, they may be stamped or marked with a number. This number stays on them all the way through the machine work. If you want to, mark all the parts yourself to be sure you get *your* parts back.

Small parts—bolts or pushrods—should be wired together to keep them from being lost. Also put small parts into a wire basket to help circulation of the cleaning solution. Don't send chrome-plated parts: Clean them by hand.

Once the parts have been hot tanked, pay close attention to the machined surfaces and oil passages. The machine shop will usually spray water through all the oil passages and water jackets of the block and then wash all the tanked parts.

All parts and surfaces that are not going to be machined right away must be protected from rust. Spray a light coat of oil on the parts after they are dry. Use a good-quality *water-dispersant* oil, such as CRC or WD-40. A water-dispersant oil displaces water—most other oils trap it underneath, against the metal's surface.

To speed up the drying, use paper towels and wipe the surfaces down. Pay particular attention to the machined surfaces to prevent rust. Keeping parts rust-free will also make it easier to paint them.

Hand Washing—If you don't want the block hot tanked you need to clean the block another way. Some car washes allow engine cleaning; some even have special spray degreasers. Before you go, prepare the block and other parts by scraping off excess dirt, grease and gasket material.

Spray the engine with degreaser. With most solutions it helps to scrub. Wait a bit and rinse it off.

Repeat the washing if necessary. When the block is clean you should be able to run your finger along a non-machined surface without picking up any grime.

Oil Passages—No matter which method you use on the block, clean the oil passages individually. Use a rifle brush or tie a rag to a wire and pull it through the galleries. Do this several times. High-pressure water or steam is great if you have it. If the block has been sitting for some time you may need solvent to assist in the cleaning.

Drying and Preserving—Once the block and passages have been cleaned, dry them. Compressed air is best, though paper towels will do. Once dry, spray the machined surfaces with CRC or WD-40. Another precaution is to spray oil in the bolt holes, but they will have to be washed out again before final assembly.

If clean block is going to sit for a while, cover it with a plastic bag. This will ensure the block will be clean and ready to go when you are.

CYLINDER-BLOCK: INSPECTION & RECONDITIONING

This is the final inspection of the cylinder block to determine what is needed to restore it to as-new or better condition. I say "better" because you will essentially be *blueprinting* the engine—bringing it up to its design specifications. You can follow this engine through its reconditioning and be sure *everything* is right—that's hard to do on an assembly line.

To make this final inspection you will need some special measuring tools. Depending on the bore of the engine, you will need 3—4-in. or 4—5-in. inside and outside micrometers. A telescoping gage—*snap gage*—can take the place of the inside micrometer.

If you have access to a dial bore gage, by all means use it. It simplifies measuring bore wear, or bore taper. You will also need 0—1-in. and 2—3-in. outside micrometers, a straightedge and feeler gages.

Checking Bore Wear—To a large degree, bore wear determines how expensive the rebuild will be. If the block needs boring, it will need new pistons: an expensive proposition.

There are four common ways to measure bore wear: with a dial bore

Check for taper by moving dial bore gage from bottom to top of bore. Reading is bore taper.

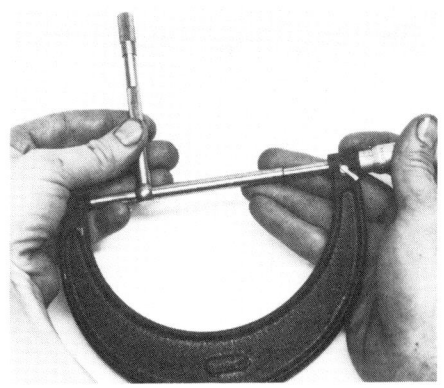
Bore wear measured with telescoping gage and micrometer is more time-consuming, but still accurate. Measure bore with telescoping gage, then check gage with micrometer. Measure in several places around and down the bore. Difference between maximum and minimum readings is bore wear, or taper.

gage; with an inside micrometer; a telescoping gage and outside micrometer; and with a piston ring and feeler gages. The last method, with a piston ring and feelers, is the least accurate but it's readily available to any rebuilder.

Bore Wear—Most bore wear occurs where the piston and its rings contact the cylinder bore. This wear is not even from top to bottom or around the circumference of the bore. Most cylinder-bore wear is caused by the compression rings. The varying load these rings apply causes varying wear in the bores.

Compression rings are designed so that pressure in the combustion chamber helps them seal by forcing them outward against the cylinder wall. The higher the pressure the better the seal.

The load on the rings is greatest immediately after the cylinder fires. As the piston travels down the bore, the load decreases rapidly. This varying load causes higher wear at the top of the bore, tapering to less wear as the piston travels down the bore.

As the crank rotates, the angle between each rod and its piston changes as the piston moves up and down the bore. This puts friction on alternate sides of the cylinder wall, particularly on the *thrust side*. The thrust side of a bore is the side the rod points toward during the power stroke—the right side on a counterclockwise-rotating engine as viewed from the rear. This friction causes each cylinder to wear in an oval shape.

This effect is not nearly as noticeable as ring-induced wear. The piston skirt's main job is to stabilize the piston in the bore—not to provide any sealing. As a result, the loads on the skirt are minor compared to those on the rings. The piston skirt is also better lubricated than the rings.

If you look at the cylinder bores in a used block, note that the crosshatching from the hone still exists at the bottom. The top of the bore is shiny. There is a *ridge* of unworn bore above where the top piston ring stops at TDC. This wear pattern is normal.

Measuring Bore Taper—Taper is the difference between the largest and the smallest diameters of each bore. The largest diameter is normally found just under the ridge, at the top of the piston-ring travel. The smallest diameter is found at the bottom of the bore where there is the least wear.

To find maximum wear, measure the bore just below the ridge. If you are using a dial bore gage or inside micrometer, set it perpendicular to the crankshaft. Compare this to the measurement at the bottom of the bore. Also measure in between to make sure you find the greatest distance. The difference between the measurements determines how much work will have to be done to the cylinder.

The cylinder with the most taper—largest diameter—determines what has to be done to the engine. If the taper in one cylinder is enough to require reboring, the rest of the cylinders must be rebored to the same size.

There is one exception: If the taper in the rest of the cylinders is acceptable, but one cylinder has been gouged, cracked or badly worn, you can sleeve the one cylinder and bore it to match the others. This saves the cost of new pistons.

You may notice when measuring the taper, the end cylinders (-1, -4, -5 and -8) will have the most wear. This is because they operate cooler than the rest. See the graph on page 72.

Cylinder-bore wear is higher at lower temperatures. This means the outside cylinders wear more. Measure the wear on the end bores, or just feel the depth of the ridges if you haven't removed them. The ridge will be highest near the outside of the engine—at the front on number-1 and -5 cylinders and at the rear on -4 and -8.

Taper alone won't tell you how much boring is required to restore a cylinder to as-new condition. This is because as a cylinder bore wears, the center line of the bore shifts.

The amount a cylinder must be bored also depends on the type of boring equipment used. If the boring machine centers on the block deck and the bore, it will consider the worn bore as on-center and remove only enough material to make the bore round again.

If the boring bar centers on the center line of the crankshaft and bore, it will remove enough material to bring the bores back to their true centers. If the center line has shifted, more metal must be removed than what is indicated by taper alone. This is why you should have the cylinders bored *before* buying new pistons.

Dial Bore Gage—Using a dial bore gage is the easiest and most-accurate way to measure taper in a bore. But the gage will only measure taper. You will need a micrometer to check the actual diameter of the bore.

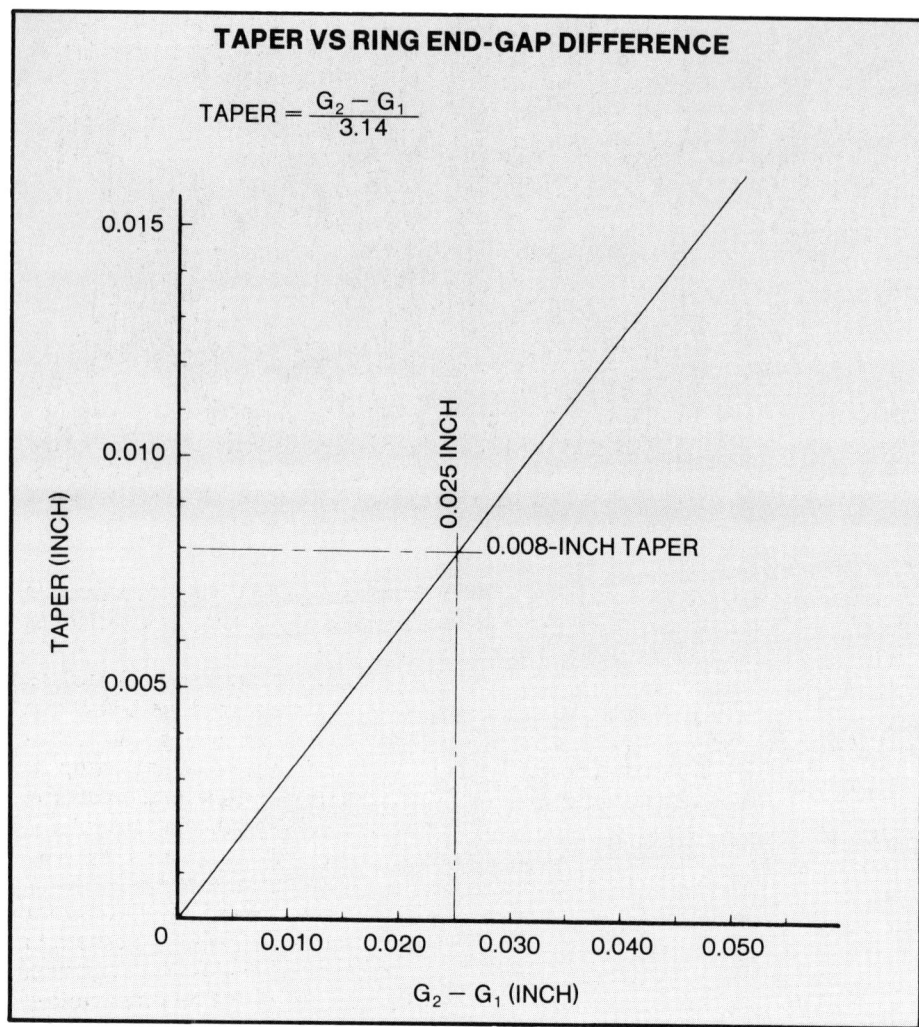

G_2-G_1 ΔG	TAPER
0.000	0.0000
0.001	0.0003
0.005	0.0016
0.010	0.0032
0.015	0.0048
0.020	0.0064
0.025	0.0080
0.030	0.0095
0.035	0.0111
0.040	0.0127
0.045	0.0143
0.050	0.0159

Approximate Taper = $0.30 \times \Delta$

Difference between ring-end gap is G_2-G_1 or $\triangle G$. Read to right for taper.

Using ring end gap to determine bore taper. Maximum end gap (G_2) is found with ring immediately below ridge. Minimum gap is measured with ring in unworn section of the bore. Square ring in bore with piston before measuring gap. After finding maximum difference between two end-gap readings, use chart or graph to approximate bore wear. Borderline figures should be rechecked with dial bore gage or telescoping gage and micrometer.

To use the gage, set the micrometer to the exact measurement the bore should be. If the bore size should be 4.0 in., set the micrometer to that size. Set the dial bore gage in the micrometer and zero the gage with its plunger compressed.

With the gage set this way, readings below zero indicate a slight undersize. Readings above zero indicate how much the cylinders are worn. Keep the plunger square in the bore and watch the dial as you *slowly* slide the gage up and down. Don't force the plunger or allow it to jump the ridge at the top.

Obtain the largest and smallest readings for each bore and record them. The difference between them is the amount of wear.

Inside or Outside Micrometer—With an inside micrometer you can take measurements directly from the bore. To measure bore taper with an outside micrometer you also need a telescoping gage.

When using an inside micrometer, first check its accuracy with an outside mike. The outside mike should be double-checked against a standard. Also, never force any measuring device and don't wedge it into place. If you are not gentle you will end up with an inaccurate micrometer.

With a telescoping gage you must check each measurement with the outside mike and then reset the gage. When measuring use a small amount of force. Pivot the gage so it self-squares and centers in the bore. Remove it carefully so you don't disturb the measurement.

Find the largest and smallest measurement in each cylinder. The difference between them is bore wear.

Ring-and-Feeler Gages—I don't recommend using this method, because it is the least accurate. You are measuring wear indirectly and taking an average, so the *best* you can get is an approximation. I think you could do better by just looking down the cylinder or using the ridge as a guide.

Another disadvantage to this method is that you can't measure the actual bore diameter. This becomes important in piston-to-bore clearance.

The only advantage to this method is convenience—all you need are feeler gages and a piston ring. If the wear on the bores is slight you may be able to determine that all you need to do is have the cylinders honed or deglazed. If you are not sure, or if the figures are borderline, have the bores checked at the machine shop.

To measure taper by this method, remove the rings from one piston. Place one of the piston rings in the cylinder bore and push the ring into place with the piston top. This squares the ring in the bore for accurate measurements.

First push the ring to the bottom of the cylinder until the end gap on the ring is the smallest. Measure this gap with a feeler gage. Then move the ring up to where the gap is the largest, immediately below the ridge. Measure this gap. Subtract the two measurements. Multiply the difference by 0.3 for the approximate taper in that cylinder, or check the nearby chart for a conversion.

How Much Taper?—No taper in a cylinder is desirable—from there its condition goes downhill. Taper not only increases piston-to-bore clearance but causes rapid ring fatigue. As

Glaze-breakers take mirror finish off bores and leave cross-hatch pattern. Spring-loaded hone is used in bores that won't be rebored.

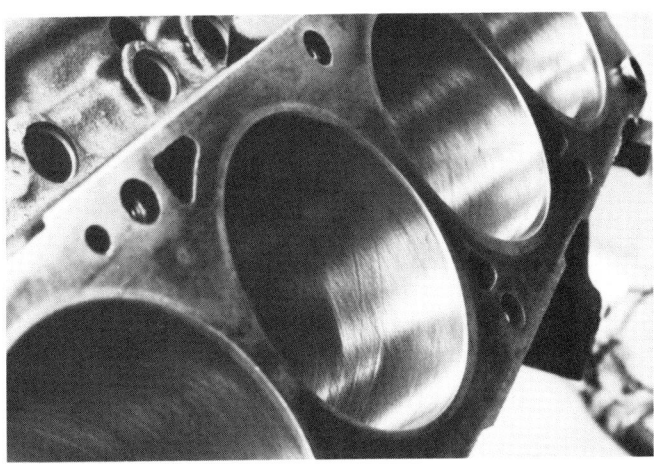
Crosshatch pattern aids oil-retention during break-in. Grade of stone used depends on ring type to be used. See ring discussion, page 77.

the piston moves up and down in the bore, the piston rings must expand and contract to match the changing bore size. This movement quickly fatigues rings and accelerates wear on ring lands and cylinder bores. Eventually the rings lose their *resilience*—ability to spring back—and can no longer seal.

The amount of taper determines how long the engine will go before needing to be rebuilt again. The more taper, the fewer miles it will last. Compression ratio, piston-to-bore clearance, ring type and engine use all affect ring life.

I've known of engines with 90,000 miles or more on them that were reringed after *glaze breaking*. These typically go another 50,000 miles or more, performing well. Glaze breaking and reringing may take care of your current needs if you don't have the money for pistons and taper isn't too bad.

How much is too much? Ford recommends a maximum taper of 0.005 in. This means honing is acceptable, but will not restore the engine to as-new condition.

The maximum taper under any circumstances is 0.006 in. measured with a dial bore gage or micrometer, or 0.003 in. with a ring and feeler gages. Engines with as much as 0.010-in. taper have been honed and reringed but end up having a very short life.

If taper is 0.005 in. or less, glaze breaking is OK. But if taper is greater than this, or you expect like-new performance from your rebuild, rebore.

Glaze Breaking—If the taper is within limits, the next step is *breaking the glaze*. Check the crosshatch pattern at the base of the bores. This is what you'll restore to the cylinder bore when you break the glaze.

A glaze-breaking hone simply resurfaces the existing bore. It will not remove taper. A glaze-breaking hone differs greatly from precision hone. A precision hone is used to make the cylinders round and straight again. It increases bore size and piston-to-bore clearance—not what you want when glaze breaking.

A glaze-breaking bore removes only enough material to restore the crosshatching on a cylinder bore. This serves two purposes: It aids ring break-in and oil retention. Consider it a requirement if you are using cast-iron or chrome rings and optional if you are using moly rings. Ring types and bore finishes are explained on page 77.

Two types of glaze-breaking hones are available—a spring-loaded-stone type and a ball type. Both types will follow the taper of the cylinder, though the ball type will follow any other irregularities in the cylinder wall. Make sure you get one to match bore diameter.

Chuck the hone in a variable-speed, 1/2-in. drill. Run the hone up and down the cylinder at about half to three-quarter drill speed. Use a little oil to keep the stone clean. Continue until there is a good crosshatch pattern the full length of the bore.

DECKING THE BLOCK
Decking the block—resurfacing the head-gasket surface—is rare on a big-block Ford. It's usually needed only because of a gouge or etching from a blown head gasket. Occasionally it is necessary because the deck surface is warped.

If you have to have the deck surface machined, it must be done before the rest of the machining. This is particularly true if the machine shop is using a boring bar that locates on the block deck. If the deck surface of the block is warped, the boring bar will shift the bore out of position.

Set a maximum cut of 0.050 in. for the deck surface—in fact, I hesitate to cut more than 0.020 in. Anything above a 0.020-in. cut requires cutting the intake manifold to fit, and also raises the compression ratio considerably. If the block is *decked,* cut both sides the same so the compression ratio will be equal for all cylinders.

If the block is etched there is one alternative to decking the block. The etching can be repaired by welding and the block given a light surfacing afterwards. If the etching is more than 0.050-in. deep, the block must be repaired this way.

Before welding, the block should be preheated. The welding is done with a special nickel-alloy rod that is compatible with the block material. Because the weld will be harder than the block, the block surface is ground with a stone instead of machined.

The process is costly and time-consuming. Before getting too deeply into this type of repair, remember that the block may still need reboring or other machine work.

Again, weigh the complete cost of restoring the block against reconditioning another used block or purchasing a new one.

Special Problems—As mentioned in

This type of boring machine locates off main-bearing bores to ensure bores are restored to their original center. No matter which type is used, main-bearing caps should be in place and their bolts torqued 95—105 ft-lb. Boring bar removes enough material to accept next over-size piston with 0.002—0.003-in. stock left for honing to size and finish.

Chapter 1, the oil galleries for the rocker-arm assemblies sometimes crack. Suspect this only if there was oil in the coolant and no sign of a blown head gasket. These can be fixed by drilling the hole and driving a tube into the passage.

It is better to let the machine shop make this repair. The tube is usually made from a pushrod from a Ford 300-CID six. The ends are cut off and the passage is drilled oversize. The pushrod is coated with sealer, driven in and machined flush with the block surface.

Another check is the lifter bores: Lifter-to-bore clearance should be 0.0005—0.0020 in. with a new lifter. Check the bores for gouges, then install a new lifter and check it for side play—you should feel no movement. If you have any doubts, check the bore and lifter with a small-hole or telescoping gage and a micrometer.

If the clearance is greater you must have the lifter bore sleeved. Check the cost of sleeving because it varies. In most cases the lifter bores will be OK, unless they have been damaged from broken valve-train components.

Another check is the trueness of the crankshaft-bearing bores. This isn't a common problem with the big-block Ford, due to the Y-block design. This check is normally left to racing engines or those that had bottom-end failure.

Align boring or honing is a machining process that aligns the bearing bores. It ensures that the bearing bores are in perfect alignment. Align boring is not a normal part of an engine rebuild, simply because it is rarely necessary. It is usually only required when non-original main-bearing caps are used or when a bearing bore is out of spec or damaged.

The simplest way to check the bearing bores is to do it after you have checked the crankshaft for runout, pages 81, 82 and 83. Install and oil the main bearings, but do not install the rear-main seal. With the crank installed and the caps torqued to specification, the crank should spin freely.

Another method of checking is to set a straightedge in the bearing bores and check each one with feeler gages. If there is more than a 0.002-in. variance between one bore and the next, have the block *align bored*.

The last thing to check for is cracks around the water jackets. If the engine is operated with insufficient antifreeze in cold climates, the coolant may freeze. The resulting ice may crack the block near the core plugs.

A visual check of the block should reveal any obvious problems, but it is possible to have a nearly invisible crack. If your block is not heading for the machine shop and a "full-dress" reconditioning, at least have it checked for cracks. Have it *Magnafluxed* or *Spotcheck* it yourself.

Magnafluxing sets up a magnetic field in the piece being checked to attract iron powder that is sprinkled over the surface. The particles group around any cracks and show up as fine, white lines.

This process will reveal even the smallest hairline crack, which could result in engine failure after the rebuild. A good machine shop will do this automatically when reconditioning a block.

Any cracks mean the block may need replacing. Even if it needs no machine work, take the block in for Magnafluxing to make sure!

Spotcheck is similar, but uses a dye to indicate cracks. Spotcheck consists of a penetrant—a liquid red dye—and a developer—a white powder. Cleaner is also included to clean the surface before and after the test. Spotcheck can also be used to check rods, heads, pistons and crankshafts.

The major advantages of Spotcheck are its convenience and the ability to check non-magnetic parts, such as pistons. Spotcheck is available from Magnaflux Corporation, 7300 West Lawrence Avenue, Chicago, IL 60656.

Another area that might crack is around the main-bearing webs or the cylinder bores. Cracks here are usually caused by a broken rod or crankshaft and will be obvious. Most of these cracks can be repaired, but the repair should be considered in the price of the block reconditioning.

CYLINDER-BORE FINISHING

When you drop the block off to have it bored or honed, you should know what type of piston rings you will use. If you already have the pistons, you can specify a final bore dimension.

Final Preparation—Before boring the block, torque the main-bearing caps. Torque them 95—105 ft-lb for all big-blocks except the cross-bolted

Some boring bars locate on deck surface. If block is bored with this type of machine, be sure deck surface is free of any high spots. Lightly file surface to remove any projections.

If possible, chamfer cylinder bores after boring but before honing. If block won't be bored, chamfer bores before glaze breaking. A 1/16-in.-wide chamfer, 60° to deck surface is sufficient. Be careful not to score cylinder wall with file.

406 and 427. On these two engines, torque the vertical main-bearing-cap bolts 105 ft-lb. Install the spacers and washers on the cross bolts and torque the bolts 20 ft-lb and then 40 ft-lb in the sequence shown on page 124.

When the bolts and caps are torqued in place they distort the shape of the cylinder block and bores. With the caps torqued in place the bores are under a load similar to that with the engine assembled. This makes them much closer to round after boring or honing and assembly.

For racing applications, another step in truing the block before the final honing is to simulate the load of the installed cylinder heads. This is accomplished with 2-in.-thick *torque plates* bolted in place of the heads. Most machine shops don't have these torque plates, though many speed shops or performance rebuilders do. It's nice to use torque plates, but most engines are bored and honed without them.

Finish—Piston rings come in three standard types: chrome, plain cast iron and moly. Each one requires a different type of bore finish.

Chrome rings are very popular and last longest, particularly in engines that inhale a lot of dust. The disadvantages for chrome are a longer break-in and heavier wear on cylinder bores.

Plain cast-iron rings break in more quickly than chrome-faced rings but wear faster. They are often used for the second compression ring, but otherwise should not be considered.

Moly rings, or *moly-coated* rings, have several advantages. They break in rapidly and cause the least wear on the cylinder walls. The moly coating is slightly porous and retains oil. This reduces wear and improves piston-ring seating.

This oil retention is important, because the oil on the cylinder walls is exposed to combustion, so much of it burns away during the power stroke. This means there is relatively little oil on the cylinder walls during the exhaust stroke.

Oil in the pores of the moly ring is shielded from the flame as the mixture burns. Chrome or plain cast-iron rings have less porosity and retain less oil, particularly after being exposed to combustion gasses.

Ring type is the major factor in honing finish. If you are using moly rings, the cylinder walls should have a smoother finish—use a 400-grit honing stone. Cylinders for chrome and cast-iron rings are finished with a 280-grit stone.

Either stone will work with any type of piston ring. But the rougher finish will make the moly wear faster. The smoother finish will increase break-in time on chrome or cast-iron rings.

The final crosshatch pattern should be 60° where the patterns cross, or

Fitting the Pistons

In most cases, when you buy new pistons the machine shop will fit them to the bores. You simply give them the block and pistons and they give you back a bored-and-honed block with pistons that fit the bores.

If you are willing to pay the extra money you can have each cylinder matched to a particular piston. You can reduce this cost somewhat if you measure the pistons and give the shop final measurements for each cylinder.

If you already have the pistons, measure each one and record this dimension on its box. You should then ask for a final bore finish with each bore sized to a specific piston and the piston-to-bore clearance you want.

When you give the machine shop the dimensions, you will get the block back with a slight difference in finished bore diameters. You can now selectively fit each piston to the bore, providing the pistons are not out of tolerance.

All measurements of pistons and bores should be checked and double-checked. If you tell the machinist the bore size you want and make a mistake it's your fault—not the machinist's. You may wind up having to rebore the block *again* and buying another set of pistons to fit.

When you measure the pistons, if you find one of them is not quite right, exchange it. That way you will have correct clearance for that bore. This attention to detail will result in a more-durable engine. Although many engine rebuilds do not warrant this extra care, the decision is up to you.

Chasing threads ensures accurate head- and main-cap-bolt torquing.

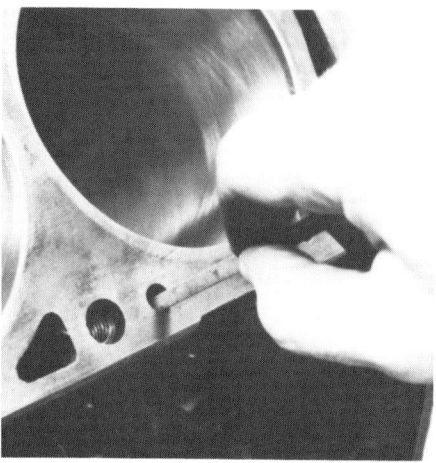
Use file to clean rust or casting slag from water-jacket ports and oil galleries.

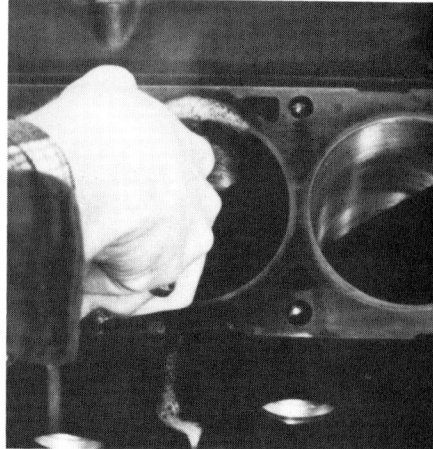
Once all machine work is done, give block thorough cleaning. Wash bores with scrub brush, detergent and hot water. White paper towel wiped through bore should come out clean.

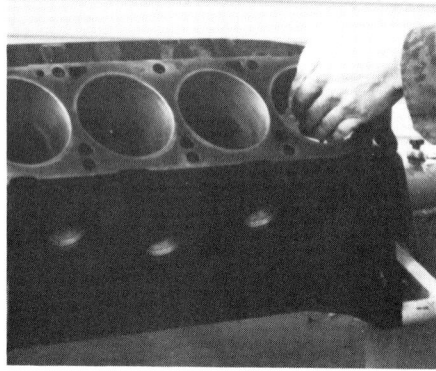
Use hose to rinse out any loose particles, then clean oil galleries with gun brushes or rags tied to a wire. *Every* part of block should be *completely* clean.

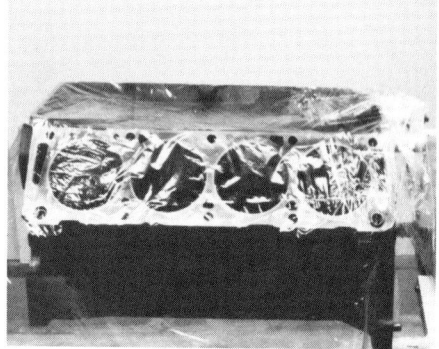
Dry block thoroughly with paper towels or compressed air. Spray machined surfaces with WD-40 or CRC 5-56, then cover block. Keep clean block covered whenever you are not working on it to keep it free of dust, dirt and moisture.

30° to the deck where the lines enter and exit the bore.

Chamfer the Bores—A chamfer is simply an angled cut to remove the sharp edge at the deck surface, or entrance of the bore. It provides a lead-in for the new piston rings when they are installed. It also removes a sharp edge that can glow red hot and cause preignition or lead to a crack forming in the deck surface.

Most machine shops chamfer the bores after boring or honing the block. If possible, chamfer the bore *after* it is bored but *before* it is honed. Chamfering is done at the machine shop with a large cone-shaped stone or other abrasive. It can also be done with a half-round or round file.

Hold the file at a 60° angle to the deck surface—30° to the cylinder bore. Cover the end of the file with your hand or a heavy layer of tape so you don't damage the opposite side of the cylinder bore. Maximum width of the chamfer should be about 1/16 in.

Many engines with High-Riser heads and all the side-oiler 427s came with bores notched for valve clearance. There was a small notch at the top of the bore by the exhaust valve and a slightly larger one near the intake. The notching was done to promote better breathing, though some blocks need it for valve-to-block clearance.

If you are installing different heads on the block, you may need or want to add these extra notches. On 427s, if the heads or block have been milled, you should check for adequate notches—particularly with the Tunnel-Port heads. See photo, page 43.

Clean it Again—After all the machining is done on the block, clean it again. The best way is to pressure wash or steam clean the block, then scrub the cylinder walls.

The block could be hot tanked again, but I don't recommend it. You should wash it anyway because shavings can still remain in the block after hot-tanking.

When you get the block home, the final step is to scrub the entire block. Use dishwashing detergent and a scrub brush with *hot* water. Pay particular attention to the cylinder bores, crank and cam journals, and the deck surface.

Use cleaning tools that will not damage machined surfaces. Use a *clean* scrub brush with a wooden or plastic handle. Don't use one that's been used around metal shavings.

Before giving the block the final rinse, chase the main-cap- and head-bolt threads one more time. Rinse the block and force water through the oil passages—I know you did it, but do it again. If possible, dry the block with compressed air. Blow out the bolt holes and oil passages.

If you don't have compressed air, you can either tip the block over to

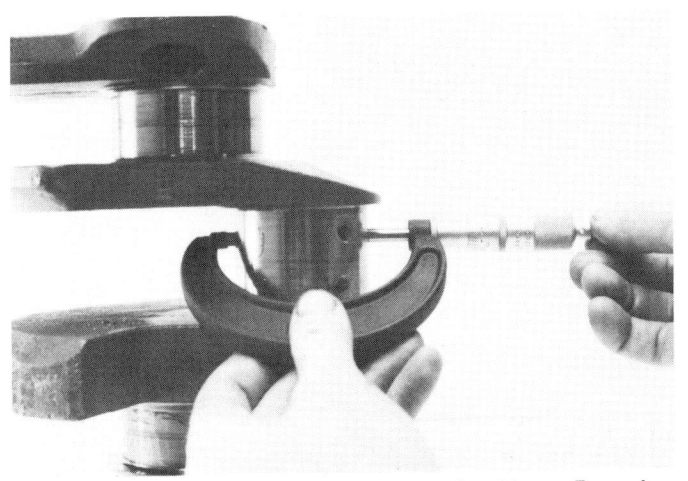

Measure crank journals for size, out-of-round and taper. Record readings. Check main- and rod-bearing journals using the worn bearing inserts as a guide. Pay special attention to journals with unevenly worn bearings.

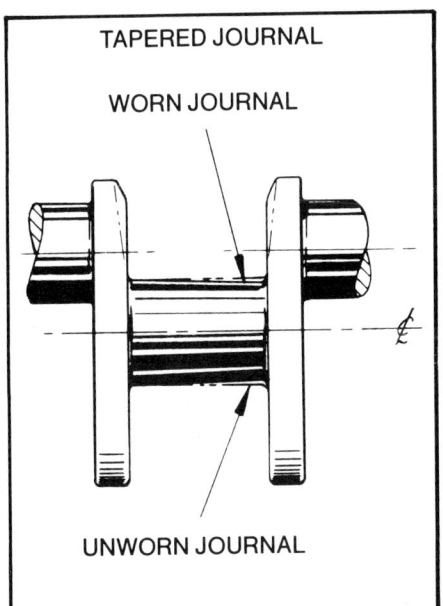
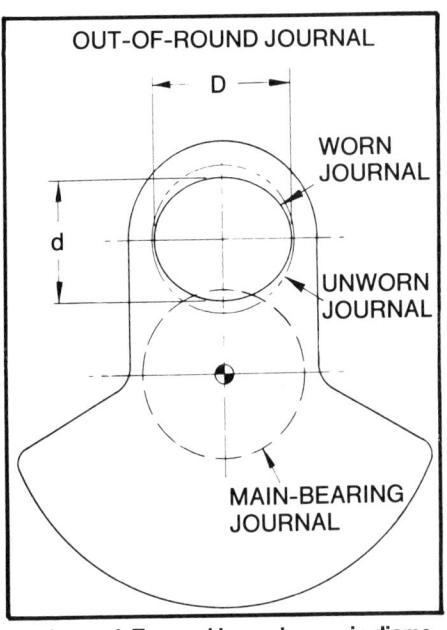

Bearing journals wear two ways: tapered and out-of-round. Tapered journals vary in diameter along their length; out-of-round journals are oval. Connecting-rod journals tend to wear out-of-round due to the way they are loaded. Drawing by Tom Monroe.

This is one dimension you'll need to determine crankshaft thrust-face width using the insert-and-feeler-gage method: width across the number-3 bearing-insert flanges.

assist drainage or blow into the holes with a straw. A hand pump or portable air tank will also work. Don't forget to dry the cam-bearing and lifter bores. Once the block is dry, spray it with WD-40 or CRC to prevent rust. Pay particular attention to the machined surfaces.

This block cleaning may seem like a lot of work, but if you want to do the job right, this is the way to do it. You can do an otherwise perfect engine rebuild and one piece of dirt can end its life or shorten it drastically.

Cover the block with a large plastic trash bag and seal off the end with a cable tie or tape to keep out dirt. If you are storing the block in your garage on the floor, place wood blocks under it to reduce condensation.

CRANKSHAFT

The crankshaft converts reciprocating motion into rotation. It is the heart of the short block. Although it is subject to heavy loads and high speeds, the crankshaft is usually a very durable part.

Crankshaft damage is usually caused by oil contamination or starvation. On standard passenger-car engines, the usual causes are infrequent oil changes or low oil level. In most cases, these cranks are serviceable and only need turning or polishing to be restored to as-new condition.

A crankshaft can be damaged beyond use by tangling with a broken connecting rod or piston. Typically this occurs when a rod bolts breaks and the big end or cap gouges the rod-bearing journal.

The first thing to check is the condition of the journals. Look for deep grooves, scratches or heavy blue marks. These will usually be on the front main or rod journal, but can be on any of them. Also check the bearing journals for cracks, *out-of-round* and *taper*. Most cracks are visible to the naked eye, but have the crank Magnafluxed to be sure.

Out-of-round is caused by uneven loads on the *circumference* of a journal; taper is caused by uneven loads on the *length* of a journal. Taper is usually caused by a bent or twisted rod or a

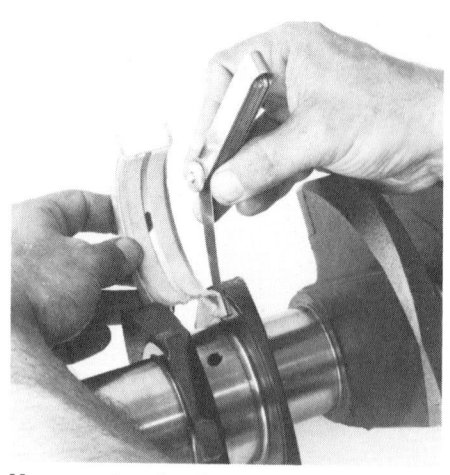
Measure bearing-insert width with micrometer. Add this to clearance between bearing insert and thrust faces. Total is thrust-face width.

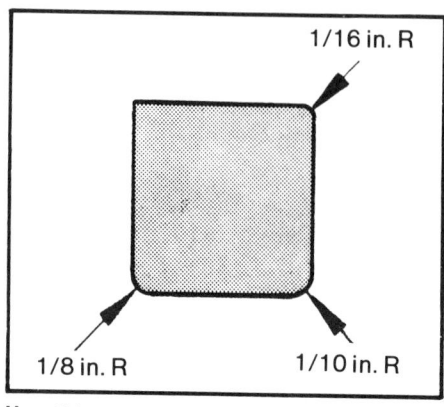
Use this pattern for a template to check bearing-journal radii. Journal radius should not be larger than 0.100 in., otherwise journal will edge ride bearing. Other radii, 1/16 and 1/8 in., are for reference.

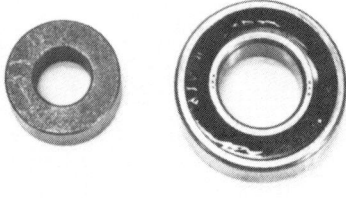
Two types of pilot bearings used on big-blocks. Most use an Oilite bushing, left. Some heavy-duty trucks use a ball-bearing, right.

poor machining job on a journal or rod-bearing bore.

Out-of-Round—To check for out-of-round, you need a 2—3-in. outside micrometer. Rod journals for all engines are 2.4380—2.4388 in. Main journals should be 2.7484—2.7492 in. for all cranks except the forged-steel FT cranks. Forged FT-crank mains should measure 2.7479—2.7487 in.

Measure around each journal until you find the smallest and largest diameters. The difference between the two is how far the journal is out-of-round. You will normally find the minimum and maximum measurements 90° apart, due to uneven loads from the piston-and-rod assembly.

If all the journals measure less than specified, the crank may have already been cut undersize. This is nothing to worry about unless the crank is 0.030-in. undersize already and needs to be cut again. Engines used only for light duty can go 0.040-in. under, but that is the limit.

Pay particular attention to the connecting-rod journals when checking for out-of-round. Rod journals are subject to heavy loads whenever the rod and piston reverse direction, particularly during the power stroke. Most of this load is on one side of the rod journal—the side closest the piston at TDC.

The maximum allowable out-of-round is 0.0004 in. for both rod and main journals. If one journal is off have *all* the journals ground. You won't save any money having only one journal ground because the main cost is setting up the grinder for machining.

Having all journals ground saves you the confusion of different bearing sizes. It is OK to have rod journals one undersize and mains another, but you don't want crankshaft journals with a random mix of standard, 0.010- and 0.020-in.-undersize bearings. Besides, parts stores won't split up bearing sets.

Taper—Crankshaft-journal taper is caused by uneven loads across length of the journal.

On FE engines, maximum taper for rod and main journals is 0.0003 in. *per inch of journal length*. Note that this is not the same as a specification for total taper. For example, suppose you measure a rod journal and find total taper is 0.0003 in. for the overall length, which seems acceptable. It's possible this same journal could have 0.0002 in. of taper in 1/2 in. of length—too much as it equals 0.0084 in. per inch.

FT cranks have a recommended taper allowance for main bearings of 0.0005 in. per inch of journal length. FT rod-bearing specifications are 0.0006 in. per inch. If you follow the FE allowance of 0.0003-in. taper per inch, the bearings will last longer and oil pressure will be higher.

Finish—Some crank damage may be visible right away, but to get a better look at the crank, clean it with solvent or detergent. Better yet, have it hot tanked if you haven't already done so. Give it a light coating of CRC or WD-40 afterwards.

Inspect the journals for scratches, nicks or discoloration—usually blue or black heat marking. If you find cracks in the crank during the inspection, throw it away. Otherwise, most other conditions can be repaired.

If the crank feels and looks smooth, polish the journals with a long piece of fine crocus or emery cloth. Remove any nicks or rough edges with a file or small stone before polishing. Finish by polishing each journal evenly all the way around. A mirror finish for crank journals is great, but don't change the journal diameter.

Check the front main-bearing journal carefully. It is particularly susceptible to damage because of the loads placed on it by accessories—alternator, power-steering pump and air-conditioning compressor. The belts for these accessories pull the crankshaft up or to the side. The front main-bearing journal is also one of the first areas to receive oil. Any dirt passed through the filter shows up here, so it's more likely to be grooved.

Thrust faces—The next area to check is the condition of the thrust surface at number-3 main-bearing journal, particularly on engines mated to manual transmissions. The thrust surface mates with the thrust-bearing faces to control crankshaft end play.

With a manual transmission, the crank is forced forward when the clutch is disengaged. Under certain circumstances an automatic-transmission torque converter can expand, or *balloon*, also pushing the crank forward. In both cases, longitudinal crankshaft movement is controlled by the thrust surfaces on the crankshaft and number-3 main bearing.

If the crankshaft has excessive end play it will not keep the rods and timing-chain sprockets in alignment. This increases wear or stress on other parts of the engine—mostly the bearings and timing chain.

Distance between thrust faces—*thrust width*—should be 1.124—1.126 in. Maximum thrust width is usually not exceeded because the bearing wears first. Maximum runout of the thrust face is 0.001 in. Also note the finish of the thrust faces—they should be smooth like the journals.

To measure crank thrust surfaces, you need an inside micrometer or telescoping gage and an outside mike. Move the mike or gage around the thrust faces and measure it in at least four places. If the width exceeds 1.126 in., replace the crank or have the area welded and reground.

You can also check the dimension between thrust surfaces with an old thrust bearing, feeler gages and a micrometer. Place the bearing insert between the thrust faces. Measure the clearance between the bearing and the thrust faces with a feeler gage. Measure the width of the bearing and add the feeler gage thickness for total width.

Crank Fillet—Each end of a crank journal, where it meets the crank throw, should have a *fillet*—radiused section. This radius at each end of the journal eliminates sharp 90° angles.

A sharp 90° angle creates a *stress riser*—a potential weak point that can cause a crack or break in the crankshaft. This potential is reduced by forming a slight radius at the ends of the journals. The larger the radius, the stronger the crank.

The radius cannot exceed a certain limit or the bearing will *edge ride* the journal. Edge riding is where the edge of the bearing actually comes in contact with the crank journal at the radius—fillet—of the crank. To prevent edge riding, the radius should be 0.10 in. and no larger. A smaller radius is OK for the bearing, but weakens the crankshaft.

To measure each fillet radius, make a template like the one shown nearby. Make one corner square, one with a 1/16-in. radius, one with a 0.10-in. (approximately 3/32-in.) radius and the last one a 1/8-in. radius. The correct radius is 0.10 in., the other corners are for reference.

If the crank has no radius or the radius is too large, it must be ground. When you pick up a reground crankshaft, recheck the radius. If it's off, return the crank and have it reground or exchanged. If there is no radius, the crank must be ground to the next undersize.

Crankshaft Kit—If the crank must be reground, consider buying a crankshaft kit. The kit includes a crankshaft and matching main and rod bearings. The advantages of the kit are: you don't have to wait to have the crank reground and the bearings *should be* included with the kit.

If you get a crank kit, ask what undersize bearings it uses. If the bearings are 0.030-in. undersize already, there is a good chance the crank can't be serviced again. Keep this in mind if you plan to have the engine around long enough for another rebuild.

If the engine has a forged crank—heavy trucks and some 427s—you may find the price of a kit to be high or a kit won't be available at all. Another rare crank is one from a 332 with its 3.30-in. stroke. With these engines you may have to regrind the original crank.

Polish & Chamfer—If you have any machine work done, check the journal surfaces when you pick up the crankshaft. They should be smooth with no rough edges around the oil holes. Some engines have oil grooves in the main journals; the edges on these should also be smooth.

Any rough edges around the oil holes can be removed with a cone-shaped stone or with a piece of emery cloth wrapped over the head of a round-head screw. Simply clamp the whole works in a small pair of Vise-Grips.

If the crank has sharp edges on the oil grooves, run the emery cloth along the edge with pressure from your fingertip. If the edge is particularly sharp, use a fine grinding stone. Carefully work your way around the groove and finish with the emery cloth.

To check the journal finish, run the edge of a copper penny across the journal. If it leaves a line of copper, the journal is too rough. Smooth the journal by polishing it with a 2-ft strip of 400-grit emery cloth.

Work your way evenly around the journals. To put on the mirror finish, use a piece of worn emery cloth. None of this should take much time if the crank was ground correctly. Be sure to work evenly around the journals to

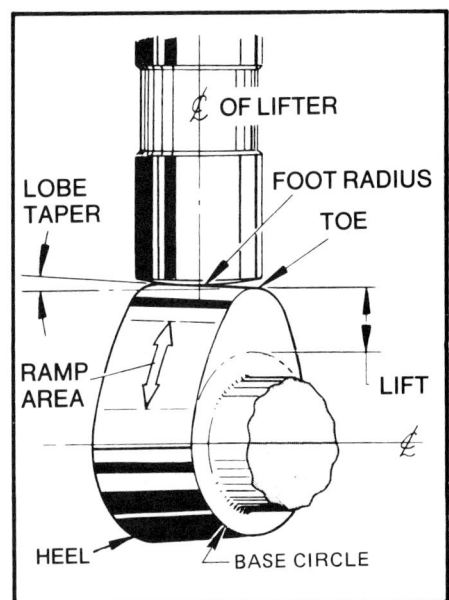

Visually inspecting camshaft lobes and lifters may tell you if the cam should be replaced. If any cam lobe is worn across its entire width of the toe and corresponding lifter foot is worn flat or concave, camshaft *and* lifters should be replaced. Drawing by Tom Monroe.

keep them round, and don't work too long in any one area.

When working around the rear-seal area of the crank, simply clean this surface and make sure there are no nicks or scratches. If you polish this surface too much, the seal will run dry and tear, causing an oil leak.

If the engine had a rear-main-seal leak, check the rear main-bearing journal closely for finish, diameter and out-of-round. Also check the slinger slots on the front sleeve. Viewed from the front of the crank, the slots should be going from left to right; if not, the oil will be pushed against the seal instead of being pushed away.

Pilot Bushing or Bearing—FE engines with manual transmissions have a pilot bushing in the rear of the crankshaft. Under most circumstances, a pilot bushing is subject to very little wear. Still, I usually replace the pilot bushing as a matter of course—it's cheap insurance.

The only time the bushing is subject to wear is when the clutch is disengaged with the engine running. At all other times the crankshaft and transmission input shaft turn at the same speed and there is no relative movement between the pilot bushing and the transmission input shaft.

Check that the bearing or bushing ID is round with no scratches or gouges. Pilot bushings wear in a cone shape, with the larger ID near the clutch. If the cone shape is visible to the eye, replace the bushing.

Another way to check for wear is to insert an old input shaft into the bushing and feel for looseness. If you feel energetic, take the crank to the transmission, brace the crank against your leg and hold it with one hand. Use your other hand to wiggle the pilot shaft up and down in the bushing to check for looseness.

To remove the pilot bushing you have a number of choices. The first is to have the machine shop do it. The second is to do it at home with rented or makeshift tools. The easiest ways to remove it are with a slide hammer or a screw-type puller.

The slide hammer has hooks that reach behind the bushing. These are connected to a shaft with a sliding weight on it. Once the bushing is hooked, slide the weight back against the end of the hammer. The impact pulls the bushing out.

The screw type is similar to a harmonic-balancer or steering-wheel puller. The tool rests on the outer edge of the crank or flywheel. A threaded rod in the center of the puller goes through the bushing hole. A hook similar to that of a slide hammer engages the back of the bushing. As the rod is turned, it pulls the bushing out of the crank.

There are still other methods you can use without buying or renting equipment. These work just as well but take more time.

One method is to pack the cavity behind the bushing with grease. Leave no air pockets. Fit a piece of round stock to the bushing-bore ID. Insert the piece of stock and hit it with a hammer. The hydraulic pressure will force the bushing out. The advantage here is that this does not damage the bushing.

The next method uses an 11/16-in. coarse-thread bolt. Thread the bolt into the bushing until it bottoms out. Keep turning until the bushing is pulled out. To make it easier, grease the hole. Place a short piece of round steel or aluminum stock inside the bore behind the bushing so the bolt will bottom on it and not the crankshaft. You don't need to tap the bushing, but grind down the first few bolt threads so the bolt will start easier.

The last method is to drill the bush-

Checking crankshaft runout using a dial indicator. Turn crank to find lowest point at number-3 main-bearing journal. Zero dial indicator, then turn crank one revolution. Runout is read directly as you turn the crank.

ing through in two places then split it with a chisel. Be sure to wear safety goggles.

Heavy trucks normally have the pilot bushing installed in the flywheel. If your engine has one installed in the crank, the ID should be 1.0 in. A ball-bearing pilot bearing is used in '75-and-later heavy-truck crankshafts. Check these by turning the inner race to check for roughness. If the bearing feels rough, replace it.

Remove the bearing with a puller, or split the outer race with a chisel or drill. *BE SURE TO WEAR SAFETY GLASSES IF YOU SPLIT THE RACE.* The hardened race can shatter very easily.

To install the new bushing, clean the crank bearing bore. Coat the bore lightly with grease. The bushing is self-lubricating bronze—Oilite. Soak it in oil before you insert it to provide additional lubrication.

Place the bushing squarely over the hole and put a large socket over the bushing. If you are installing a ball bearing, tap *only* on the outer race. Hitting the inner race or the ball bearings will ruin the bearing.

Drive the bearing or bushing in until it bottoms in the bore. If the bushing cocks, remove it and start over. If you accidentally distort the bushing, get a new one—the cost is small.

Clean the Crankshaft—With all the machining and polishing done, clean the crankshaft oil galleries. Even if you had the crank hot tanked, there may be remaining debris lurking in the oil passages.

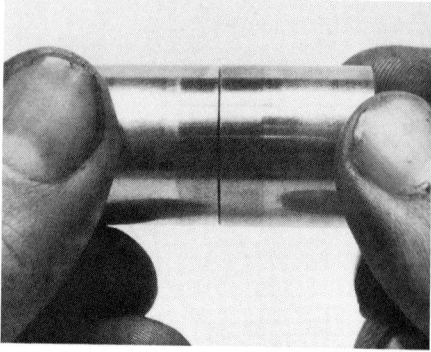
Placing two lifters foot to foot is simplest way to check for wear. Radius is so small—about 30 in.—that it can be felt more easily than seen. One worn lifter indicates a worn lobe, so if one lifter fails this test, cam and lifters should be replaced.

If you didn't have it hot tanked and the entire crankshaft is full of carbon or varnish, clean the outside of the crank first. Otherwise you will get dirt in the oil galleries.

Clean the counterweights and cheeks of the crank. Don't scrape the crank with anything sharper than a piece of wood and be careful around the bearing journals. Use caution not to distort pressed-in plugs or gouge the retaining-ring grooves.

If you have a forged-steel crank from a 427, remove the plugs from the crank throws near the rod journals. Production cranks and early service cranks were fitted with pressed-in plugs, held by retaining rings. Late service cranks had screw-in plugs.

Clean the oil passages with solvent and a rifle brush or bits of paper towel tied to a piece of wire. When a clean paper towel remains clean when dragged through the gallery, it's OK. Squirt motor oil into the galleries once they are clean.

To install the pressed-in plugs, inspect the retaining-ring groove in the crank to make sure it is clean. Use new plugs and a *small* amount of sealer on the edges. Make sure the sealer doesn't get into the oil passages. Seal the threads on the screw-in plugs with Teflon tape. Tighten the screw-in plugs. Oil the crank and store it in a plastic bag.

CRANKSHAFT RUNOUT

Crank runout refers to misalignment of the main-bearing journals. It's a polite way of referring to how far the crank is bent.

There are two common methods

Lobe lift can be checked by comparing measurements made across base, and heel and toe. Difference is lobe lift. This will not work on long-duration cams where ramps extend farther into base circle. You can also check cam by comparing lobe to lobe, but be sure you compare intake to intake and exhaust to exhaust. All similar lobes should be within 0.005 in.

Lifters worn as badly as one at right should mate to equally worn cam lobe. Cam lobe in photo on page 84 mates to worn lifter.

for checking crankshaft runout. The first is to turn the crank in the block and measure runout directly with a dial indicator.

The second method is far simpler—install the crank and main bearings and *feel* for drag. This method won't tell you exactly how much runout there is, but if the crank turns easily, runout is acceptable.

By Dial Indicator—Place the upper bearing inserts in the number-1 and -5 bearing bores. Oil the bearing inserts and lay the crankshaft in the block.

Mount the dial indicator with the plunger on the center main-bearing journal. You will probably need a tip extension for the indicator so the crank throws don't hit it. Offset the plunger to miss the oil holes.

Rotate the crank until the indicator is at the lowest reading. Zero it at that point. Rotate the crank again until the indicator is at its highest reading. The highest reading is maximum runout.

Rotate the crank again to make sure the indicator shows zero and rotate it back to the high reading to confirm your earlier reading. You can also check the other journals before putting the indicator away.

Maximum runout is 0.004 in. If the crank exceeds this don't panic. Place the high spot up—away from the block—and install the main-bearing cap with the bearing insert. Torque the cap bolts to specification. Recheck the crank in a day or two. If the crank is still out, have it reground or exchange it for a crank kit.

By Installation—To check runout without an indicator you must install the crankshaft. Wait until you are ready to assemble the engine. Oil all the bearing halves, install the bearings, crank and main-bearing caps. Don't install the rear-main seal. Torque the caps to specification.

Rotate the crank by hand and feel for any drag. If the crank turns freely, it's OK. If the drag seems excessive, recheck the crankshaft with a dial indicator.

CAMSHAFT & LIFTERS

A camshaft lobe controls valve opening: how soon, how fast, how far and how long. Although it rotates at only half crankshaft speed, the camshaft is much more likely to be worn.

Camshaft lobes are subject to extremely high contact pressure at the lifters. This is particularly true with racing or high-performance cams. These are subject to even higher loads because of the high lifts, solid lifters and high spring pressures.

Camshaft and Lifter Design—A quick course in cam and lifter design will help you understand why these areas are so critical. See drawing, page 81.

Notice the cam-lobe *profile* is egg-shaped as viewed from either end of the cam. As the cam lobe rotates, the *opening ramp* raises the lifter in its bore, opening the valve through the pushrod and rocker arm. The lifter and valve reach maximum travel when the lifter is on the *toe* of the lobe. The valve returns to its closed position as the cam continues to rotate and the lifter rides down the *closing ramp* to the *base circle*, or *heel* of the lobe.

Changing the shape of the lobe changes the rate at which the valve opens. It also changes the length of time—*duration*—the valve is open or closed. Not only can the duration be changed, but the distance the valve opens—*lift*—can also be changed. Changing duration or lift affects the amount of power and where it occurs in the rpm range.

When the valve is closed, the lifter is on the *base circle* or *heel* of the lobe. The lifter rides up the *ramp* to the *toe* and opens the valve through the pushrod and rocker. The cam lobe determines the cam's *lift at the lifter*, not at the valve. Valve lift depends on lobe lift *and* the mechanical advantage, or ratio, of the rocker arm—1.73:1 or 1.76:1 for big-block Fords.

A cam-lobe surface is not ground parallel to the cam center line, but at a *rake angle*—usually 1°. The lifter base or *foot* that contacts the lobe has a convex *spherical radius*, like a magnifying glass.

Lobe rake angle and lifter-foot radius ensure tolerable contact pressure with good lubrication. Each lifter is also placed slightly off-center to its lobe. This spins the lifter as the cam rotates, which increases cam and lifter life.

CAMSHAFT & LIFTER INSPECTION

Camshafts can last as long as the rest of the engine, if the engine is cared for and the cam isn't subject to greater-than-normal loads. Still, my

Note rounded edges on lobe (arrows). Toe on this lobe is completely gone.

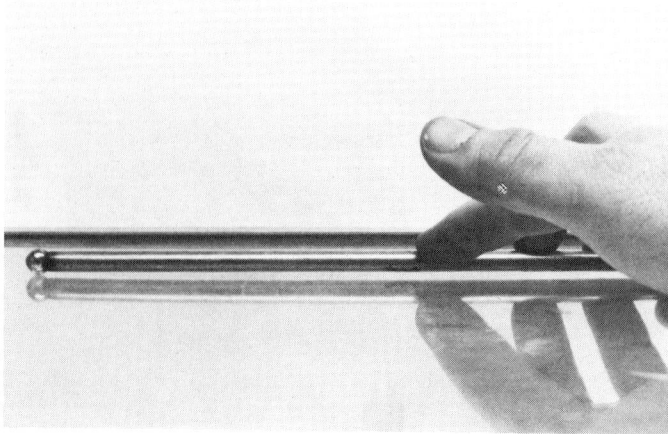
Rolling pushrods on plate glass is simple way to check for straightness. Check bent ones at midpoint with feeler gages. Replace any bent more than 0.020 in.

own preference is to install new lifters in any rebuilt engine and either replace the camshaft or have it reground.

No matter what you do, do not install the old lifters out of order, and **never install old lifters on a new cam.** A worn lifter will not mate with a different cam lobe. The mismatching will wear the lobe away very quickly.

Check for Lobe Wear—Before you get out the micrometer or caliper, check the wear pattern on the lobes. Each lifter will leave a worn area where it rides on the lobe. This area will usually be darker or shinier than the rest of the lobe.

Normal wear patterns vary from a thin line to a band about 1/2-in. wide. If the wear pattern extends all the way across the toe, the lobe is worn out. If it's not completely gone already, it will be shortly.

To measure the cam lobes you will need a micrometer, dial indicator or vernier caliper. Measure the height of the lobe at the highest point and compare this to the base circle. The difference is lobe lift. Compare this measurement to the original lobe lift. See the chart on page 85. Lift should be within 0.005 in. of specification.

Another method of checking lobe lift is to place the camshaft back into the engine with a lifter—cam bearings must be installed. Set a dial indicator so the plunger rests on the lifter. Turn the camshaft until the lifter is at its lowest point: zero the indicator. Rotate the camshaft again and read lobe lift directly.

If you just want to check lobe wear, you can check each lobe with a micrometer or vernier caliper. Compare the measurement of each lobe to the others. Be sure to compare intake to intake and exhaust to exhaust. From either end of the cam, the lobes are EEIIEEIIIIEEIIEE. All the lobes should measure within 0.005 in. of one another.

As you measure the lobes, note that some lobes taper toward the front of the cam and others to the rear. This is so the cam will not be thrust to one end of the block more than the other.

Also check for pitting or chipping of the cam lobes. Pitting will lead to very rapid lobe and lifter wear. So, if you find any pitting, replace the cam.

Cam Journals—Unlike the lobes, cam journals are almost never worn or damaged. They are lightly loaded in comparison to the lobes. They also have a comparatively large bearing surface.

Journal diameters are all the same at 2.1238—2.1248 in. The out-of-round allowance is 0.0010 in. and the maximum runout is 0.0005 in. Out-of-round is checked by measuring each journal at several points with a micrometer or caliper.

To check runout, the cam must be placed in some kind of fixture, either a lathe or on V-blocks or rollers. Rotate the cam while checking the various journals with a dial indicator or feeler gages.

If the bearings are installed, install the cam and check rotation. If the cam rotates freely, the runout is acceptable. If the cam rotated OK before and won't rotate now, the cam bearings may have been installed wrong, the cam may have been bent or the journals are varnished or dirty.

If the runout exceeds specification, the camshaft can be straightened by a crude, but effective, way. Place the camshaft in a pair of blocks with the high-side *down*. Place a chisel between the journal and one of the lobes. Give it a sharp blow with a hammer.

I use an impact tool with a chisel attachment. This way the cam can be straightened without bombarding the cam with heavy blows—which may bend it too far. If you decide to use this method, use a light hammer and chisel.

Once the cam is given a sharp impact the metal is displaced around the chisel and *draws* the cam toward the low side. It seems backwards, but try it. Afterwards measure the cam at all journals and repeat the process until the cam is in tolerance.

Once you've checked the cam, keep it covered with a light coat of grease. Store it in a plastic bag where it won't be damaged.

Lifters—The only use I have for used lifters is as an additional check for camshaft condition. This will only help if the lifters were kept in order during removal.

As I noted, cam lobes and lifters wear to match one another. If you find a worn lifter, it's a sure bet the matching lobe is worn.

The lifter base, or *foot,* has a convex shape with a 30-in. radius. If this radius is gone, the camshaft lobe will be worn in proportion—the more wear on the lifter, the more on the lobe. If you found a lifter with the base worn concave, you definitely should have found a badly worn cam

CAMSHAFT LIFT (Inches)

Year	Engine	INTAKE at lifter	INTAKE at valve†	EXHAUST at lifter	EXHAUST at valve†	Comments
*58	332	0.241	0.401	0.244	0.404	adj. rockers
58	332	0.232	0.408	0.232	0.408	adj. rockers, hyd. lifters
59	332	0.232	0.401	0.232	0.401	includes 58–59 Edsel
*58	352	0.241	0.401	0.244	0.404	adj. rockers
58	352	0.232	0.408	0.232	0.408	adj. rockers, hyd. lifters
58–60	352 Police	0.254	0.440	0.254	0.440	
59–66	352	0.232	0.401	0.232	0.401	includes 60 Edsel
*60	352HP	0.286	0.480	0.286	0.480	
*65–68	352 Marine	0.264	0.440	0.264	0.440	reverse rotation
67	352	0.232	0.401	0.232	0.401	trucks w/AIR, trucks w/o AIR to serial # A84001
67	352	0.247	0.427	0.249	0.431	trucks w/o AIR from serial # A84001
68–76	360	0.247	0.427	0.249	0.431	
58–59	361	0.232	0.401	0.232	0.401	
61	390	0.232	0.401	0.232	0.401	includes T-bird
*61	390 Police	0.264	0.440	0.264	0.440	
*61	390 Special	0.286	0.480	0.286	0.480	
62	390	0.232	0.401	0.232	0.401	includes 4V T-bird
62	390 6V T-bird	0.258	0.446	0.258	0.446	
*62	390 Police	0.264	0.440	0.264	0.440	
*62	390 Special	0.298	0.499	0.298	0.499	
63–65	390	0.232	0.401	0.232	0.401	includes 63 6V T-bird
*63–65	390 Police	0.264	0.440	0.264	0.440	
*65–68	390 Marine	0.264	0.440	0.264	0.440	reverse rotation
66	390 2V & 315 HP 4V	0.253	0.438	0.253	0.438	
66	390 335-hp & 4V GT	0.278	0.481	0.283	0.490	
67–68	390 270-hp 2V	0.253	0.438	0.253	0.438	
67–68	390 280-hp & 320 HP 4V	0.247	0.427	0.249	0.431	
67–68	390 335-hp 4V GT	0.278	0.481	0.283	0.490	
68–76	390 2V & 4V	0.247	0.427	0.249	0.431	pickup
69	390 270-hp 2V	0.253	0.438	0.253	0.438	
69	390 280-hp 2V & 4V GT	0.247	0.427	0.249	0.431	
70–71	390 2V	0.247	0.427	0.249	0.431	
*62–63	406	0.298	0.499	0.298	0.499	
66–67	410	0.253	0.438	0.253	0.438	
*63–67	427	0.298	0.499	0.298	0.499	
*65–68	427 Marine	0.264	0.440	0.264	0.440	reverse rotation
68	427	0.278	0.481	0.283	0.490	
*66	428 Police	0.298	0.499	0.298	0.499	
66–68	428	0.253	0.438	0.253	0.438	
66–70	428 Police	0.278	0.481	0.283	0.490	
68–70	428 CJ	0.278	0.481	0.283	0.490	
69–70	428 SCJ	0.278	0.481	0.283	0.490	
64–78	330MD & HD	0.245	0.423	0.233	0.403	
73–74	359	0.233	0.403	0.233	0.403	F600 Rental
64–78	361, 389, 391	0.245	0.424	0.245	0.424	

*Mechanical cam (solid lifters)
†Valve lift includes 0.025-in. valve lash (clearance)

lobe.

If the bottom of a lifter is only worn slightly you may not be able to see it without some help. Check for the foot radius by placing two lifters—preferably one new one—base to base. Try to rock the lifters against each other. They should rock noticeably. When placed together, you should be able to see the radius on both lifters.

If you find that all lifter feet are in good condition and the camshaft checked out OK, you can reuse the old cam and lifters. Remember that the old lifters *must* be installed in their original order. If you have *any* doubts about any of the lobes, lifters or their order, get new lifters and have the cam reground or replace it. Replacing the cam and its lifters once the engine has been installed in a vehicle is a hassle at best.

Clean the Lifters—If you decide to use the old lifters, give them a good cleaning—particularly the outside. If the engine oil was fairly dirty or the engine had varnish or sludge buildup, clean the inside of the lifters too. You should disassemble each lifter to clean it, so unless the lifters are in really good condition, it is hard to justify the cleaning time.

To clean the outside of the lifters, just about any solvent will do. Carburetor cleaner is fine. Clean the outside until it is shiny and there are no deposits left. Use a plastic scouring pad such as Scotch-Brite to remove any heavy buildup.

Cleaning the inside can sometimes be done without taking the lifter apart. Soak the lifters in solvent or cleaning fluid and work the plungers up and down in their bores with a pushrod. Be *extremely* careful not to scratch or score the lifter while cleaning it. If you do, it must be replaced.

If the plunger sticks at the bottom of the lifter when you push it down, hold the lifter upside down and give it a sharp rap against a piece of wood. If a few sharp raps won't budge it, soak the lifter in solvent and try again. If a lifter is this dirty, I advise replacing it or at least taking it apart to clean it.

Continue working the lifters until the plungers work freely in their bores. If they were free from the beginning they were probably clean at the start.

Taking a lifter aprt to clean it is much slower, but lets you do the best job. Lifter parts are matched, so when you disassemble a lifter keep its parts together. The simplest way to ensure this is to clean the lifters one at a time.

Remove the retaining ring with needle-nose pliers. Release the pressure on it by pushing down on the plunger with a pushrod. Once the retainer is out the next items out are the pushrod cup, plunger assembly and spring.

If the plunger doesn't fall out, work it up and down with a pushrod or force it out by shooting oil through the metering hole in the lifter body with a squirt can. There is also a special tool that locks into the inside of the plunger to pull it out.

Once the plunger is out, carefully pry the retainer cap off the bottom of the plunger. Remove the spring and valve disc. Now you can clean all the parts to almost brand-new condition. I say *almost* because cleaning won't fix a weak spring, restore clearance between a lifter body and plunger, or repair a bad retainer, cap or weak spring.

If it sounds as if I'm trying to convince you to replace the cam and lifters, you're right. If you still feel confident that the old the cam and lifters are reusable, assemble the lifters in reverse order, oil them, and keep them in order.

PISTONS & CONNECTING RODS

In most cases you will need to replace the pistons. The only exception is if the bores have little taper and only need glaze breaking and the pistons are in good shape. The odds are against you, unless the engine has low mileage.

Pistons are expensive. However, if you plan to keep the engine a long time, the money is worth spending. Also, if the bores checked out OK but the pistons don't, have the block rebored anyway. This will remove all taper and restore the bores to as-new condition.

Piston Operation—A piston truly goes through torture. It must withstand extremely high temperatures and pressures, yet be lightweight and transfer heat well. The piston must be strong enough to transfer the pressure in the combustion chamber and pass it to the crankshaft through the connecting rod and its bushings and bearings. In addition, the piston undergoes a wide and quick change in temperatures, from below zero in some climates to over 900F (480C) under some operating conditions.

The piston must maintain close tolerances to seal combustion-chamber pressure during compression and power strokes. Overall, the piston is one of the hardest-worked parts in an engine.

Removing the Rings—Carefully remove the rings from the piston, using either a ring expander or your hands. Be careful not to scratch the sides of the piston. Remove the rings up over the top, not down over the skirt.

If doing this by hand, spread the ring far enough to pull it completely out of the groove. Carefully guide it over the top of the piston to avoid gouging the skirt or ring land. Note that there is much less tension in the old rings than in new ones.

If you have a ring expander, by all means use it. A ring exander is more important during installation, where the rings have plenty of tension and you don't want to twist or break them. During removal the expander makes it easy to keep the ring from damaging the piston.

Piston Inspection—Unless part of the piston is missing, the most-obvious damage is a crack or break. The most-likely places for this type of damage are at the bottom of the piston thrust surface, or *skirt,* and the top of the piston, the *dome.*

The skirt is the thinnest section of a

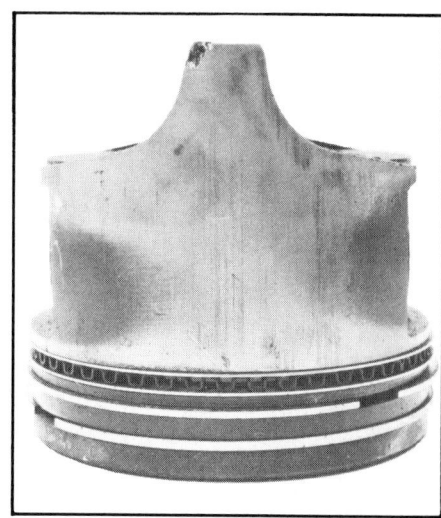

Piston wear pattern should be symmetrical—mirror image—across thrust faces. Asymetric pattern like this indicates a bent rod. Turn to page 90 to read about connecting rods.

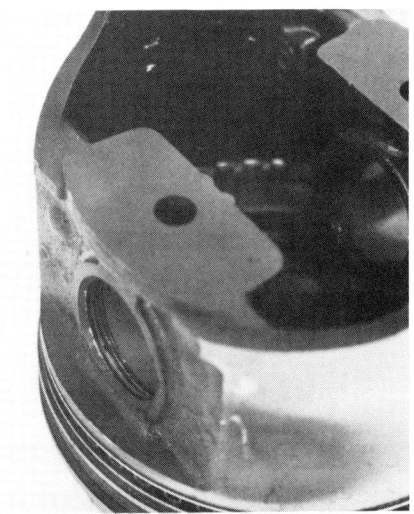
Underside view of retaining-ring groove shows damage by wrist pin. Bent rod loaded pin against retaining ring, resulting in retaining-ring groove being pounded out.

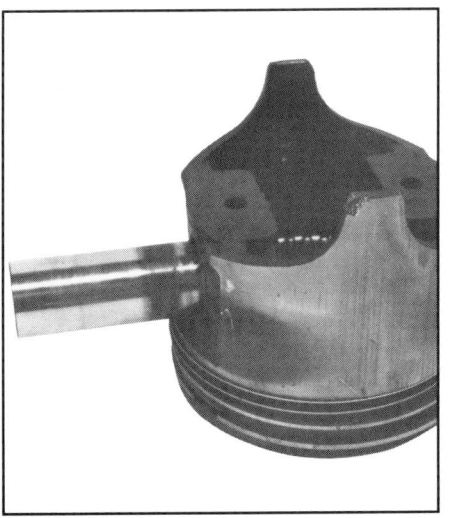
Further evidence of bent rod. Note offset wear pattern on wrist pin. Wear pattern ranges from dark-blue to black; result of oil film being squeezed from between wrist pin and rod.

Textbook failure of a piston. Preignition caused piston to overheat, scuff, then seize in bore.

piston. Its main job is to stabilize the piston in the bore. As piston-to-bore clearance increases, the piston is "slapped" harder against the cylinder wall. Eventually this can collapse or crack the skirt.

The dome and ring lands have a tough life under normal circumstances. Conditions such as detonation, preignition and carbon buildup are sources of added heat and pressure. If one of those things doesn't ruin a piston, contact with a valve probably will. Any of these items can punch a hole right through a piston.

Ring lands are also subject to breakage, mostly from preignition, detonation or broken rings. This type of damage is normally visible to the eye—especially after the piston is clean. If you want to check further, pistons must be X-rayed or checked with a dye-penetrant like Spotcheck. Pistons cannot be Magnafluxed.

The last area to check is the wrist-pin bore. You should inspect the area around the pin bore for cracks or galling both on the inside and outside of the piston. Also check each retaining-ring groove to see if it is wider than the ring.

Wrist pins are a problem area on the big-block Ford, mostly with heavy-duty trucks. The problem can be insufficient piston-to-bore clearance, which causes the pin bore to deflect more and puts added pressure on the wrist pin. More than likely, the problem is a result of a bent connecting rod, incorrect pin-to-bushing clearance or too little wrist-pin end play.

A bent rod may cause the wrist pin to shift and place more load on one of the retaining rings. This can eventually pound the retaining-ring grooves wider, or cause the ring to seize and spin in the piston.

If the pin-to-pin-bore or pin-to-bushing clearance is wrong, the pin can wobble. Eventually the pin can push or spin the retainers out if the pin seizes in the rod bushing. These are important clearances to check, especially if you are building the engine for heavy-duty or racing use.

Scuffing or Scratching—Scuffing is fairly common on engines with high mileage. Scuffing is the result of excessive heat. When the engine is overheated, the pistons may expand to the point that piston-to-bore clearance becomes insufficient. This forces the oil film out from between the cylinder wall and the piston so that the piston contacts the cylinder wall. Without sufficient lubrication, the piston will eventually seize in the bore, an expensive proposition.

Scratching can be caused by the same problems as scuffing, plus a few others—dirt in the oil, carbon particles, a broken ring or a gasoline- or ether-washed cylinder wall.

Wear—High-mileage, abuse or poor maintenance causes piston wear. The main thing to check is piston diameter, especially at the piston *skirt*—the section below the wrist-pin bore.

A piston is not a true cylinder but is flared—it's wider at the skirt. The piston should be at least 0.0005-in. wider near the bottom than at the wrist pin. Both measurements should be made across the thrust surfaces—90° to the pin bore.

Depending on the amount of wear, you may find the piston is the same size or smaller at the skirt. If this is the case, the piston should be discarded. Once the piston skirt has started to collapse, the piston rocks—slaps against the cylinder wall. Eventually the piston skirt cracks or breaks.

Pin Bore—The next areas to check are the wrist pin and pin bore. Before disassembling the piston and connecting rod, you can *feel* for excessive wear.

Place the piston on its top and grab the connecting-rod big end. Twist the connecting rod back and forth. Then try to tilt the connecting rod 90° to its normal movement. Feel for any movement.

If it feels tight, soak the piston head in some solvent to remove any oil, which will cushion the movement. Try the test again with the pin and pin bore dry.

If the piston passes this test, re-oil the pin, pin bore and bushing so you won't forget before engine assembly. Squirt oil in the hole at the top of the connecting rod and around the edges of the retaining rings. Hold the piston upside down and squirt more oil

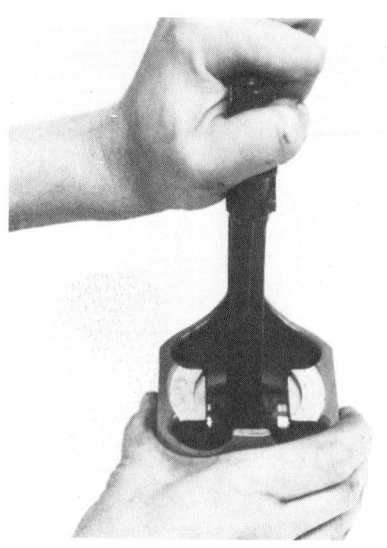
To check for pin-bore wear, soak piston and rod in solvent to remove any oil. Try to twist the connecting rod 90° to the pin, they try rocking the rod sideways to its normal rotation. If any movement is felt, the pin or bore is worn. Disassemble the rod-and-piston assembly and check them with micrometer and telescoping gage.

Pin boss cracked when piston, shown on page 87, seized in bore. Connecting rod and cap parted company shortly afterwards.

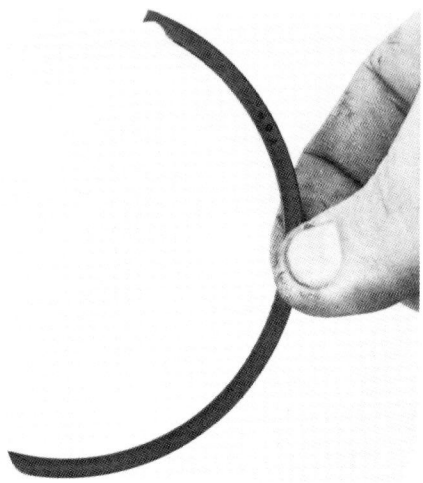
If you don't have a ring-groove cleaner, break an old ring in half and grind one end as shown. Use it to clean the grooves. Photo by Don Taylor.

Wrist-Pin-to-Piston Clearance (inches)		
Application	Clearance	
standard FE engines	0.0001–0.0003	0.0008 maximum
FT engines	0.0003–0.0005	0.0008 maximum
racing FE engines	0.0007–0.0009	

directly on the wrist pin. Slide the rod back and forth to work the oil in.

The last way to check wrist-pin clearance is to disassemble the piston and rod and measure it directly. To disassemble the piston-and-rod assembly, turn to page 93. Once you have the piston-and-rod disassembled, measure the pin and bore.

Measure the wrist-pin bore in the piston with a micrometer and a telescoping gage. Do the same for the rod bushing. Take several readings in the pin bore and rod bushing to be sure they are round.

Next measure the diameter of the pin. Subtract the pin diameter from the other two measurements for the clearance. Pay particular attention to the rod bushing—it gets the most wear.

When checking these assemblies, be sure all the parts are at the same temperature. If you hold the wrist pin in your hand long enough, the pin can warm up to the point where it will show no clearance—try it and see.

The best way to do these measurements is to have all the parts warm when you measure them—80F (25C) or more is great. Allow all the parts to warm up to that temperature.

If you find the piston is in good shape except for out-of-round or too-large pin bores, they can be reamed and an oversize pin installed. Pins are available in 0.001 and 0.002-in. oversize. If you do this, the rod bushing must be honed to correct clearance.

Clean the Ring Grooves—If the pistons have checked out OK this far, the next thing to do is clean and check the ring grooves.

Before starting this job you may want to compare the value of your time to the cost of pistons, particularly if the pistons are borderline. Ring-groove cleaning is tedious and you have to be careful not to damage the piston. If you decide to clean the ring grooves, pull up a chair. It's going to take awhile.

Ring-groove cleaners are available at automotive shops or rental stores. Most cleaners have a V-shaped bar at one end to stabilize the tool and a cutter to remove the carbon. The tool is adjustable for piston diameter and ring-groove width. One of these will save you a lot of time and lessen the chance of damaging the pistons.

You can also use an old piston ring. Break off a little less than half and grind it as shown in the photo. I have done it this way a few times and found it less than enjoyable, depending on your patience and the amount of carbon buildup.

To clean ring grooves, scrape all the carbon away without removing the metal. Be sure you don't scratch or gouge the ring lands or backs of the groove, or the outer surface of the piston. When cleaning the ring grooves, the piston-and-rod assembly should be protected from damage.

I prefer to do this while holding the assembly in my hand. The piston-and-rod assembly can also be placed in a vise with wood blocks over the jaws. Clamp the vise on the rod, not the piston, and gently rest the piston skirt on the top of the jaws. These pieces are tough, but shouldn't be abused.

Measure Ring-Groove Width—With the pistons clean, check for ring-groove damage and wear.

You may see *steps* in the grooves, particularly if the engine had extreme bore taper. As a piston moves up and down in a bore, the rings move in an out to follow the taper. This constant movement wears the ring grooves, leaving a step in the bottom of the groove.

If you see this step, quit—buy new pistons and have the block rebored.

PISTON-RING WIDTH (IN.)

ENGINE	YEAR	#1 COMP	#2 COMP	#3 COMP	OIL
332	58—59	5/64	3/32		3/16
352	58—67	5/64	3/32		3/16
360	68—76	5/64	3/32		3/16
361	58—59	5/64	3/32		3/16
390	61—65	5/64	3/32		3/16
	66—67	5/64	5/64		3/16
	68—76	5/64	3/32		3/16
406	62—63	5/64	3/32		3/16
427*	63 early—65	5/64	3/32		3/16*
†	late-65—68	1/16	1/16		1/8†
428	66—70	5/64	5/64		3/16
330 MD	64—77	5/64	3/32		3/16
330 HD	64—66	5/64	3/32		1/4
	67 3 ring	5/64	3/32		1/4
	4 ring	5/64	5/64	5/64	3/16
	68—72	5/64	5/64	5/64	3/16
	73—78	5/64	3/32		3/16
359 & 361	64—66	5/64	3/32		1/4
	67 3 ring	5/64	3/32		1/4
	4 ring	5/64	5/64	5/64	3/16
	68—69	5/64	5/64	5/64	3/16
	70—73	3/32	5/64	5/64	3/16
	74 to ser. U40001	3/32	5/64	5/64	3/16
	from ser. U40001	3/32	5/64		3/16
	75—78	3/32	5/64		3/16
389 & 391	64—66	5/64	3/32		1/4
	67 3 ring	5/64	3/32		1/4
	4 ring	5/64	5/64	5/64	3/16
	68—69	5/64	5/64	5/64	3/16
	70—73	3/32	5/64	5/64	3/16
	74 to ser. W00001	3/32	5/64	5/64	3/16
	from ser. W00001	3/32	5/64		3/16
	75—78	3/32	5/64		3/16

NOTE: Some pistons had a spacer or expander behind top compression ring.
*except High-Risers
†includes 63—65 High-Risers

If pistons pass preliminary checks, give them a thorough cleaning. Dull screwdriver works well for cleaning tops.

PISTON-TO-BORE CLEARANCE

FE exc. 427	0.0015—0.0023
427 street*	0.0030—0.0038
427 racing*	0.0060—0.0070
FT	0.0025—0.0033

*forged pistons. With aftermarket forged pistons, use manufacturer's recommendation.

The groove can be machined wider and groove inserts installed to take up the additional width. Considering that used pistons won't last as long as new ones, it's hard to justify this expense.

When measuring ring-groove width, keep in mind that too much clearance will allow the new rings to move up and down in the grooves as the piston moves. This will increase ring and land wear, possibly leading to a broken ring or land.

See above chart for the ring-groove widths for your engine. Ideally, you should have new piston rings to check the grooves. If you have new rings, all you have to do is place them in the groove and then check the side clearance.

You can check the grooves with old rings, though it's more complicated. Measure an old ring with a micrometer. Insert the ring into the groove and check the clearance with a feeler gage. Be sure to mike the width of the rings and be sure the rings and grooves are clean.

To find the width of the grooves, add the clearance to the ring width. When measuring old pistons, check each groove in several places to be sure you find the widest point. Pay particular attention to the top compression-ring groove.

The top compression ring is the main barrier against combustion heat and pressure, so it takes the most punishment. It's more likely to be damaged than any other ring groove.

Oil-ring grooves are not subject to the same heat and pressure as the compression-ring grooves, and normally have a more-than-adequate supply of oil. If they fail, it is usually due to tapered cylinder walls. In either case, the top compression rings and grooves are most likely to be worn or damaged.

You can usually just look at an oil-ring groove to check its condition. In most cases it will look new compared to the compression-ring grooves.

To measure the oil-ring grooves, place two compression rings together and measure their width. Then slip the rings into the oil-ring groove and check clearance with a feeler gage.

Note: A number of the production 427s used 1/16-in.-wide compression rings. On these pistons, you must use *three* compression rings to check oil-ring grooves.

Piston-to-Bore Clearance—One of the major factors affecting engine performance is piston-to-bore clearance. Even if the old pistons and bores are acceptable, if this clearance is too great you must buy new pistons.

There are two ways to measure piston-to-bore clearance. In both cases, make sure both the piston and bore are at the same temperature. Also be sure you compare each piston to *its* bore.

The first way to check piston-to-bore clearance is to measure the diameter of each piston and its bore. The difference is the clearance.

Before you measure the pistons, refer to the drawing on page 90. Mea-

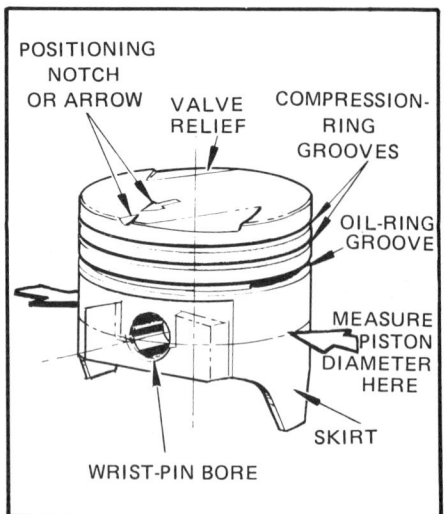

Pistons are not cylindrical, but are flared wider at the skirt. When measuring pistons for piston-to-bore clearance, measure in plane of wrist-pin bore as shown.

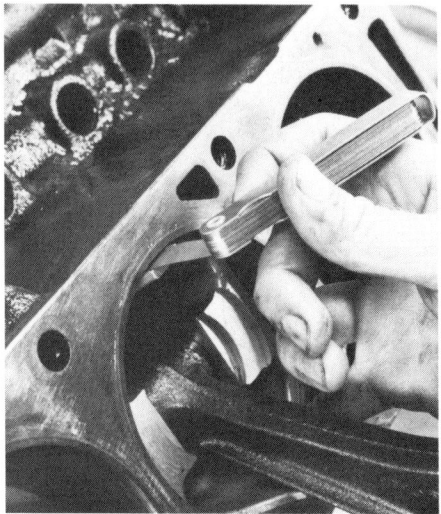
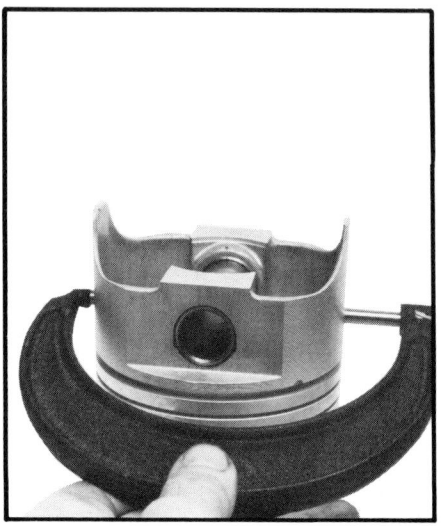

Piston-to-bore clearance can be determined by measuring width across piston skirt and subtracting from bore diameter. Clearance can also be measured directly with piston in bore using feeler gage.

sure across the thrust surfaces in the same plane as the wrist pin, 90° to the wrist-pin axis. To measure the bore, measure where the thrust faces of the piston touch the cylinder walls at TDC, which is 90° to the crankshaft center line, about 2 in. below the head-gasket surface.

The next way is to measure piston-to-bore clearance directly. Remove the rings from the piston. Place the piston in its bore right side up and close to TDC. Measure the clearance between the thrust face of the piston and cylinder wall with feeler gages.

This measurement will give you a direct reading. Be careful when inserting the piston into the bore that you don't damage it.

Piston-to-bore clearance is greater on racing and heavy-duty truck engines to help maintain minimum clearance. Pistons in these engines expand more due to increased heat from increased combustion loads. They simply run hotter when they work harder.

The 427 pistons are forged aluminum—other big-blocks use cast pistons. The denser aluminum in the forged piston is stronger, but it also expands more. Cast pistons use a cast-in steel strut to control heat expansion. Forged pistons don't have this strut, so they expand more. Consequently, initial clearance must be greater with forged pistons.

Knurling—If the piston-to-bore clearance is too great, there is a solution that's cheaper than a rebore and new pistons. The clearance can be taken up by a process called *knurling*. Knurling is a temporary, stopgap measure; don't use it unless you want the cheapest way out.

Knurling is a process that makes impressions in the thrust surfaces of a piston. These impressions cause the surrounding material to raise up, effectively increasing piston diameter.

These impressions also retain oil to aid lubrication, but the raised portions

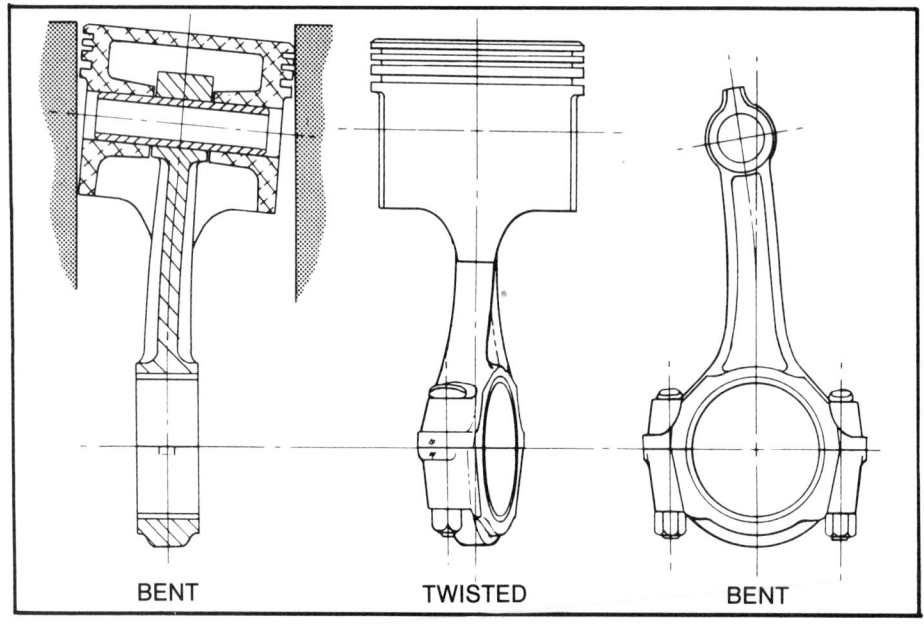

First two rods, bent and twisted as shown in these exaggerated drawings, cause uneven piston-skirt wear as pistons are tilted in their bores. First rod tilts its piston most at TDC and BDC. Twisted rod does its tilting progressively toward the mid-point of its stroke. A connecting rod bent like third one shown can be reinstalled unless the bend is obvious to the eye or rod is used in an all-out performance engine. Drawing by Tom Monroe.

of the piston do not have much surface area. Consequently they wear rapidly and you're back to square one—needing new pistons.

CONNECTING RODS

As the link between the pistons and crankshaft, connecting rods are subjected to heavy loads. Although they are tough, connecting rods frequently get out of tolerance. Fortunately connecting rods are usually repairable.

I found when disassembling the

This may look crude but it works. Rod is bent slightly past being straight. It then springs back into position.

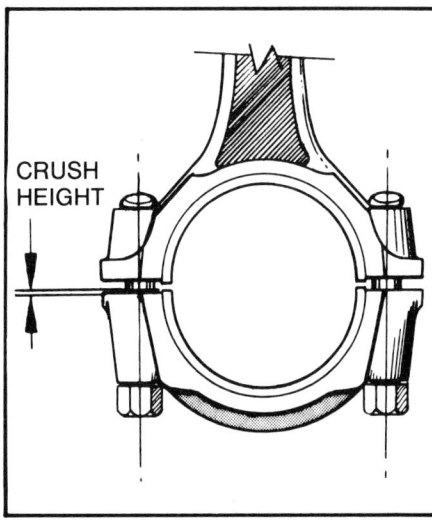

How much a bearing insert projects above its bearing-bore parting line is called *crush height*, or *crush*. This should be about 0.001 in. As the bearing cap is tightened down, insert ends contact first, forcing the insert to conform to the size and shape of its bore. This locks the insert into place and ensures good heat transfer from the bearing to the rod and cap. Drawing by Tom Monroe.

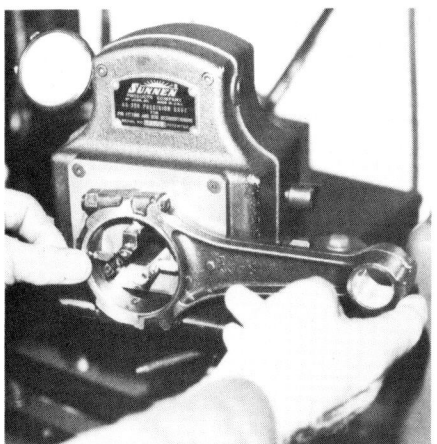 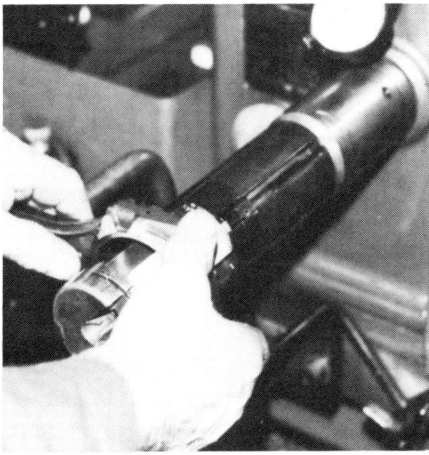

Crank end of rod is checked for size and out-of-round. Specified diameter ensures correct bearing crush. If diameter is too large, rod and cap parting faces are ground to reduce size of opening. Rod is then honed to the correct diameter.

428, pictured in this book, that one of the piston skirts was cracked and the wrist-pin bushing was breaking away. A check at the machine shop turned up a badly bent rod, the probable cause of the problems.

The major areas to check on connecting rods are bending or misalignment, out-of-round bearing bores or bores damaged by a spun bearing, and wrist-pin bore too large or damaged from a failed bushing.

Bent or Twisted—It's a good idea to have all the rods checked by a machine shop. If you don't want to spend the money, take a good look at the piston and rod-bearing wear patterns.

The wear on the piston should be symmetrical across the thrust surface. That is, one side of the wear pattern should be a mirror image of the other. See photo, page 86.

The connecting-rod bearings and journals should also have even wear. A bent rod places uneven loads on the bearings and journals. Be sure to check any rods that were on tapered journals.

A tapered crankshaft-bearing journal will give a similar bearing wear pattern, but the wear will be on the same side—front or rear—of *both* inserts. A bent rod will wear *opposite* sides of the bearing inserts.

For example, if a rod is bent, the *upper* insert will be worn on the *rear* portion while the *lower* insert will be worn on the *front*, or vice versa. Double-check against the crank.

If you suspect a rod is bent or twisted, take the rod to a machine shop and have it checked or straightened. Machine shops have a fixture to check connecting-rod alignment. The rod is checked with the piston on or off, depending on the type of fixture.

A bent or twisted rod can be straightened. Of course, if a rod is bent or twisted 90° it's a little ridiculous to have it straightened. Consider a bend of 15° maximum for standard engines and 5° maximum on performance engines. Any more means you should replace the rod.

Out-of-Round Bearing Bores—Both big- and small-end bores should be checked for size and roundness. You've given the wrist-pin bores a preliminary check, so pay particular attention to each big-end bearing bore.

To obtain the proper *bearing crush*, a rod bore must be slightly smaller than the OD of the bearing halves. The slight undersize of the bore causes the bearings to *crush*, which locks them in place and makes the bearings conform to the shape of the rod bore.

Bearings need this tight fit to lock them in place. Engine bearings depend on tight contact with the bearing bore to transfer heat away from the bearing surface.

If the bore is too small, there will be too-much crush and the bearing parting faces will bend in towards the journal. This causes the bearing to wear at the parting faces.

If the bore is too big, there will be too-little bearing crush. This will cause the bearing to overheat, and also increase the chance of the bearing spinning in the bore. Either of these conditions can damage the rod, bearing or crankshaft.

If the wrist-pin bore is too large, the wrist pin will knock. This will destroy the bushing, wrist pin and finally the rod. If it is too small, the wrist pin will seize and score the bore in the piston and destroy the retaining rings or piston.

To measure the connecting-rod bores, you will need an inside micrometer, a dial bore gage, or a telescoping gage and outside micrometer. Before checking the big-end bore, assemble the rod and cap. Set the big end of the rod in a vise between two blocks of wood. Install the cap and nuts or bolts, then torque them to specification as specified on page 130.

Beginning with the big end, mike the opening for diameter, then check for out-of-round. If you have the Sunnen-type indicator, rotate the rod to check out-of-round. With an inside mike or telescoping gage you must reset after each measurement. Write down the measurements for each rod.

If the bore at the big end of the rod is out-of-round or oversize, the mating surfaces of the cap must be ground down, and the bore honed out to the correct dimension. If the bore is too small it only needs to be honed to size.

Big-end bore diameter should be 2.5907—2.5915 in. Maximum out-of-round is 0.0004 in. for FE engines and 0.0006 in. for FT engines. Maximum bore taper is also 0.0004 in.

If you have to replace the wrist-pin bushings, check the bushing bore after you have removed the bushing. If the bushing-bore diameter is too big, the rod must be discarded. Usually the bushing-bore diameter is OK, unless the bushing came out of the bore and the wrist pin hammered the small end out of shape.

The diameter of the small end without the bushing should be 1.028—1.038 in. With the bushing installed, bushing ID should be 0.9752—0.9755 in. Pin-to-bushing clearance is 0.0001—0.0005 in. for standard FE engines. For FT and racing FE engines, pin-to-bushing clearance is 0.0002—0.0005 in. Maximum clearance is 0.0010 in. for all engines.

All honing on a connecting rod should be done on a precision hone—so the big- and small-end bores will be in alignment. The fit between the wrist pin and rod bushing should be a *light press fit*—you should be able to push the wrist pin through with your thumb with little effort.

Test the fit with the parts at room temperature—70F (20C)—or more. Remember not to hold onto one part too long. Warming them unequally will change the clearance. Before testing the wrist pin in a new bushing, chamfer the lead-in edge at both ends of the bushing—see the section on rod-bushing replacement on the following page.

Inspect Rod Bolts—Visually check the rod bolts for cracks. Better yet, have them *Magnafluxed*—magnetic-particle inspected. These bolts are highly stressed and are subject to fatiguing and cracking.

The LeMans or NASCAR rod bolts thread into the rods. These simply unscrew. Removing the bolts causes no problem with rod-to-cap alignment. The cap aligns on concentric dowels, not the rod bolts.

This is not true with standard rods. Bolts from the other rods must be pressed or knocked out with a soft hammer. If the rods use pressed-in bolts, the big ends must be checked or honed for alignment after the bolts are reinstalled.

To remove the pressed-in bolts, place the big end of the rod in a vise. Clamp the rod between two pieces of wood. Place a brass or hardwood drift against the threaded end of the bolt and drive it out. To reinstall, press the bolt back in or install the cap and draw the bolt in by torquing the nut. Be sure you have the big end held securely in a rod vise or a bench vise between two pieces of wood. Also check that

Wrist-Pin-to-Rod-Bushing Clearance (inches)		
Clearance	Maximum	Application
0.0001—0.0005	0.0010-in.	standard FE engines
0.00015—0.00045		FT engines
0.0003—0.0005		racing FE engines

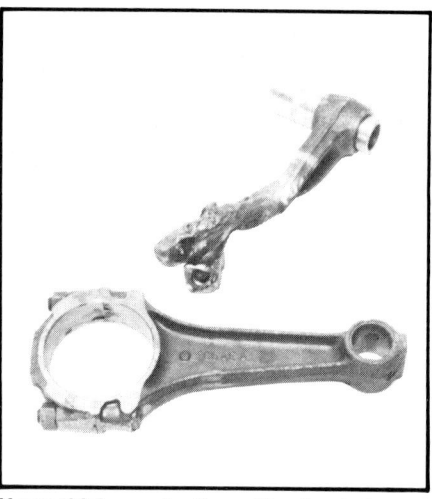

If you think you don't need to check the rod bolts, think again. Rod at top was doing just fine until rod bolt said, "goodbye." A broken rod bolt can be an expensive failure. Not only is the wayward rod destroyed, it often takes the crankshaft and cylinder block with it.

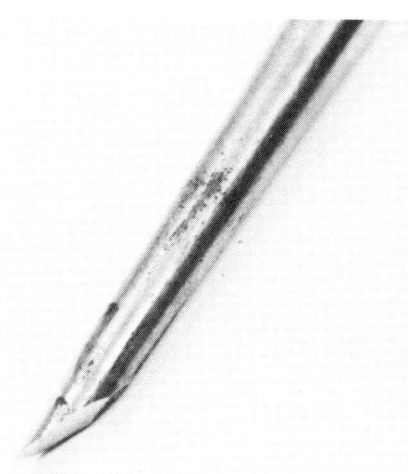

If you don't have a cape chisel, a sharp-pointed punch can be used to remove rod bushings.

New rod bushings can be solid or split, with or without an oil hole. If they each have an oil hole, make sure you align the hole in the bushing with the hole in the rod before pressing in bushing. If new bushing has no hole, you will need to drill it after bushing is pressed in place.

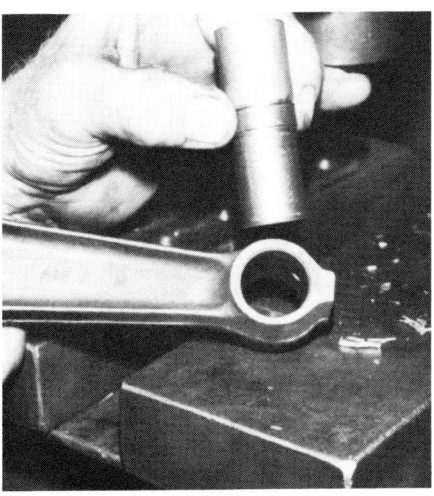

Here is special tool Paul has machined to press out old bushing and install new one simultaneously. Close-fitting mandrel keeps bushing from collapsing while it is pressed into place. Also note bushing has been chamfered to provide lead-in to assist with the installation.

the square edge of the bolt lines up with the flat side of the rod.

On the LeMans and NASCAR rods, chase the rod-bolt threads with a 7/16-20 bottoming tap. This ensures accurate torquing of connecting-rod bolts.

The stock LeMans rod uses a 7/16-in. bolt with a necked-down shank. Although these bolts were originally intended to minimize bolt breakage, the necked-down bolts would stretch before they would break. Ford discovered the bolts caused more problems than they prevented.

Replace these bolts with an aftermarket bolt with a straight shank or bolts from the NASCAR rods. Rebalance the rods with the new bolts installed.

DISASSEMBLING & ASSEMBLING PISTONS & CONNECTING RODS

Before disassembling the rod-and-piston assemblies, mark the parts so you can match them up again, or disassemble, inspect and reassemble them one at a time. Keep the parts clean and at the same temperature when making measurements, preferably room temperature, 70F (20C) or more.

Wrist Pin—To remove the wrist pin, first remove the retaining—"snap"—rings at both ends of the pins. Use a pair of inside-retaining-ring pliers.

Next drive the wrist pin from the piston-and-rod assembly. Often you can remove the pin by pushing it through with your thumb. Hold the piston in your hand while removing the pin. A spray of WD-40 or CRC on the exposed part of the wrist pin under the piston will help removal. Be careful not to drop the piston, rod or pin when they separate.

If the pin won't come out with thumb pressure, soak the assembly in carburetor cleaner to remove any varnish buildup. If the pin still won't come out, you'll have to use a special pin-driving tool. You can either buy one or make one.

The tool is made of round steel stock, slightly smaller in diameter than the wrist pin. A short shank protruding from the end pilots in the center of the wrist pin. The pilot keeps the tool from scratching or gouging the pin bore.

If you can fabricate one of these out of brass or steel, great. If not, a hardwood dowel will work. Be sure to sand the sides smooth. The diameter must be smaller than the wrist-pin bore.

Rod Bushing—To remove the rod bushing, you can either peel it out with a special chisel or press it out on a press. If the rod bushing must be replaced, leave the work to the machine shop—the bushing must be honed to size after being installed anyway.

A special driver is used to press out the bushing. The driver pilots in the small end of the rod and has a small —*very small*—shoulder to butt against the outer diameter of the bushing. The rod must be supported squarely from the other side when the bushing is pressed out. This is another reason to leave the job to a machine shop.

To remove the bushing with a homemade tool, cut a long 3/8-in. or larger bolt off above the threads. Grind the end of the bolt until you have a round-bottomed chisel—shaped similar to a crescent moon.

Carefully work the point under the bushing and peel the bushing out. Use care not to scrape the rod surface during removal or it will distort the bushing when it is installed. A split bushing will peel easily by the seam. On a seamless bushing, work from one side and drive the chisel between the bushing and pin bore. Collapse the bushing and push it out.

To install the new bushing, clean the rod bore and check the replacement bushings for a chamfer. New bushings should come with the OD at one end chamfered. This helps start the bushing in the rod.

If there is no chamfer, cut one with a knife or file. The *lead-in* will ease installation and prevent the bushing from being crushed.

Use a vise or press to install the new bushings. On a press, use the same tool you used for removal. If you are using a vise, protect the bushing end and the rod with soft jaw covers. The bushing should fit completely inside the small-end bore, but doesn't have to be perfectly centered side to side.

If the bushing comes with a predrilled oil hole, line it up before pressing in the bushing. If there is no oil hole press the bushing into place, then drill the hole. Place a piece of wood in the bushing to protect the far wall. Carefully drill a hole in the bushing the same size as the hole in the rod.

The bushing must now be honed to size. The bushings are reamed to match a particular wrist pin, so once the reaming is done, *keep the wrist pin with the rod*. Wash the bushing out before you check the fit so the cuttings don't score the pin or bushing.

Once the bore has been honed, recheck the oil hole to make sure it is fully open. Bronze from the bushing

Pin bushing can be pressed into place with a press or with a simple vise. Regardless of method used, chamfer leading edge of bushing and use smooth, parallel surfaces against rod and bushing when pressing it into place.

Once bushing is installed, it must be honed to size. Wrist pins are often selectively fitted to rods, so keep rods and pins together if this method is used.

When installing wrist-pin retaining clip, compress clip only enough to fit it into its groove. If distorted, clip won't fit tightly. This could result in severe engine damage.

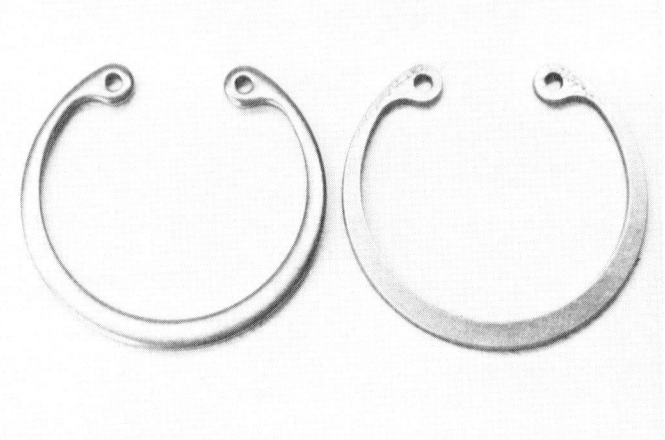

Retaining rings have a front and a back. Rounded edge, left, installs against wrist pin. Sharp edge at right install away from pin.

may have been dragged over the hole during honing. If the hole is blocked, reopen it with a drill bit turned by hand. Again, don't hit the side opposite the hole.

Finally, chamfer both ends of the bushing with a bearing scraper or pocket knife. The chamfer can be relatively small—about 1/32-in.—enough to provide a lead-in for the wrist pin. Again, be careful not to scratch or bump the inside of the bushing.

New Pistons—If you are using new pistons, they should come with new wrist pins already fitted. Check the pin-to-bore clearance. As previously discussed, the clearance is so small that you can check it by feel. You should be able to insert the pin by hand, but shouldn't be able to wiggle it in the bore. A more accurate check, outlined on pages 87 and 88, gives exact clearances.

Another clearance, often not checked, is wrist-pin end clearance. Install the oiled wrist pin in the piston with the retaining rings.

Install the rings with their rounded edges toward the wrist pin. See above photo. Install one ring, then the wrist pin and the other retaining ring. Push the wrist pin against one of the retaining rings and measure the gap between the pin and the other ring.

Wrist-pin end clearance should be 0.001—0.009-in. for passenger-car or light-truck engines. On race engines or heavy trucks give them a bit more clearance, 0.005—0.012-in. If the clearance is less than specified, try switching the pin from piston to piston to obtain the right clearance. Otherwise, the pin must be shortened.

If the pin needs to be shortened, it must be held in a fixture like that used for surfacing the rod faces. This will hold the pin perpendicular to the grinding surface. Sufficient clearance might be gained by simply polishing both ends of the pin.

Assembling the Pistons and Rods—Once all the pistons, rods and wrist pins have been checked or recondi-

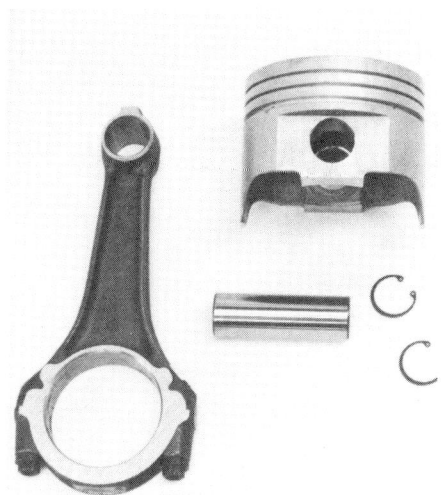

Piston and rod ready for assembly. If pistons have been matched to specific bores, be sure you match them with the correct rods.

Oiling-hole alignment is important. All rods have holes for wrist-pin oiling (outline arrows). Standard-performance rods have groove cut in cap-to-rod parting faces to aid cylinder-wall oiling (solid arrows). LeMans rod at right does not. All oiling holes must be correctly oriented when rod is assembled to piston; see text.

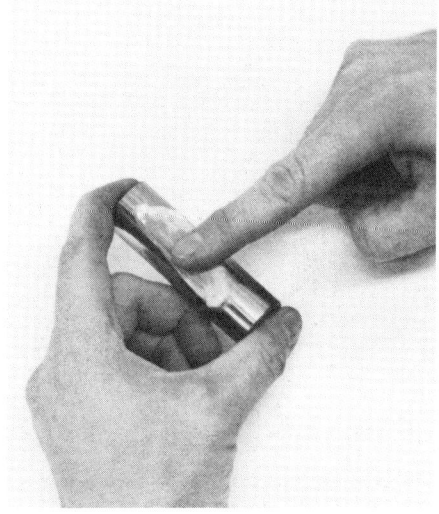

Thoroughly lubricate wrist pin before installing.

tioned they are ready for assembly. All the parts should be clean and their contact surfaces oiled. Pistons, pins and rods are matched to one another, so keep them together.

Although pistons can be installed on their rods two ways, only one way is *usually* right. Some heavy-truck pistons can be installed two ways. Most pistons have an offset to reduce piston slap. Most also have either a dome or valve reliefs, which must fit the combustion-chamber and valves.

Pistons will have a notch or arrow on top. This mark should point to the front of the engine when the piston-and-rod assembly is installed.

The connecting rods must also be installed in a certain direction. A chamfer on the big end of each rod must fit the *crankshaft fillet*—the radius at the joint between the crankshaft journal and cheek.

On all except the LeMans and NASCAR rods, there is an oiling hole at the rod-to-cap mating surface. This hole should point toward the opposite bank when the piston-and-rod assembly is installed. That is, the oiling holes on rods 1, 2, 3, and 4 should point toward cylinders 5, 6, 7, and 8 and vice versa.

To line up the rods, stand them with the small end up and the wrist-pin and crank-journal bores facing you. In this position, the rods for the right bank (1, 2, 3 and 4) should have the bearing locks and rod numbers on the right, and the wrist-pin oil hole and cylinder-wall oiling hole on the left.

NASCAR and LeMans rods are positioned the same way. Even though they have no cylinder-wall oiling hole, the bearing locks and rod numbers for the right bank face right and the wrist-pin oil holes face left.

A quick double-check for the pistons on the right bank is to hold the rod with the wrist-pin oil hole away from you. The notch for the piston should be to the right.

Pistons and rods for the left bank are assembled exactly the opposite. With the small end up and the wrist-pin and crank-journal bores facing you, the rods for the left bank should have the bearing locks and rod numbers on the left. The wrist-pin and cylinder-wall oil holes should be on the right.

To double-check the left-bank pistons, hold each rod with the oil hole away from you. The notch should be to the left on the piston.

Because the wrist-pin clearance is so small in both the wrist-pin bushing and the piston, assembling rods and pistons should be done at room temperature. If you try to insert a warm wrist pin into a cold piston it won't go. If you have trouble assembling the parts, heat the piston and rod with a light bulb.

Lubricate the wrist pin, rod bushing and pin bores of the piston with moly or white grease—I use Lubriplate #105. After all the parts are thoroughly lubricated, assemble them.

Again, the rounded edge goes against the pin and the sharp edge to the outside of the piston.

Install one retaining ring to keep the wrist pin from sliding through. Hold the piston-and-rod assembly together. Double-check the relationship between the piston and rod. Slide the pin through the piston and rod and install the other retaining ring. Put the piston-and-rod assemblies aside until installation time.

TIMING CHAIN & SPROCKETS

Timing chains and sprockets are discussed in Chapter 3, but there are a few things to consider when buying these parts.

There are two crankshaft sprockets for FT engines, which all use the heavy-duty roller chain. The first, C4TZ-6306-A, is for forged-steel cranks; the second, C4TZ-6306-C, is for cast-iron cranks.

FE engines also use two crankshaft sprockets. One, B8A-6306-A, uses a 3/16-in.-wide Woodruff key while C4AZ-6306-A uses a 1/4-in.-wide key. The 3/16-in. key is used on '58—'63 cranks and the 1/4-in. key is used from '64 on. There is one catch—'63 390s used both keys, so you'll have to compare new and old parts.

The next area to watch on FE en-

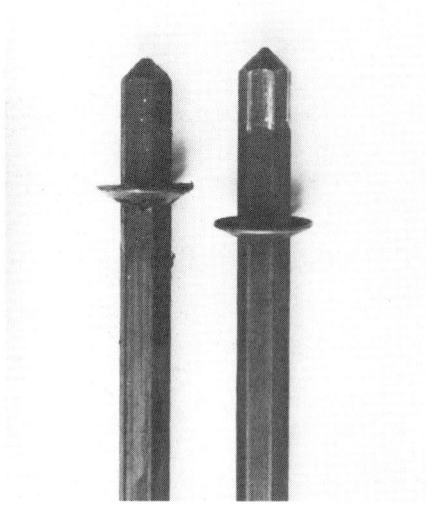

Engine depends on small-diameter drive shaft for its life's blood. New oil-pump drive shaft is cheap insurance against failure. Note rounded edges on used shaft, right.

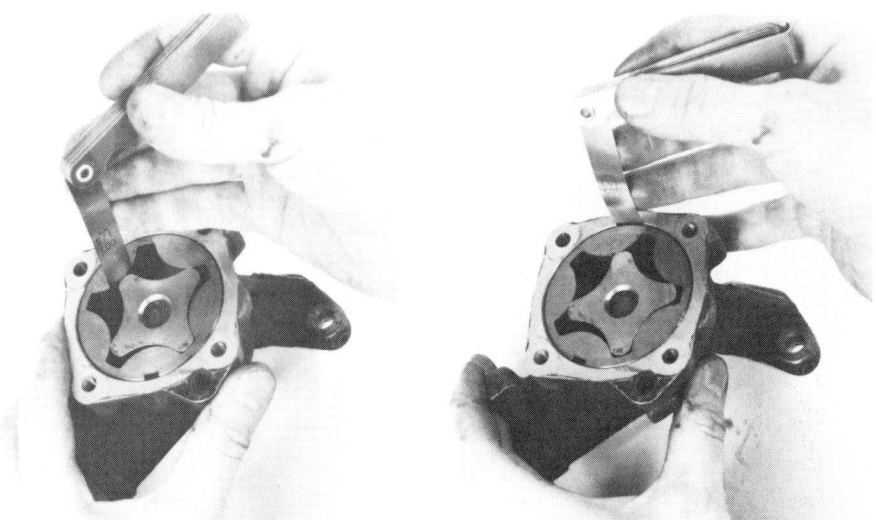

Remove bottom plate from oil pump to inspect condition of inner and outer rotors and pump housing. Check for scores or pits. Then measure rotor clearances: Clearance between rotor tips should be less than 0.015 in., with 0.006 in. desired. Outer-rotor-to-housing clearance should be less than 0.012 in., also with 0.006 in. desired.

gines is the camshaft sprocket. If camshaft thrust is controlled by a thrust button, the cam uses a 1-1/4-in.-long bolt and a 7/8-in.-long dowel. If the camshaft is held in place by a thrust plate, the bolt is 1-7/8-in. long and the dowel is 1-1/2-in. long.

From '63—'69, FE camshaft sprockets with a thrust plate used a separate spacer between the sprocket and thrust plate. On '70-and-later FE engines, the spacer is part of the sprocket. Most replacement sprockets will have one built-in—so don't try to install a separate spacer.

Some FE engines use camshaft sprockets with nylon teeth. These work well unless the engine is subject to extreme heat. If it is, install a cast-iron sprocket. The truck roller chain can also be used. It requires the truck cam sprocket, and the crank sprocket for cast-iron crankshafts.

Many people think nylon-tooth cam sprockets have no business in a performance or racing car—not so. Nylon-tooth sprockets absorb heavy shock loads, so they are gaining popularity in drag and NASCAR racing. In fact, the first nylon-tooth cam sprockets for big-blocks were introduced on the 427.

For other types of racing or high speeds, consider a roller-type timing chain and sprockets like those available from Cloyes. They will wear longer than the standard silent-chain type and maintain cam timing longer.

OIL PUMP & DRIVE SHAFT

Start the oil-pump inspection with the oil-pump drive shaft—replace it! If you have any doubts about throwing the old shaft away, remember that the engine depends on that small shaft for its life's blood. If that shaft breaks, there will be *no* oil pressure.

It is easier—and cheaper—to buy a new shaft than it is to replace the crank, rods and bearings when the old shaft fails. Use an aftermarket chrome-moly shaft if you plan to race the engine—it is stronger and longer wearing.

One of the major reasons an oil pump wears is because it picks up *unfiltered* oil. Whatever passes through the pickup screen will be munched in the oil pump. Metal shavings, carbon deposits or plain dirty oil severely shorten the life of an oil pump.

The next common problem with the pump is the relief valve. Sometimes it will stick open, either because the spring gets weak or the valve sticks in its bore due to dirt or varnish buildup. A valve sticking is more common on a high-mileage engine—or pump. I have taken engines apart that were supposedly worn out and found the only problem to be a stuck oil-pump relief valve.

If the relief valve sticks, you can replace the spring and plunger. Do this only if the rest of the pump is OK, otherwise buy a new pump.

You can test the pump by immersing the pickup in solvent. Use the drive shaft to turn the pump a few times. The solvent should squirt out forcefully each time you turn the shaft.

Pump Clearances—Open the pump to inspect it by removing three of the four cover-plate bolts. Loosen the remaining one. Rotate the cover to the side and check for damage to the rotor, outer-rotor housing and pump housing. Check for scratches, pitting or burnt marks caused by oil starvation.

Pull the rotors up and look at the inside of the housing for any scoring or other signs of wear. Also check the underside of the cover plate.

If the rotors look OK, check clearances with feeler gages. The clearance between the inner- and outer-rotor tips should be a maximum of 0.015 in.—0.006 in. is best. Clearance between the outer rotor and the pump housing should be 0.006 in., with a maximum of 0.012 in.

The last clearance to check is be-

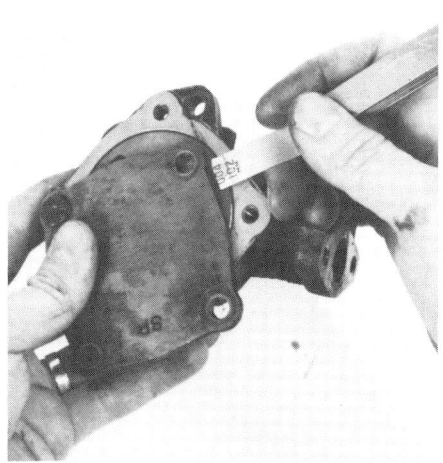
Oil-pump end play should be 0.001—0.004 in., with 0.002 in. desired.

If clearances are correct, remove and inspect rotors. Oil pump ingests unfiltered oil, so it is much more likely to be damaged than other machined parts. Replace pump if it's severely scored.

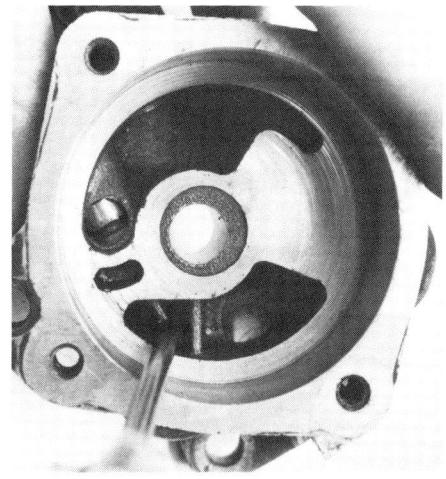
If pressure-relief valve works smoothly, reassemble pump. In most cases, a sticky valve can be freed by soaking the pump in solvent and working the valve back and forth.

tween the rotors and the cover plate. With the rotors installed, check clearance between the plate and rotors as shown above. Rotor end clearance should be 0.001—0.004 in., with 0.002 in. desirable.

Assemble the Pump—Clean and lubricate the parts and assemble the pump. Be careful with the outer rotor. It is tapered, and can be installed upside down with the taper the wrong way. Look for an indent on one face, similar to the one on the inner rotor. This mark should go away from the cover. Install and torque the cover bolts 6—9 ft-lb.

Oil-Pump Drive Shafts—The 1/4-in. oil-pump drive shaft is used on all FE engines, all 330MD and '73—'78 330HD engines with velocity governors. The 5/16-in. drive shaft is used on all '64—'72 330HD, '73—'78 330HD with Centrivac governors and all 361 and 391 engines.

FE and FT engines using the 1/4-in. shaft use the same drive shaft, B8AZ-6A618-A—5/16-in. shaft for the FT is C4TZ-6A618-A.

Shafts alone cannot be switched because the holes in the distributor and oil pump are different. To switch the entire system—distributor, oil pump and drive shaft—refer to Chapter 3 for block modifications.

When reassembling pump, note small indentations on inner and outer rotors (arrows). Indents go toward housing, away from cover.

CHAPTER 6
Cylinder-head reconditioning

Most machine work—including cylinder-head work—is usually done by an engine machine shop. You simply take the assembled heads in and all the work is done for you: teardown, inspection, reconditioning and assembly. This is OK, but it is important to be well informed when the machine shop asks you a question or when you want a particular thing done to the heads.

Few people have the tools needed for cylinder-head machine work—it just isn't practical for a now-and-then rebuild. If you have access to the equipment and can do the work yourself, by all means do so. Otherwise this chapter will help you understand what is done by the machine shop and help you make decisions when the need arises.

Cylinder-head work can go from relatively inexpensive—valve and seat grinding—to very expensive—new valves, seat inserts, springs and retainers, guides, welding or other repairs. And work must be done in order. For instance, don't grind valve seats before you do valve-guide work.

DISASSEMBLY
Keep The Valves In Order—You need some way to protect the valves and keep them *in order*. This is a must if the valve guides don't need reconditioning. Like other machined parts in the engine, valves and guides have mated to one another.

You can either drill holes in a board or yardstick, or punch holes in the bottom of a cardboard box. Mark the storage device as to what valves came from which guide in which cylinder head.

Also be sure you have the cylinder heads marked RIGHT and LEFT. Because machining or hot tanking may remove most marks, use a punch or scribe to mark the heads. If you can't

Standard big-blocks found in cars and trucks have logged many trouble-free miles for their owners. Excellent power, quiet operation and relatively good fuel economy have given this engine a good reputation. Photo courtesy of Ford Motor Co.

tell which is which, just call one RIGHT and one LEFT. The heads will interchange left to right.

This extra preparation is unnecessary if you need new valves or guides, but trying to save the valves is a good idea. Some valves—those in some 427s, for example—cost more than $50 apiece.

Valve-Spring Compressors—You need a *valve-spring compressor* to remove the valves. There are two standard valve-spring compressors: a *C-clamp* type and *hook* type.

The C-clamp type is a large adjustable C-clamp with adjustable fingers at one end and a pad at the other. The fingers fit around the valve-spring retainer while the pad butts against the valve head. C-clamp compressors are

used to remove the valves when the head is off the engine.

If the head is off the engine, you can also use the hook type for disassembly. Hook-type spring compressors can be used to remove the valve springs and retainers with the heads installed. The rocker shaft—less rocker arms—is installed on the head. A set of fingers slip under the rocker shaft—the shaft is used to lever under to compress the valve spring.

With the C-clamp compressor, adjust the fingers to fit the retainer. Place the C-clamp over the head and center the pad against the valve face. Center the fingers on the retainer and compress the spring. Adjust the travel of the compressor so the valve-spring retainer clears the keepers with the

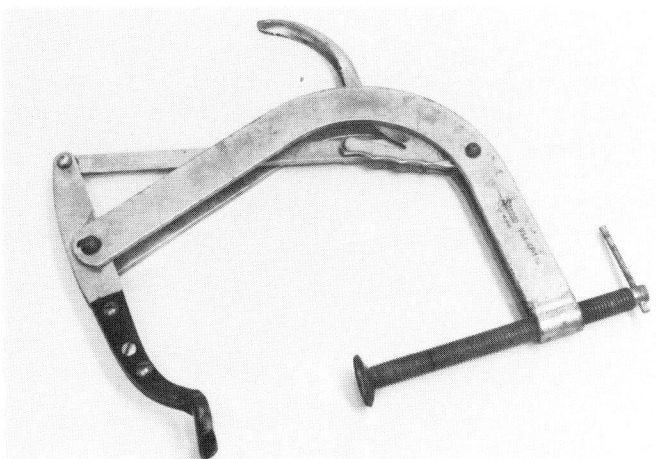
Though heads can be disassembled without it, you'll need a valve-spring compressor for teardown and assembly. C-clamp-type compressor is common. If you don't own one, rent one.

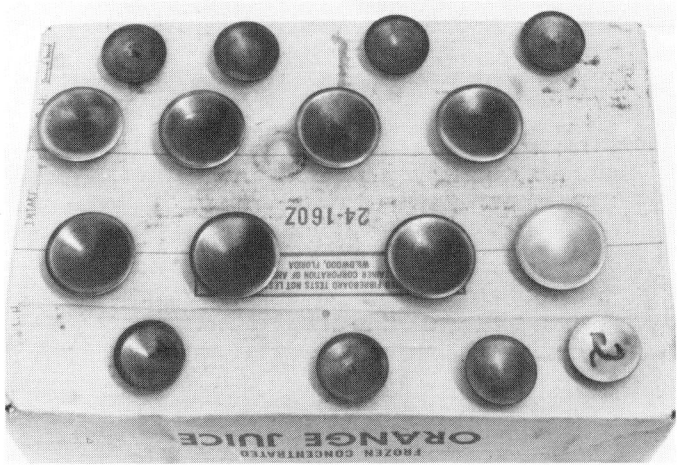
You'll need some way to keep valves in order once they've been removed. Cardboard box with holes punched in it and labeled works well.

Sharp blow to spring retainer loosens sticky keepers. This will often dislodge keepers so retainer and spring can be removed without using a spring compressor.

Valve-spring compressor should be adjusted only far enough to release keepers.

Remove keepers, then slowly release compressor so spring and retainer do not become projectiles.

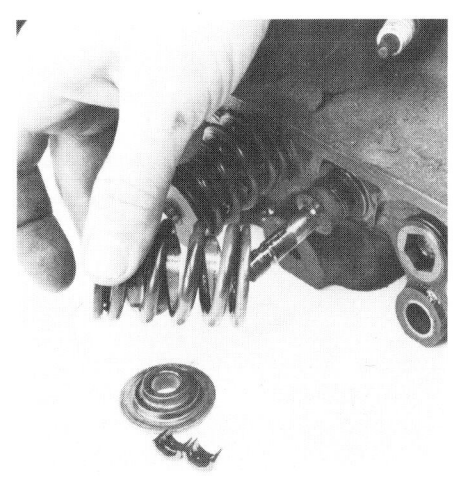
Remove valve spring/s. Most high-performance engines use dual valve springs or single spring and flat-wound damper as shown here.

Scraping cylinder heads prior to hot tanking will ensure good cleaning.

Rotary wire brush is excellent for final gasket-surface and combustion-chamber cleaning. When removing carbon use care not to gouge machined surfaces.

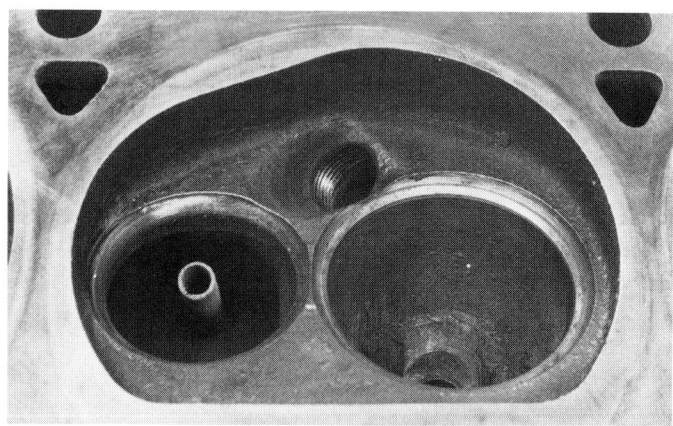
Air-injection (AIR) tube in exhaust port supplies air to complete combustion inside exhaust manifold. Burned exhaust valves are more common on engines with AIR because combustion continues in exhaust manifold.

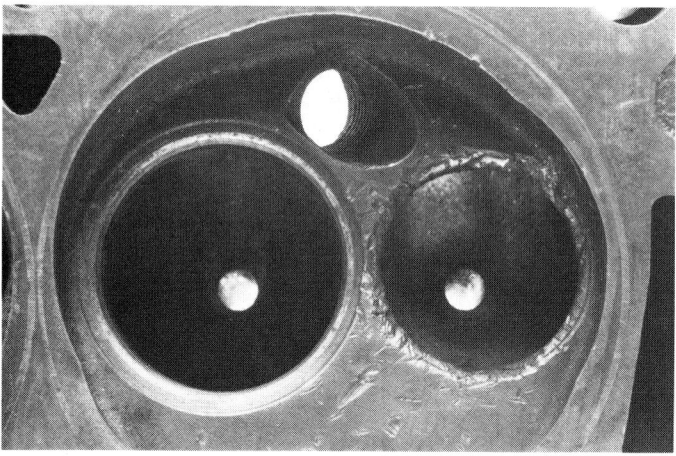
Exhaust-valve seat can be replaced. This kind of repair is expensive, so unless heads are very valuable or rare, consider new or used heads if old ones need extensive work.

Checking cylinder head for flatness. Feeler gage with a precision straight edge is normally used for this, but a precision carpenter's square or metal ruler will also work.

compressor drawn down and locked. *Do not compress the valve spring more than necessary.*

Once the spring is compressed, the keepers should fall off, or can be easily picked off with a magnet or screwdriver. Release the compressor *slowly,* so parts won't fly all over.

Stem-To-Guide Wiggle Test—The retainer, valve spring/s, seal, spacer and, possibly, the spring seat can now be removed. Before removing the valve, *feel* for valve stem-to-guide clearance by wiggling the valve. This method is not accurate, but will give you an idea of wear.

Pull each valve about 1/2 in. off its seat and wiggle it back and forth. At best, this *wiggle test* is a ballpark check, particularly on racing and truck engines. However, if the valve seems loose, the guides will probably need attention. If an engine has a lot of miles on it, you can count on reconditioning the guides.

The only accurate way to check stem-and-guide wear by this method is to use a dial indicator, or compare each valve and guide to a *new* valve and *new* valve guide. You should feel *very little movement* with a new valve and guide.

For more accurate ways to check valve-guide wear, turn to page 102.

CLEANING THE HEADS

Remove any gasket material, carbon or heavy varnish from the heads before having them hot tanked. Scrape carbon from the combustion chambers with a worn flat-tipped screwdriver. Be careful not to gouge the valve seats or other machined surfaces of the head.

Don't forget to scrape the carbon out of the ports. Be careful not to damage the valve seats or the bottom end of the valve-guides—they protrude into the ports.

Now is a good time to remove any casting slag or rough edges in the water passages and ports. Remove these from the water passages with a small round file—use a flat file for the ports. Chase the bolt holes for the intake and exhaust manifolds with a 3/8-16 tap. The heads should now be ready for hot tanking.

CYLINDER-HEAD INSPECTION

Before you deliver the heads to the machine shop, check for cracks. Cracks usually show up around the valve seats, particularly on racing engines or those used for pulling heavy loads.

If you want to be sure there are no cracks, have the heads Magnafluxed or Spotchecked, page 76.

Cracks commonly found in Tunnel-Port heads are rarely found in standard FE or FT heads. Magnafluxing exposed crack in this head.

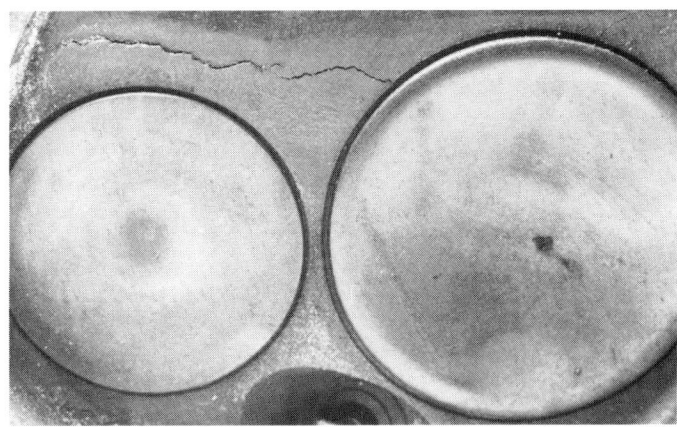
No need for Magnafluxing to see this crack. Crack can be pinned and welded.

Cracks—Cracks can be repaired and damaged valve seats can be replaced, but these are expensive repairs. Cracked valve seats can be *pinned*, welded and then cut to accept seat inserts. Other cracks can be repaired by welding or pinning.

Pinning: The crack is drilled at its ends to stop the crack, then a piece of steel stock is threaded into the holes. The excess is trimmed off and the remainder of the crack is either welded or pinned again.

Crack repair is expensive. If a lot of repair is required, weigh the costs against the value of the head. With standard heads, it may be less expensive to buy a new or used head than to have the cracks repaired or seats replaced.

The Tunnel-Port heads in the photos were brought in to be checked for cracks in one combustion chamber. A valve had broken and pounded dents in one chamber. No cracks were found there, but six of the eight combustion chambers had cracks through the intake-valve seats. Two cracks went the entire length of the sparkplug holes.

The point is this: The Tunnel-Port heads needed several intake-valve and exhaust seats replaced, plus six cracks welded. Add the cost of any guides, valves, retainers or springs that might be needed and the total repair cost really starts to climb. If these were standard heads, they would have been on their way to the scrap pile.

Don't expect to find cracks; just be aware of the possibility. Tunnel-Port heads are more susceptible to cracking than other big-block heads. The 2.25-in. intake and 1.73-in. exhaust valves and large ports make these heads fairly thin in some areas. In addition, Tunnel-Port heads are not usually pampered.

Head-Surface Flatness—Despite the aforementioned problems with the Tunnel-Port heads, big-block-Ford heads are pretty tough. They have little trouble with warping. Warped heads are usually caused by extreme heat.

Check head-surface flatness with a long straightedge, the same way you checked the block. Set the straightedge diagonally across the head, then the length of the head surface parallel to the intake and exhaust ports. Check the head with feeler gages in at least three places each time.

The amount of surface warping or irregularity allowed is 0.003 in. in any 6-in. length and 0.007 in. for the full length of the head.

You can also check for low spots with a 6-in. straightedge using the 0.003-in. tolerance. The long straightedge should extend the full length of the head.

If you are using thin, shim-type head gaskets—0.020-in. thick or less—limit the maximum warp to 0.003-in. overall. The thin gasket can't comply with the irregularites as well as a *composite-type* gasket.

Cylinder-Head Milling—Cylinder heads can be milled up to 0.050-in. but don't take them this far. Removing this much material can reduce combustion-chamber volumes up to 12cc, depending on the cylinder head.

A reduction this large can raise the compression ratio significantly—nearly two full points on some engines. Most would increase one full point. An engine with a 10.5:1 compression would go at least to 11.5:1 compression with the heads cut that much. This compression ratio would be impossible to run on the best pump gas.

The next area of concern is matching the intake-manifold ports to the heads. If you remove more than 0.020 in. from either the head or the cylinder block, you must have the intake manifold machined. See the chart, page 102.

In most cases you shouldn't need to mill a head more than 0.010 in. to correct a surface irregularity. On some 427s, valve-to-piston clearance, can be a problem if the heads are milled more than 0.020 in.: Valve-to-piston clearance will decrease by the same amount the heads are milled.

If the heads are milled more than 0.010 in. or have been milled before, check the rocker-shaft oil hole. The machined channel in the gasket surface of each head connects the oil hole from the number -2 or -4 cam bearing in the block to the oil hole in the head. Be sure this channel is at least 0.18—0.20-in. deep to provide rocker-shaft lubrication.

Also check that the rocker-shaft oil passage in each head is clean. Use the rifle brush or wire and rag to clean it.

VALVE GUIDES AND STEMS

Valve guides are important. A guide's primary function is to guide a valve onto its seat each time it closes. This is necessary for valve sealing. The guide also provides oil control at the valve stem.

Valve guides in the cast-iron heads are cast as part of the head, then machined. Aluminum heads use separate, aluminum/bronze guides.

Special tools are needed to check

Cylinder heads milled more than 0.020 in. will require machining of intake manifold.

INTAKE-MANIFOLD VERSUS CYLINDER-HEAD MILLING REQUIREMENTS (INCH)		
Removed from Cylinder Heads	Remove from Manifold Bottom	*Remove from Manifold Sides
0.020	0.028	0.020
0.025	0.035	0.025
0.030	0.042	0.030
0.035	0.049	0.035
0.040	0.056	0.040
0.045	0.063	0.045
0.050	0.070	0.050
0.055	0.078	0.055
0.060	0.084	0.060

*Double amount when milling one side of manifold.

VALVE-TO-PISTON CLEARANCE

For a number of reasons, cylinder-head or block-deck milling should only be done to correct etching or warping. First, it can raise the compression ratio to a point where detonation can become a serious problem, particularly with today's low-octane gasolines.

There is another reason to avoid milling the heads or block. On certain high-performance FE engines, milling the heads or block decks can cause insufficient valve-to-piston clearance. This is particularly a problem on the High-Riser 427, but it should be checked on any of the high-compression engines when the block is decked or heads milled more than 0.020 in.

Valve-to-piston clearance must be checked with the lower end assembled. It is easiest to make this check before installing the heads. Unfortunately, if the clearance is insufficient the pistons must be removed and relieved—machined for valve clearance.

The most-common method used to check valve-to-piston clearance is to start by placing a thin strip of clay—about 1/4-in. thick—across the piston valve pockets, or reliefs. A thin coating of oil on the valve heads will keep the clay from sticking to the valves. Then, using old gaskets, install the assembled heads—less sparkplugs—and the valve train. Torque the head bolts 80—90 ft-lb using the proper torque sequence, page 133.

Position the camshaft so both valves for the cylinder being check will be closed when the pushrods and rocker shaft are installed. Install the pushrods for that cylinder only, then install the rocker shaft. There should be no load on the pushrods, so you so you don't need to worry about bending the rocker shaft. Torque the rocker-shaft bolts 40—45 ft-lb, then adjust valve lash to 0.002 inch.

Slowly rotate the crankshaft. If the crankshaft starts to bind, **STOP!** A valve may be hitting the piston, so don't force it. Only a light force should be required to turn the crank. So remove the head and check valve-to-piston clearance immediately.

If no resistance is felt rotate the crankshaft two full revolutions, then remove the head. Examine the clay. With a sharp knife, slice the clay across the depression made by the valve. Peel half of the clay off the piston.

Measure the thinnest section of the clay with a micrometer, vernier caliper or check it on the piston with feeler gages. This is the piston-to-valve clearance. Minimum clearance should be 0.080 in. for the intake valve, 0.125 in. for the exhaust.

If there is insufficient clearance, the pistons must be removed and the notches or eyebrows deepened. Leave this work to a machine shop.

Checking valve-to-piston clearance is time-consuming. But it's much easier—and cheaper—than tearing the engine down and replacing a set of valves bent from contact with the pistons.

---

valve-guide and stem wear accurately. You will need a 0—1-in. outside micrometer and a *small-hole gage* or dial indicator and a new valve.

Measuring Valve-Guide Wear—The most-accurate way to measure stem-to-guide clearance is with a small-hole gage and a micrometer. A small-hole gage works similar to an inside mike, however, it doesn't read out directly. It expands to fit the hole you're measuring. The gage is then withdrawn from the hole and measured with an outside mike to determine hole size. With the small-hole gage, measure around the guide at its top and bottom to find its maximum diameters. Remove the gage and measure the ball end with an outside micrometer.

Mike the valve stem and subtract this measurement from the guide measurement to find stem-to-guide clearance. If the clearance limit is exceeded—see chart, page 103—the guide should be reconditioned.

To check guide wear with a dial indicator—probably the second-most -accurate method—install a valve in the guide about 1/8 in. off the seat. Mount the dial indicator 90° to the valve-stem tip in the direction you want to measure wear. The dial-indicator plunger should contact the valve stem as close to the top of the guide as possible.

Pull the valve away from the tip of the indicator and zero the dial. Now, push the valve toward the indicator and read the clearance directly.

As I already mentioned, the ball-park way to determine wear is also the

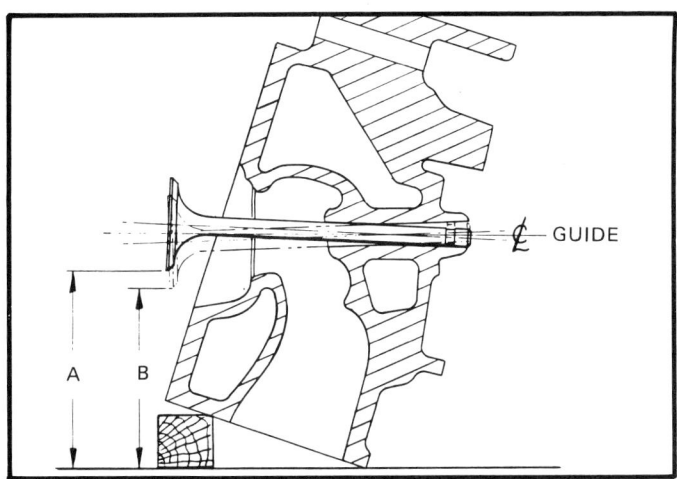

"Precision" wiggle test. Difference between A and B can be measured with dial indicator or vernier calipers. Dividing by 3.5 gives approximate valve-stem-to-guide clearance. Drawing by Tom Monroe.

VALVE STEM-TO-GUIDE CLEARANCE

INTAKE
FE engines
0.0010—0.0024 in. 0.0045 in. maximum

FT engines
0.0010—0.0027 in. 0.0055 in. maximum

EXHAUST
FE engines, exc. 427
0.0015—0.0032 in. 0.0055 in. maximum

427
0.0020—0.0034 in. 0.0055 in. maximum

FT engines
0.0020—0.0040 in. 0.0055 in. maximum

Engines used for heavy-duty or high-performance require greater exhaust-valve-guide clearance because of higher operating temperatures.

One can be saved, other can't. Carbon buildup on back of valve can be removed and valve reground. Burned exhaust valve is scrap.

Cast-iron valve-guide inserts have spiral grooves for better valve-stem lubrication.

simplest: Try wiggling the valve back and forth. If it wiggles more than a hair, the guide is worn! Other methods are more sophisticated and accurate, but this is the most popular.

The accepted *wiggle test* is to place the valve in its guide until the tip of the stem is flush with the top of the guide. This uses an unworn portion of the valve stem, so the measurement reflects guide wear, not stem-to-guide clearance. The valve is then moved—wiggled—from side to side.

Movement from one extreme to the other is measured with a dial indicator mounted 90° to the valve head, or the depth-gage end of a vernier caliper. This measurement is then divided by 3.5 and the result is the approximate valve-guide clearance. As an example, if the limit of wiggle is 0.0085 in., dividing this by 3.5 gives 0.0024 in.

This checking gives you *approximate* clearance. If the figures are borderline, double-check with one of the other methods, or play it safe and consider guide wear excessive. With the wiggle method and all others, measure clearance in two directions 90° to one another—front to rear and side to side.

Stem-to-guide clearance should be a function of the intended use of the engine. If the engine will be used to pull heavy loads or for racing, set the clearance to the higher side of the limits. If the engine is going to be used for a passenger car, set clearances near the low or tighter side of the limit.

Do not use the maximum for a reference. This is what is known as a *service limit*. It is the maximum clearance allowed on a used engine.

Note that the recommended stem-to-guide clearance for intake valves is about the same for all big-blocks. But exhaust-valve clearance is greater on the 427 race engine and heavy trucks. This is because exhaust valves run hotter on these engines, causing them to expand in their guides. Standard clearance would result in the exhaust valves seizing in their guides.

At hotter-than-normal temperatures, stem-to-guide clearance of a racing engine is actually less than that of a standard, passenger-car engine operating under normal conditions. If you build the engine with large clearances, then operate it under light loads as in a passenger-car, the clearance will be excessive. Oil consumption and wear will increase and performance will decrease.

Excessive stem-to-guide clearance causes a valve to wiggle in its guide and bounce off its seat before closing. This problem is aggravated by larger-diameter valves and high rpm. This can damage valve seats, bend or break valves and cause lost power because

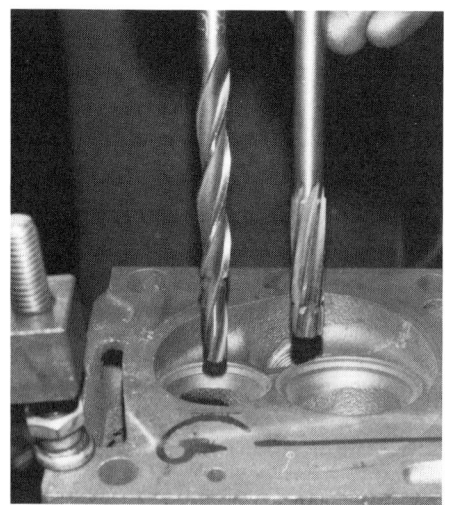

Reamer at left enlarges guide for new guide inserts. Bit at right reams opening to exact undersize for interference fit. Smaller portion at bottom of each reamer pilots in guide.

Old guide is reamed for 0.001-in. interference fit for guide insert.

New guide is driven or pressed into place.

Head is releveled because setting may have been disturbed by pounding. Excess guide is then trimmed off.

Guide opening is chamfered at both ends, then reamed to size.

of incorrect valve sealing.

Again, before setting up the engine with additional stem-to-guide clearance, take a *close*—and in the case of racing engines, *honest*—look at the intended use. A high-performance street engine is *not* a racing engine.

In drag racing valve stems don't need additional clearance because of the short duration of the race. If the engine is going to be run at extended high speeds most of its life—as in a stock car or power boat—build in the extra clearance. Heavy-truck engines use the additional clearance because sustained high loads raise the exhaust-valve temperature.

VALVE-GUIDE RECONDITIONING

When restoring valve stem-to-guide clearance, there are several considerations: Are new valves needed? What is the intended use of the engine? How long do you intend to keep the engine?

I will assume that you plan to keep the engine for a long time, but I cannot know the intended use. Durability and price are prime considerations, because what you do to the valve guides affects engine durability, performance and oil consumption. Correct stem-to-guide clearance extends the life of valves, valve seats and valve-stem seals.

There are three ways to restore valve stem-to-guide clearance: guide knurling, guide inserts and new valves with oversize stems. Results range from fair to better-than-new.

Guide Knurling—To knurl valve guides, a knurling tool or arbor is inserted in the old guide. The arbor is turned to form a groove in the guide. The material displaced by the knurling tool to form the groove makes the guide effectively smaller.

The guide is then reamed to size with a fluted reamer. An added advantage of knurled guides, besides low cost, is that the grooves retain oil to lubricate the valve stems.

For this reason you can use less stem-to-guide clearance—up to half that of a new guide. Another advantage is that a knurled guide retains more oil for startup than a standard cast-iron guide.

The disadvantage of a knurled guide is that it reduces valve-guide surface area to support the valve, so wear is accelerated. A knurled guide simply won't last as long as the original guide. Also, the valve stem runs against the raised portion of the guide and with less cushioning from the oil. Finally, Teflon valve-stem seals cannot be used with knurled valve guides, because they will starve the guide of oil and cause excessive wear.

Knurled guides *may* work fine for you, and problems may never show up during the life of the engine. In most cases, however, it's a stopgap solution.

Valve-Guide Inserts—There are many different kinds of valve-guide inserts on the market. Cost and performance vary. They all work well, though some work better than others.

Two basic materials are used: cast

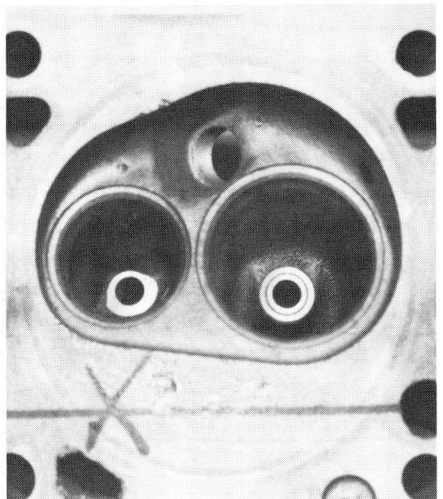
With new insert installed, valve seat can now be cut.

Final chamfer removes sharp edge at guide opening, easing valve installation.

Unless valve stem is worn excessively, guide insert will restore correct stem-to-guide clearance.

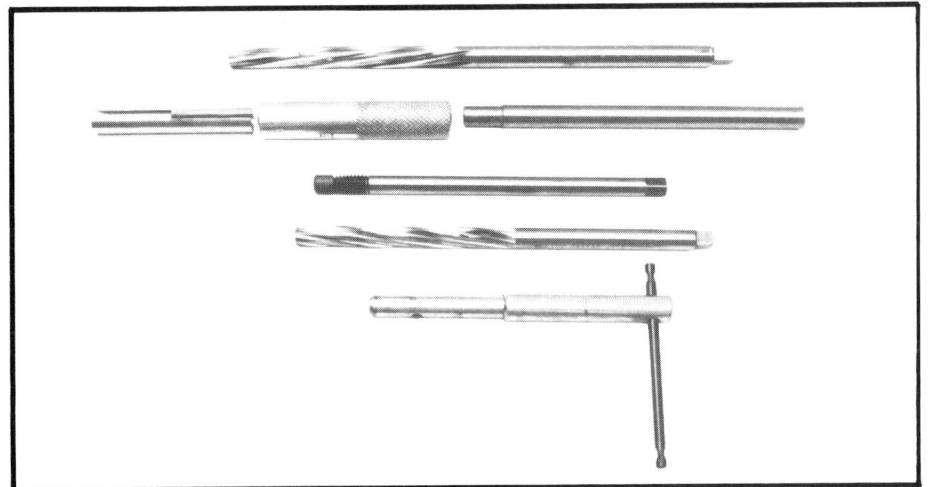
Valve-guide inserts can be installed by hand. Tools are simpler and cheaper, but method is more time-consuming. Just the thing for the occasional rebuilder.

iron and bronze. Cast-iron replacement guides are similar to the original guides and will last about as long. Bronze inserts last longer than the original guides or cast-iron replacement guides.

The cast-iron insert is pressed in with an *interference fit*. The old guide is reamed so that its ID is 0.0001—0.0002-in. smaller than the insert OD.

Bronze inserts install in two basic ways: press-in, similar to cast-iron guides, or thread-in, like a *Heli-Coil*—a replaceable thread.

To install a thread-in insert, the existing guide is tapped and an insert threaded into it. Good insert-to-guide contact is important for stability, longevity and heat transfer, so a sharp tap is a must.

After the insert is installed, excess thread is trimmed off. The insert is then locked into place so the action of the valve can't loosen the insert. The guide is then reamed to size.

Thin-wall bronze inserts are driven or pressed into place. Then a knurling tool is run through the guide to expand the insert, lock it in place, and ensure good heat transfer. The guide is then trimmed to length and reamed to size.

Oversize Valves—Ford manufactures valves with oversize stems: 0.003-, 0.015- and 0.030-in. oversize. Some valves are available in more sizes.

To install oversize valves, the original guides are simply reamed to the correct oversize and the new valves are installed.

This is an acceptable method, especially if new valves are needed. Considering the price of new valves it is expensive. If excess stem-to-guide clearance is due to worn guides, it's cheaper to keep the valves and recondition the guides and valves.

VALVE INSPECTION AND RECONDITIONING

The next step is to inspect the valves. If you see obvious damage—the valve is visibly bent or has a burned or pitted face—replace it. To inspect further you need a 0—1-in. micrometer. You must also check the valve-stem tips for wear.

To check valve-stem wear, compare the worn and unworn parts of the stem. Measure the valve where wear isn't evident, such as the area just below the valve-keeper groove. This area should have the original dimension. Compare this to the area with the most wear—the shiny section at the valve-tip end of the stem.

The amount of allowable wear, like cylinder-bore wear, depends on the job you want. You can have a cobbled job, one better-than-new or something between.

Compare the maximum valve-guide clearance to the smallest stem diameter of the valve. When using new guides and old valves, don't forget to take into account the different diameters of the old valve stem where it contacts the guide. The unworn part of the stem is larger, so don't use the minimum diameter of the valve stem to obtain the clearance. The larger part of the valve stem may then have

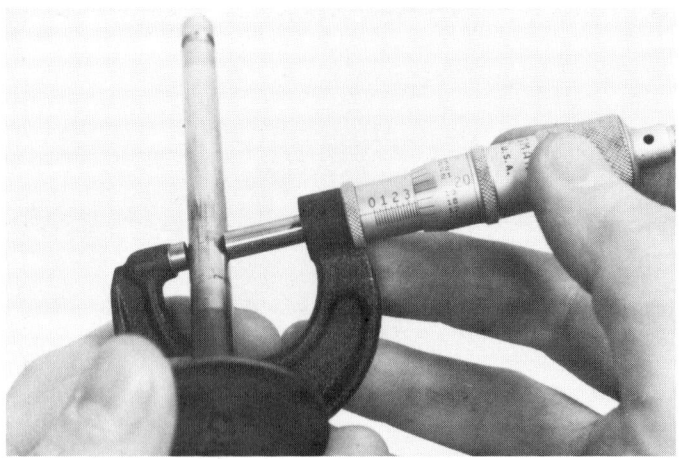
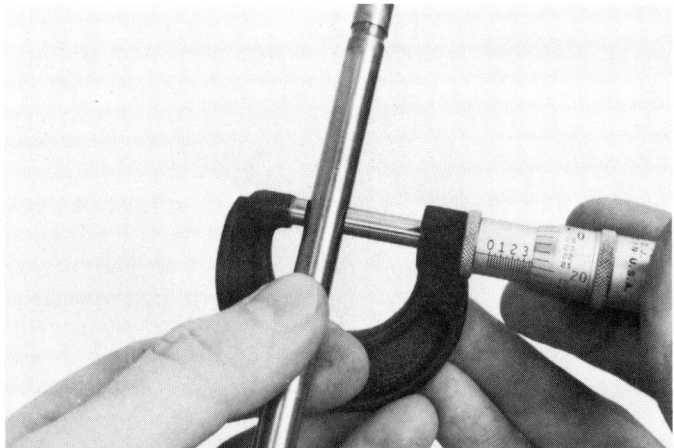

Comparing worn and unworn parts of valve stem will give you an indication of valve-stem wear. If difference is more than 0.002 in. replace valve.

Valve face is ground at 44° or 29° angle, depending on application. Flow of oil keeps valve face cool and free of grindings.

VALVE-STEM DIAMETER (inches)

INTAKE	
all engines	0.3711–0.3718
EXHAUST	
FE engines	
all except 427	0.3706–0.3713
427	0.3701–0.3708
FT engines	
all except 330 MD	0.4338–0.4348
330MD	0.3701–0.3708

All valves are available with 0.003-, 0.015- and 0.030-in. oversize stems.

Unworn diameters of valve stems should match these specifications.

---

too-little clearance.

This is one reason you may have to do a complete job and replace the valves *and* guides. Guidelines are the clearance between the valve stems and guides and the clearance between the walls of your wallet.

Finally, if the valve stem is worn 0.002 in. or more, replace the valve. Many machine shops use 0.001 in. as a throwaway dimension.

Another check is to see that there is an even-width wear pattern on the valve face. If the valve guides are worn excessively, the valve may have been hitting *off-center*—not closing squarely on its seat. This can severely weaken the valve stem. An uneven wear pattern on the valve face indicates the valve should not be reused.

Reconditioning The Valves—Regrinding the valves involves squaring the tips and truing the face. The face is the most-important part of a valve. A valve face must be ground at the correct angle and on a given diameter, which requires precision machining.

The valve face determines the valve's sealing ability. Both valves must seal the combustion chamber during the compression and power strokes. The intake valve must also seal on the exhaust stroke and the exhaust must also seal on the intake stroke.

Valves for the big-block are ground at 44°, except for the intake valves on certain engines. The exceptions are: '58–'59 332, 352 and 361; '60 352HP and '61–'62 390HP; '62–'63 406, '61–'63 Thunderbird Special 390; all 390 and 428 Police Interceptor engines, and all 427, 428CJ and 428SCJ. On these engines, the intake valves are ground at 29°; seats at 30°.

Before the seat is ground, the valve tip is refinished, or squared and chamfered. A 90° cut is made across the valve-stem tip to square it. The tip is then beveled, or *chamfered*, to remove the sharp edge. The chamfer minimizes wear on the rocker arm and helps prevent damage to the valve-stem seal during installation.

If a valve tip has been ground more than 0.010 in., check the tip for hardness. Run a fine-tooth file across it—if the file cuts, the valve must be rehardened or replaced.

Valve grinding is done by placing a valve in a valve-grinding machine. The machine has a collet-type chuck to rotate the valve while a grinding wheel passes over the valve head to *cut* the face. The chuck and grinding wheel turn in opposite directions as the cut is made.

When the valve face is ground, the machine is adjusted so the valve face

Valve tip being squared. If tip is ground 0.010 in. or more it may need to be rehardened.

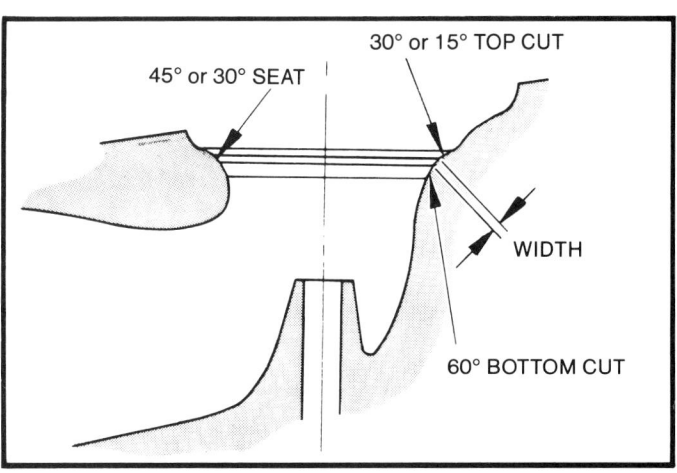
Top and bottom cuts establish seat width. Drawing by Tom Monroe.

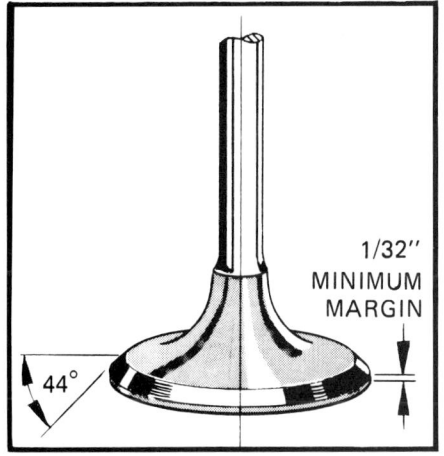
Minimum margin for exhaust valves is 1/32 in.; 1/16 in. for intake valves. 1° difference between valve face and seat angles provides edge contact to ensure valves seal immediately.

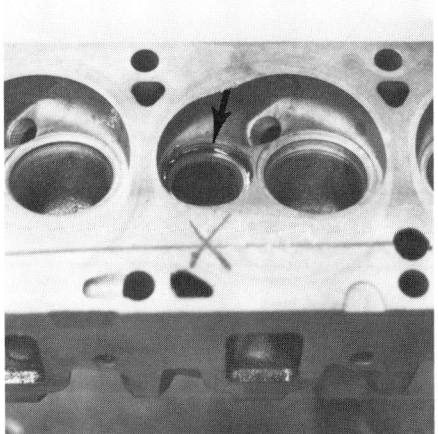
Begin valve-seat-insert installation by boring out damaged seat. This will provide a counterbore for the insert (arrow).

Seat insert is driven into place. This operation is easier if head is warmed and seat *and* installation tool are cooled.

is held at the correct angle to the grinding wheel. The valve is cooled with oil during the grinding. The oil also washes away abrasive grit and metal fragments.

Only enough material is removed to produce a concentric face with the correct width. The cutting must leave the correct *margin*—the thickness of the valve at its outer edge. A sharp edge will cause the valve to burn.

For exhaust valves the minimum margin should be 0.030 in.—about 1/32 in.—to keep the valve edges from warping or burning. Exhaust-valve durability can be helped by using a wider margin, but breathing will suffer somewhat. Intake valves should be kept near 0.030 in., but can go down to 0.015 in. because they operate much cooler.

If you replace the guides, there is no reason to keep the valves in order once they are ground. But, if the old guides are acceptable, or you have reamed some of the guides for oversize valves, keep each valve with its guide.

Valve-Seat Reconditioning—Valve seats are reconditioned the same way the valves are—by grinding them to the correct width and concentricity. The only difference is when the seats are cracked, pitted or cut too deep—they are not easy to repair. They have to be machined and valve-seat inserts installed. This can only be if the damage is not too extensive. The repair can be very expensive.

Remember, head work must be done in sequence. Before the seats are cut, be sure the valve guides are in good condition—straight and round. The seat grinder *pilots*—centers—in the valve guide.

Seats are cut or ground at a 45° or 30° angle to provide a 1° *edge-contact* with the 44° or 29° valve faces at their outer edges. This ensures that hot combustion gases will be sealed the first time the valve closes. Leaks at either valve due to poor seating can warp the valve or burn both seat and face.

Valve seats are actually a series of three cuts: On standard engines, the *top cut* above the seat—combustion-chamber side—is cut at 30°. The actual *seat* is cut at 45°, and the *bottom cut*—port side of the seat—is cut at 60°.

Exceptions to this are the intake-valve seats for the engines listed on page 106. On these engines, the top cut is 15° and the seat is 30°. The bottom cut remains at 60°.

Seat is ground to correct angles and width. Because valve-grinder pilot centers in valve guide, guides must be reconditioned before valve seats.

VALVE-SEAT WIDTHS (inches)	
INTAKE	**EXHAUST**
FE engines	FE engines
standard 0.060—0.080	standard 0.070—0.090
racing 0.035	racing 0.050
FT engines 0.070—0.090	FT engines 0.090—0.110

The top cut should form a small —approximately 1/32-in.—overhang for the valve face. This determines the outside diameter—OD—of the seat. Overhang is where the valve face extends past the seat. The overhang ensures correct airflow characteristics as the valve opens. The valve depends on seat contact to transfer heat from the face, so if the overhang is too large the valve will burn.

The bottom cut—also referred to as the *throat* or *port cut*—is done to obtain the correct seat ID. The 60° angle provides a good lead-in or -out for the fuel mixture or exhaust gases.

Be careful not to grind too far into the head, *sinking* the valves. This can ruin the flow characteristics of the valve and port. Sinking the valves can also change combustion-chamber volume. The geometry of the rocker arm, pushrod and valve-stem tip may have to be adjusted with a shorter pushrod.

Dykem or machinist's blue dye spread over the cutting area will make the cut or ground valve-seat highly visible. Use a divider and scale to measure seat width. You will also need a special dial indicator to check seat concentricity, or *runout*. Maximum valve-seat runout should be no more than 0.002 in.

If valve seats are too deep, or are cracked, pitted or etched, you must install valve-seat inserts or replace the heads. Seat inserts are standard on some FT engines: intake- and exhaust-valve seats on 361 and 391 engines and exhaust-valve seats of the 330HD engine. Some 427 heads— later Tunnel-Ports, for example—also have exhaust-valve-seat inserts.

Hand Lapping Valves—Hand lapping is basically an old process used to finish the faces on valves and seats. Lapping involves rotating a valve against *its* valve seat with light pressure after *lapping paste* is applied to the valve face. After lapping, valves should be kept in order.

With a correctly machined valve and seat, lapping does little to no good. There are better ways to use your time, so don't bother.

Hand lapping is a double-check at best. It will show how evenly the valve face is mating with the valve seat. If the valve job is done right you should have an even pattern all the way around the valve face and seat. Otherwise, it's a waste of time.

Valve Sealing—A better way to check valve sealing is with a leak test. To do this check you must assemble the valves, valve springs, retainers and keepers—or rotators in the case of some truck engines.

You may wait until after you have finished checking the springs and retainers before doing this test. This way you can completely assemble the heads, and won't need to take them apart again unless one of the valves leaks. Assembly is covered beginning on page 112.

Install the sparkplugs. Turn the assembled head upside down, with the combustion chamber up. Fill each chamber with solvent, kerosene or fuel oil. Check the intake and exhaust ports for any leaks. If there are none, the valves and seats are OK.

If you find moisture in one of the ports, either the valve or the seat must be reground. To check, swap the valve with one from another chamber and reinstall both valves. If the same port still leaks, the seat is bad. If the other port leaks, it's the valve.

VALVE SPRINGS

The final phase of valve-related cylinder-head inspection and reconditioning is spring inspection. A valve-spring tester is helpful but not necessary. Mostly, you should understand the importance of maintaining certain minimum standards and know the terminology. This includes: *valve float, spring rate, free height, solid height, load at installed height and load at open height.*

Valve Float—Valve springs don't just close valves. After the cam lobe pushed the lifter that pushed the pushrod that pushed the rocker that opened the valve against the force of the spring—whew—the valve spring returns the compliment and puts all of it back where it came from. The return side of the cam lobe helps by making sure everything is eased into place. But the valve spring must hold the lifter in contact with the lobe on *both* sides of the lobe.

Valve float—better termed *valve-train float*—occurs when engine rpm is so high the springs can't get everything back in time. Lifters momentarily lose contact with the cam lobes and the valve train momentarily *floats* away from the cam lobes. Instead of being controlled by the camshaft, the valve train is momentarily controlled by inertia, spring force and a bit of physics called *harmonics*.

It all gets a bit too involved to explain fully here, but the results are sheer chaos—the valves bounce when they close, they close late, the lifters

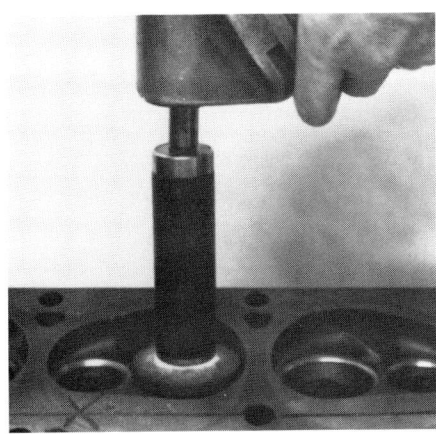

Valve seat consists of three cuts: top cut and bottom cut with seat in between. Seat forms the actual sealing surface that contacts the valve face. Bottom and top cuts determine seat width and provides increased air/fuel-mixture or exhaust-gas flow into or out of the combustion chamber.

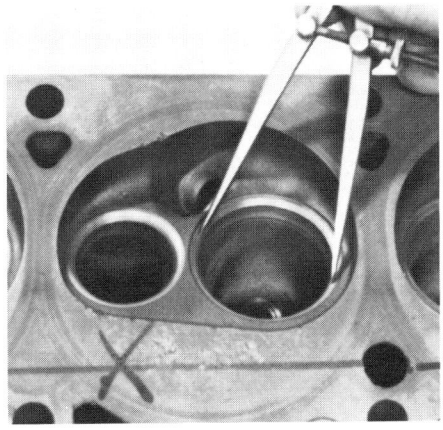

Calipers are used to check seat diameter.

Seat, top and bottom cuts are made simultaneously with this tool. Tool reduces seat-finishing time and allows accurate setting of valve height in head.

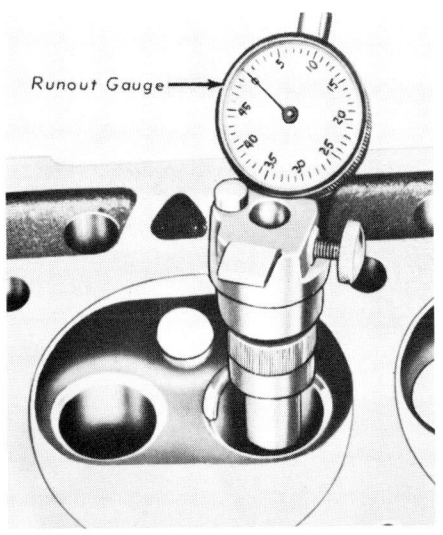

Once valve seat is ground, its concentricity or runout in relation to guide is checked. Special gage indicates runout. Photo courtesy of Ford Motor Co.

Valve spring being checked for load at installed and open heights. Both readings should be within 10% of specification. Check following chart for specific engines.

can't get back where they're supposed to be. The pushrods and rockers are somewhere in between.

To prevent valve float—at least to the desired engine rpm—the valve springs must provide sufficient force to overcome the inertia of the valve train to keep the lifters against the cam lobe.

Spring Rate—A spring requires a given amount of force to be compressed a given amount, usually expressed as pounds per inch of deflection. Spring rates relate to intended engine rpm. The higher the rpm, the higher the valve-seat load—*installed load*—must be to control valve float.

Free Height—Free height is the unloaded length of the spring. If free height is too long or too short, the spring will be compressed too much or too little by the spring retainer. The force exerted by the spring when installed—*load at installed height*—will be incorrect.

Load at Installed Height—This is the load exerted by the spring when compressed to its *installed height*. Installed height is the distance from the cylinder-head spring seat to the underside of the spring retainer with the valve closed.

A typical load-at-installed-height specification might be 78—88 lb at 1-11/16 in. This means that it should take between 78 and 88 lb of force to compress a spring to a height of 1-11/16 in.

The *absolute minimum limit* is 10% less than the minimum standard limit—about 70 lb in this example. If a spring does not meet or exceed the minimum it should be replaced.

Load at Open Height—The next specification is spring load when the valve is fully open. The specification might be 255—280 lb at 1.320 in. As with the load at the installed height, a spring must fall within the 10% limit or be replaced. In this example the

VALVE-SPRING SPECIFICATIONS

Engine		Year	INSTALLED Height (in.)	INSTALLED Load (lb)		OPEN Height (in.)	OPEN Load (lb)		Free Height (in.)	Comments
FE										
332		58–59	1.82	94–104	85	1.42	180–198	162	2.26	w/ or w/o damper*
352		58–66	1.82	94–104	85	1.42	180–198	162	2.26	w/ or w/o damper*
	to ser A84001	67	1.82	94–104	85	1.42	180–198	162	2.26	w/ or w/o damper*
	from ser A84001	67	1.82	85–95	77	1.38	209–231	186	2.12	w/o damper
352HP	early	60	1.82	94–104	85	1.42	180–198	162	2.26	w/damper
	late	60	1.82	80–90	72	1.32	255–280	230	2.06	w/damper
360		68–76	1.82	85–95	77	1.38	209–231	186	2.12	w/damper
	F100 exhaust	72–76	1.67	76–84	69	1.24	175–194	157	2.00	w/valve rotator
361		58–59	1.82	94–104	85	1.42	180–198	162	2.26	w/damper
390		61–65	1.82	74–84	67	1.42	190–208	171	2.15	w/damper
	(includes 410)	66–67	1.82	80–90	72	1.38	233–257	210	2.15	w/damper
		68–76	1.82	85–95	77	1.38	209–231	186	2.12	w/damper
	F100 exhaust	72–76	1.67	76–84	69	1.24	175–194	157	2.00	w/valve rotator
	police	61–65	1.82	80–90	72	1.32	255–280	230	2.06	w/damper
	T-bird Specials	61–63	1.82	80–90	72	1.32	255–280	230	2.06	w/damper
	4V Comet, Fairlane	66	1.82	80–90	72	1.32	255–280	230	2.06	w/damper
	4V Comet, Cougar, Fairlane, Mustang	early-67 to 10-11-66	1.82	80–90	72	1.32	255–280	230	2.06	w/damper
	4V Comet, Cougar, Fairlane, Mustang	late-67 from 10-11-66	1.82	80–90	72	1.32	244–268	220		w/o damper
	4V Comet, Cougar, Fairlane, Mustang	68	1.82	80–90	72	1.32	255–280	230	2.06	w/damper
390HP	dual spring, inner	61–62	1.72	28–32	25	1.22	91–99	81	1.88	damper btw springs**
	outer	61–62	1.82	92–98	78	1.32	186–194	159	2.28	
	single spring	61–62	1.82	80–90	72	1.32	255–280	230	2.06	w/damper
406	dual spring, inner	62–63	1.72	28–32	25	1.22	91–99	81	1.88	damper btw springs**
	outer	62–63	1.82	92–98	78	1.32	186–194	159	2.28	
	single spring	62–63	1.82	80–90	72	1.32	255–280	230	2.06	w/damper
427	dual spring, inner	63	1.72	28–32	25	1.22	91–99	81	1.88	damper btw springs**
	outer	63	1.82	92–98	78	1.32	186–194	159	2.28	
	single spring	63–68	1.82	80–90	72	1.32	255–280	230	2.06	w/damper
428	Ford, T-bird	66–67	1.82	80–90	72	1.38	233–257	210	2.15	w/damper
	Ford, T-bird	68	1.82	85–95	77	1.38	209–231	186	2.12	w/damper
	police	66	1.82	80–90	72	1.32	255–280	230	2.06	w/damper
	police	early-67 to 10-11-66	1.82	80–90	72	1.32	255–280	230	2.06	w/damper
	police	late-67 from 10-11-66	1.82	80–90	72	1.32	244–268	220		w/o damper
	police	68–70	1.82	80–90	72	1.32	255–280	230	2.06	w/damper
	CJ	68	1.82	80–90	72	1.32	255–280	230	2.06	w/damper
	CJ & SCJ	69–70	1.82	86–96	76	1.32	271–299	245	1.95	w/damper
FT										
330MD	intake	64–77	1.82	94–104	85	1.42	180–198	162	2.26	w/o damper
330HD, 359, 361, 389, 391										
	intake	64–75	1.67	76–84	69	1.24	175–194	157	2.00	w/valve rotator
	intake to serial number C10001	early-76	1.67	76–84	69	1.24	175–194	157	2.00	w/valve rotator
	intake from serial number C10001	late-76	1.82	85–95	77	1.38	209–231	186	2.12	
	intake	77–78	1.82	85–95	77	1.38	209–231	186	2.12	
	exhaust, all	64–78	1.67	76–84	69	1.24	175–194	157	2.00	w/valve rotator

* most 4V's and police engines had dampers
** dual springs were special order

minimum is 230 lb.

Solid Height—Solid height occurs when a spring is totally compressed, or *coil bound*. Here, each coil touches the adjacent coils. Valve springs should never be compressed this far in normal engine operation. Obviously, if the spring were totally compressed before maximum valve lift was reached, the valve couldn't be opened fully. The first time the engine cranked over, something would give—probably a pushrod would bend.

CHECKING THE SPRINGS

Free Height—The first and easiest thing to check is free height. Line all the springs up to check their heights—they should all be within 1/8 in. of specification. Check the above table.

One that is 1/16-in. shorter than specification might be used for light service, though you should replace it. It is starting to fatigue and won't get better.

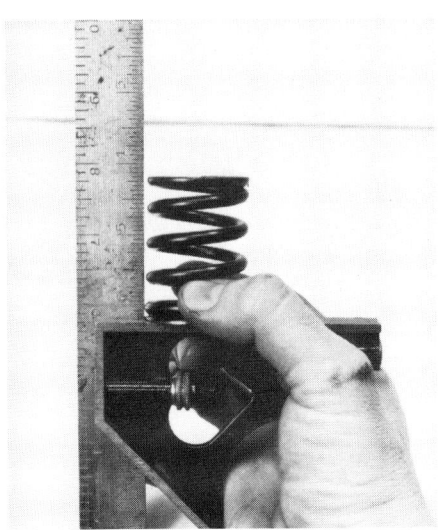
Checking valve spring for squareness. Rotate spring against inside corner of square. Largest gap to appear at top of spring should be no more than 5/64 in.

If engine uses separate valve-spring seats, they must be installed when checking valve-spring installed height.

Install valve retainer and keepers. A dab of grease will hold keepers in place.

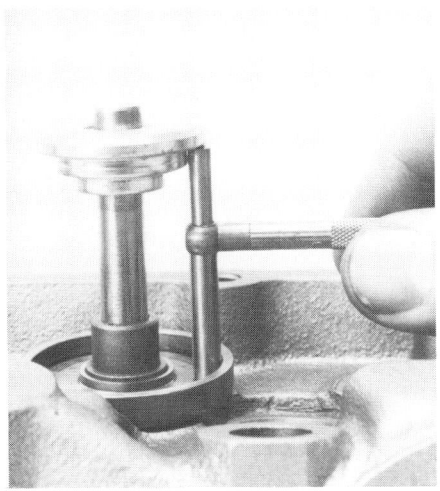
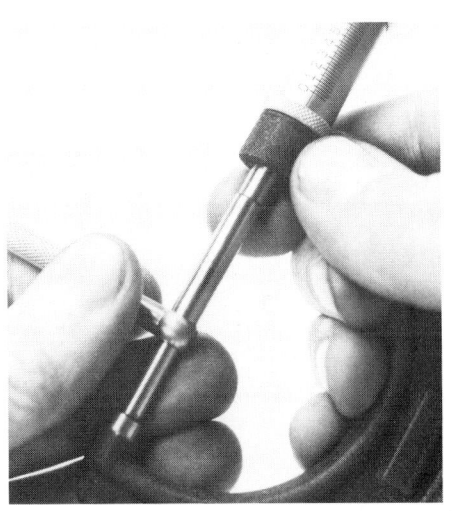
Valve-spring installed height is checked with telescoping gage. Gage is then measured with micrometer. A piece of heavy wire or welding rod cut and ground to length of installed height works well as a checking gage.

Installed height is adjusted with shims.

Squareness—For this test you need a square. A carpenter's square works fine. Position the spring against the square as pictured above. Check that the coil is flat against the surface. The end of the coil may be burred or a new spring may have a lump of paint on it. If necessary, dress the spring with a grinding stone. Slide the bottom coil of the spring against the square.

Rotate the spring against the square until you find the largest distance between the top coil and the square. Measure this distance with feeler gages. It should be no more than 5/64 in. (0.078 in.). If the distance is more than this, replace the spring.

An out-of-square spring will side load the valve stem, causing excessive wear on the valve guide and stem. This side loading and resulting wear will eventually cause the valve to seat off-center.

Checking Spring Load—Valve-spring load is measured in a special test fixture made for that purpose. You also need the load and height specification: 255—280 lb at 1.320 in., for example. See chart at left.

Insert the spring in the fixture and compress it to the specified height. Note the force at that measurement. It should be within 10% of the *lower* value given in the chart. In the above example this is 255 less 10%, or 230 lb. Discard any springs that don't meet the specification.

Next check the open height for correct force. Also be sure that only the end coils are touching one another—none in between should touch. If any of the center coils are touching when the spring is at open height the spring has collapsed and should be replaced.

Also check that the spring is not *coil-bound*—compressed to the point that all the coils are touching one another. Coil binding at open height shouldn't be a problem with the stock springs, rocker arms and camshaft. But it should be checked at the new open height if any modifications are made—a higher-lift camshaft, for instance. Coil binding can also be a problem if a fatigued spring is shimmed to bring it up to the correct load at installed height.

If some springs fall below specification you have two choices: You can either replace them or shim them to

Valve-springs, shims, valve, keepers and retainer must be installed in the same guide in which they were checked.

Install spring seat, slide valve into place, then install shims.

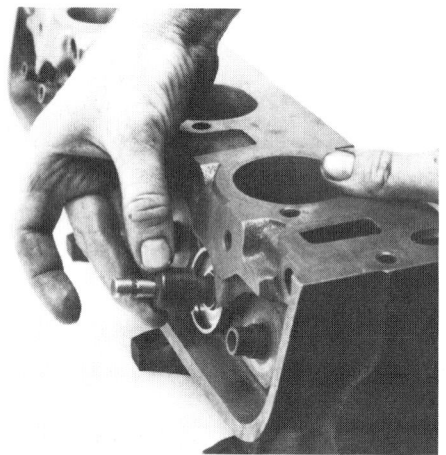

For Teflon seals, use cellophane boot over valve end to protect new seal as it is installed. Push valve seal into place and compress spring and retainer assembly. Set valve-spring compressor same as for disassembly—so springs are not over-compressed.

Spring-and-valve assembly installed. Do this seven more times and store head in plastic bag.

specification. Unless you're trying for the world's cheapest rebuild, replace weak springs.

A spring is designed to be compressed a given amount—the difference between its free height and open height. A spring compressed past its open height will be over-stressed, become fatigued and quickly lose its load-producing capacity.

If you shim a weak spring, recheck it with the valve-spring tester with the shims installed. Be sure the spring is not coil bound when compressed to its open height with any shims installed.

Remember that open height changes with different camshafts and rocker arms. If you change the cam you will need to use the correct springs and spring specifications to match the camshaft.

ASSEMBLE CYLINDER HEADS

Now for the finishing touch. At this point, all the parts for the cylinder head have been inspected, reconditioned or replaced. Now it's simply a matter of putting them all back together.

Valve-Spring Installed Height—A slight but important change has been made to the cylinder head. When the valve seats were ground, material was removed from each seat.

This removal reduced the distance from the valve seat to the spring pad, so each valve seats "deeper" in the head. The valve stem and keeper grooves extend farther out the top of the valve guide.

As a result, valve-spring installed height—the distance from the spring pad to the bottom of the retainer—has increased. The increase must be compensated for by adding shims between the valve spring and its seat, or pad.

This is a necessary step to correct valve-spring installed height. It should not be confused with shimming a weak or partially collapsed spring.

Valve-spring installed height can be checked by partially assembling the head or after the head is completely assembled.

To do the check with the head partially assembled, install the valve, retainer and keepers without the springs. Because the spring is not holding the retainer and keeper up, you have to hold the valve against the seat. To measure installed height use an inside micrometer, a telescoping gage and outside mike, a vernier caliper or a ruler.

With the telescoping gage or inside mike, simply measure between the spring seat and underside of the spring retainer. You can easily check installed height by grinding a bolt or welding rod to the correct length. Use it with a set of feeler gages to check

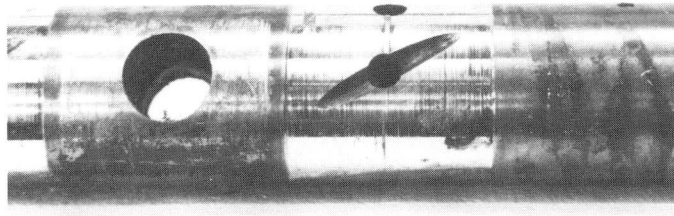
If engine was full of sludge or carbon, you will find rocker-shaft oil passages dirty or plugged. Passages can be cleaned, but restricted oil flow usually means shafts are worn like this and need replacing.

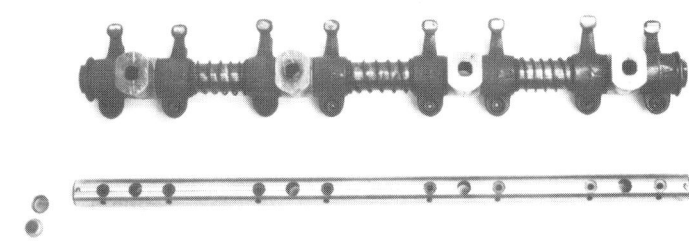
Rocker-arm shaft before and after disassembly, cleanup and inspection. Many rebuilders skip this step: Don't you.

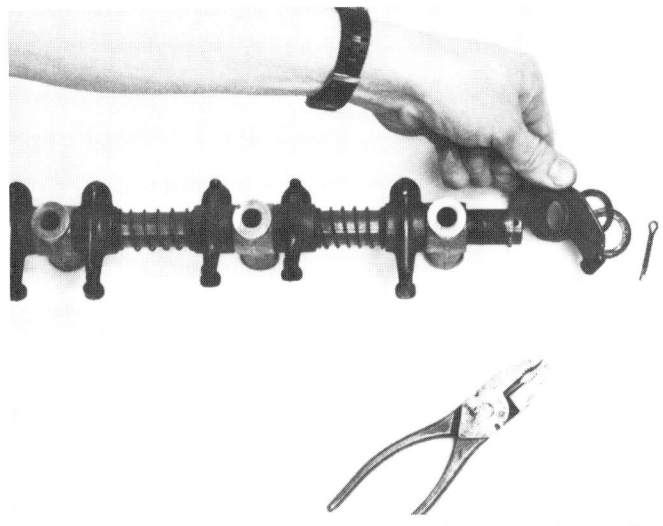
First part is easy. Pull cotter pin out, slide flat and spring washers off, and remove rocker arm.

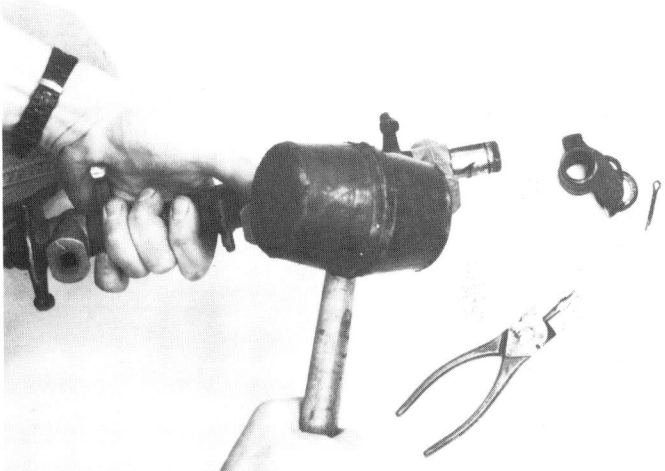
Cast-iron stands will now slide off. Aluminum ones can be driven off with rubber or hard-plastic mallet.

the height. When taking the measurements, be sure to keep the retainer square with the spring seat.

Once you have the measurement, record it so you will know what thickness shims are needed to bring installed height to specification. Shims are available in various thicknesses: 0.015, 0.030 and 0.060 in. Keep the valve, retainer, keepers and shims together, and mark them for the cylinder they match.

Installing the Valves—Have all the parts matched to their cylinders and lined up in order of assembly. Make sure the head is clean. Have a clean place to assemble the parts and some assembly lube or oil in a squirt can.

Position the cylinder head on its exhaust-port face. Squirt some oil into the valve guides and onto the valve stems. Slide each valve into its guide and push the oil seal down over the top of the stem. The oil seal will keep the valve from falling out while you assemble the head.

If installing valve seals other than the plastic umbrella type, follow the installation procedure recommended

---

RETAINERS & KEEPERS

Standard retainers and keepers are good for anything but all-out racing. If you use high-rate springs with stock retainers, but run the engine at high rpm, the stock keepers and retainers will not hold up. The valves will eventually pull through the keepers or the retainers and keepers will break.

Aftermarket companies make hardened retainers out of steel, aluminum and titanium. If you will be using the engine primarily for street driving and occasional high-rpm bashes, use steel retainers. They are less expensive, wear better than aluminum or titanium and are less subject to cracking and breaking.

Aluminum and titanium should be reserved for racing, where the engine undergoes frequent, thorough inspections. Use the keepers that match the retainers and replace them at every rebuild.

Valve Rotators—Positive valve rotators are used on the exhaust valves on the 330HD, 361 and 391 truck engines, and on the 360 and 390 in '72–'76 F100 pickups. If all the parts are together, and there are no stress cracks in the retainer, use the rotators again. If you have any doubts, replace them or have them Magnafluxed or Spotchecked. Most rotators are good for at least one rebuild.

---

by the seal manufacturer. If you are installing Teflon seals, be sure to use the cellophane boot over the valve stem. It protects the seal from damage

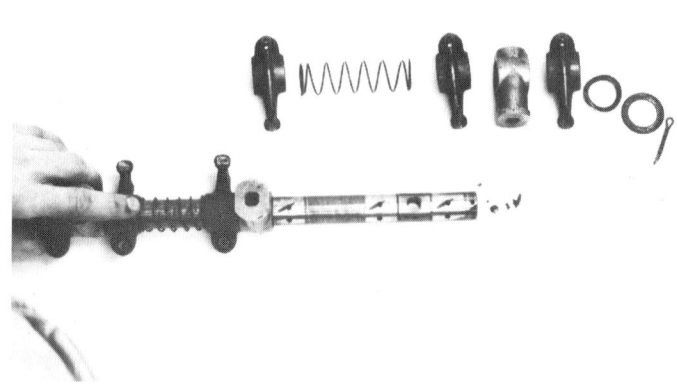

Simplest way to keep parts in order is to line them in order of disassembly.

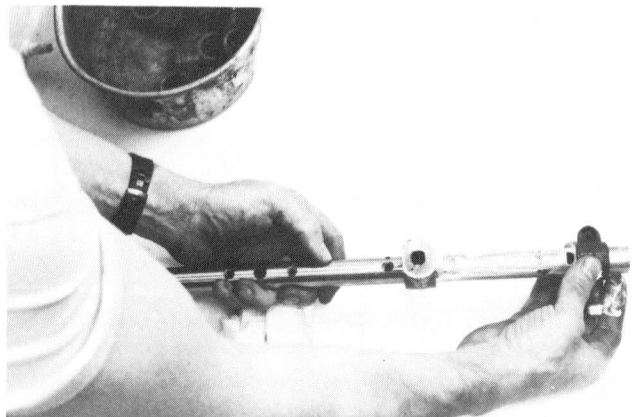

To ease assembly, shafts were left in freezer and stands placed in boiling water. Shaft is lubricated and one stand installed, followed by a rocker arm.

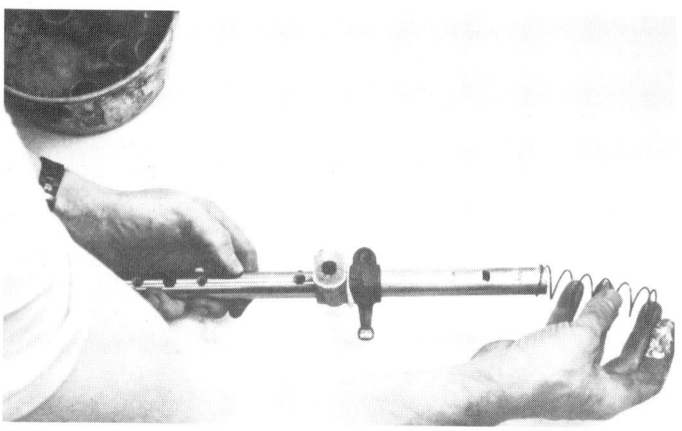

After stand and rocker arm, spring is next.

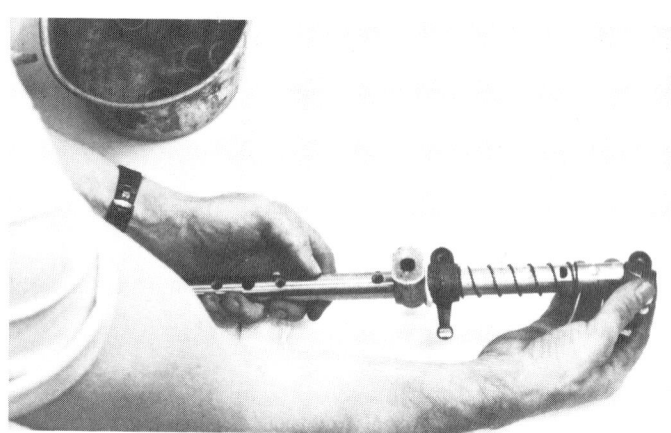

Then another rocker arm.

by the valve tip or keeper grooves. If the boot is torn or missing, use a piece of cellophane tape over the tip and grooves. Next to be installed are any shims, then spring/s, damper spring and finally the retainer.

Note: With the exception of the '68 engines, all 427 heads use removable valve-spring seats. If you are assembling one of these heads, you must install the seats before the seals.

Adjust the compressor so the valve spring is compressed only far enough to install the keepers. Compress the assembly carefully. *Do not compress the spring until it coil binds.* A dab of grease on the keeper grooves will hold the keepers in place. Insert the keepers and *slowly* release the compressor. You don't want a keeper to pop loose and hit you in the face!

Once the heads are assembled, spray them with a light coat of WD-40 or CRC and store them in plastic bags until installation.

ROCKER-ARM ASSEMBLIES

To prevent any delays when assembling the engine, inspect the rocker-arm assemblies. Be sure they are clean and in good condition. It's a good idea to check their clearances. During a rebuild, most rocker-arm assemblies get dipped in solvent at best—let alone cleaned or inspected. But take the time to do it right and check them thoroughly for wear.

Inspect—The first thing is to check the shafts for carbon or varnish buildup. If the rocker arms are spaced apart by springs, slide the rockers apart far enough to expose the shaft.

The shaft should have a polished look, but should not be scratched or gouged. Don't be alarmed if you see a spiral groove in the bottom of the shaft. Some hydraulic-lifter shafts were grooved to improve rocker-arm lubrication.

The most-common damage is scoring, caused by either a lack of oil, dirty oil or overheating. If you find this condition, disassemble the rocker shaft completely. Inspect the rocker arms and replace damaged parts.

If the rocker-arm assemblies look clean and the shaft is polished it's still wise to take them apart to check the oil passages. If none of the rocker arms feel excessively loose on the shaft and the assembly is fairly clean, clean the assembly in solvent and re-oil it.

If the assembly has a varnish buildup or you are unsure of its condition, disassemble it. It's more time-consuming, but much more thorough.

Disassembly & Cleaning—Before taking the rocker-shaft assembly apart, mark the rocker arms for their positions on the shaft. To disassemble the shafts, remove the cotter pin at each end. The flat and spring washers then will slide off along with the end rocker arms. If the shafts are fairly clean, the remaining parts can also be removed. Otherwise the varnish buildup must be scrubbed off before

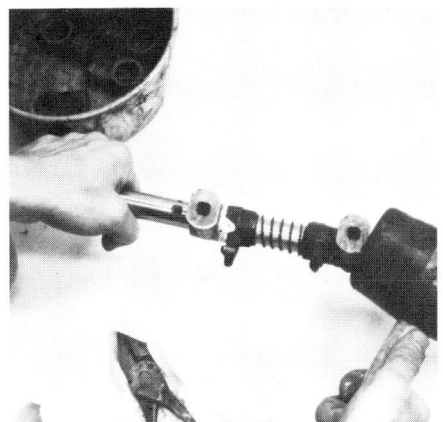

Next stand is installed to hold rockers and spring in place. Light rap with rubber mallet persuades stubborn stand.

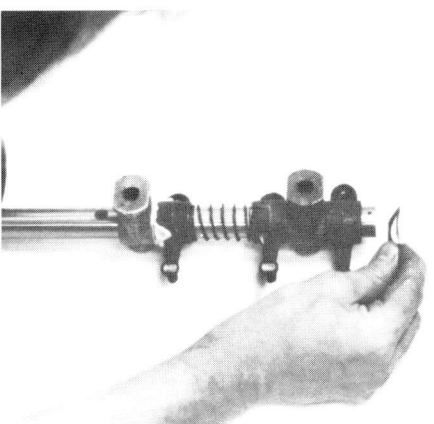

End rocker arms are held in place by spring washers rather than coil springs. Flat washer is next, then cotter pin.

the shaft can be disassembled further.

It will definitely help if the shafts are at 70F (20C) or more and you use some penetrating oil. If you still have problems, use a plastic or rubber mallet to knock the stands off the shafts. Most stands are not split and will be *glued* to the shaft by varnish. Split stands should slide off easily.

The last parts to remove are the cup plugs in the shaft ends. Drill a hole in one of the plugs. Don't use a punch; you'll probably just drive the plug farther into the shaft. Either run a long rod through the hole in the plug and shaft and knock the plug out at the opposite end, or run a sheet-metal screw into the plug and lever it out with a pair of pliers. Remove the remaining plug the same way.

The shafts and rocker arms can now be hot tanked or scrubbed. In most cases, if the shafts are really dirty, they will be badly worn where the rocker arms contacted the shafts, especially on the bottom side. Check each shaft with a micrometer—or your fingernail—before you waste time cleaning it.

If the shafts are in good condition, clean their IDs with a rifle or shotgun brush. Make sure the oil-feed and bolt holes are clean.

Checking for Wear—To measure rocker-arm-to-shaft clearance, you will need a 0—1-in. outside micrometer and a telescoping gage.

Shaft OD should be 0.839—0.840 in. Rocker-arm ID should be 0.843—0.844 in. Clearance between the two should be 0.0030—0.0055 in., with a wear limit of 0.0060 in.

If you are using the rocker arms on a race engine, limit rocker-arm-to-shaft clearance to 0.0015—0.0030 in.

This will require some selective fitting or honing of the rocker arms to match the shaft. It is possible to have the rockers reamed and have bushings installed, though this can be extremely expensive.

Usually the shafts will be severely grooved or scored if they are worn past specification. Most of the time the rockers can be saved, along with the springs. The stands see no wear and can almost always be reused.

Before you replace a complete rocker-shaft assembly, check with a machine shop. Some aftermarket manufacturers have oversize rocker shafts for the big-block Ford. These allow you to reuse worn rockers by reaming them and the stands slightly oversize.

The last thing to check is *breakaway torque* for the adjusting screw in adjustable rocker arms. Breakaway torque is simply the amount of torque necessary to *start* turning a threaded fastener in this case, the adjusting screw. This *interference-fit* screw should require at least 7 ft-lb to turn. If breakaway torque is less than this, replace the rocker arm or install an oversize adjusting screw.

Rocker-Shaft Assembly—With all the parts cleaned and inspected you are ready for assembly. Use a lot of molybdenum or white grease when assembling the parts.

Start the assembly by installing new plugs in the ends of the shaft. Use a large drift or pin punch, or a bolt ground flat on the end to drive the plug into place. Make sure whatever you use fits into the center of the plug.

If the plug cocks to one side during installation, remove it and insert a new one. A small amount of sealer can be used on the plug, but be sure none gets into the oil passages. Drive the plug in until it bottoms in the shaft.

Before proceeding, you must determine which "side" of the shaft is *bottom* and which "side" goes toward the *pushrods*. Most shafts have a notch at one end. On the installed shaft, this notch should be down and to the *front* of the engine when installed on the *right* bank. The notch should be on the bottom on the *rear* when installed on the *left* bank. In case you forgot, *right* and *left* refer to the orientation with you sitting behind the steering wheel.

Notches ensure that the rocker shafts are installed so that the rocker arms are correctly lubricated. With the valve closed there is no pressure on the rocker arm. As the rocker arm opens the valve, pressure on the bottom surfaces of the rocker arm and its shaft increase. This pressure peaks when the valve is fully open, then decreases as the valve closes.

For this reason, oil is supplied under pressure through holes in the bottom of the shaft. Most shafts have an additional, smaller hole or a groove. This additional hole or groove should be toward the pushrods or intake manifold with the shaft installed. The hole feeds oil to the rocker as it operates.

Correct orientation of the oiling holes is the reason for the notches on the shaft. With this in mind you should be able to keep things in the right orientation when assembling the shafts.

When assembling the shafts have the parts at room temperature—at least 70F (20C). Lubricate the shaft.

The rocker shaft is lubricated through the rocker stands, so pay attention to the stands during installation. Most rocker stands have a square hole for the bolts. Oil-hole alignment is no problem on these stands.

On the split, cast-iron rocker stands, the bolt hole is grooved on one side. This groove should align with the oil hole in the head. Although all split cast-iron stands are grooved, oil-hole alignment is only critical on the *third stand from the front* on the right head, and the *second stand from the front* on the left head. Got that?

It's easiest if you work from the *inside out* when assembling the rocker shafts. Slide one of the stands to one

Cleaning carbon buildup from underside of intake-manifold heat passage requires removing baffle. If baffle is retained by rivets, use thin chisel to start rivet out for removal.

A few scrapes with chisel and, presto! Carbon is off.

Although some people reuse the old rivets, I prefer to drill and tap the holes and install bolts.

Baffle bolted in place.

Be sure baffle drain is open.

of the two inner positions. Be sure the oil holes—and notch—are at the bottom of the shaft. Install a bolt and washer through the stand and shaft to keep the stand from turning. Next, from the *long end* of the shaft, install a rocker arm, spring, then another rocker arm and the next stand.

Install the next set of rockers, springs and stands and their bolts. The final rockers are installed at the ends of the stands. The spring washer is installed next with the flat part of the washer toward the rocker arm. Next comes the flat washer, then the cotter pin with the eye up. After you have them both assembled, store both rocker-shaft assemblies in a plastic bag so they'll remain dry and clean.

INTAKE MANIFOLD

The outside of the intake manifold is easy to clean, but there is a carburetor heat passage underneath. In most cases it is sealed by a stamped-steel shield. Some High-Riser and Tunnel-Port intakes didn't have heat-riser passages. Some other high-performance intakes have core plugs for cleaning the underside of the heat passage.

On some early models the heat shield is bolted on. To remove these just unbolt them. Later models use rivets to retain the shield. The rivets have a spiral thread on them and are more difficult to remove. To remove these rivets, use a 1/4—1/2-in.-wide chisel.

Wedge the chisel under the edge of the rivet head. *Gently* tap the chisel just enough to lift the rivet head. You are trying to *lever* it up, not *shear* it off, so angle the chisel so it will miss the shank. Stay to the side of the rivet once it starts lifting up. Once the rivet comes out far enough, clamp onto the head with a pair of Vise-Grips. Pull the rivet up gently, twisting it counterclockwise to *unscrew* it.

With the cover off, remove the carbon buildup from the bottom of the intake manifold. You can use a chisel or old screwdriver, but be careful, particularly with an aluminum manifold. It's easy to damage an aluminum casting. Clean the manifold until there is no carbon buildup left. Use a wire brush or screwdriver. A rotary wire brush works well, but be sure to wear safety glasses when using one.

When you finish cleaning, rotate the baffle back into place and reinstall the rivet. Use a pair of needle-nose pliers to hold the rivet while you start it in. Finish by driving the rivet back into place. A couple of good whacks will seat the other rivet back against the baffle.

It's relatively easy to break a rivet, so don't be discouraged if you do. Simply drill out the rivet and tap the hole so you can install a bolt. In fact, I prefer to attach the baffle this way.

If you break a rivet, first finish the cleaning. Center punch the broken rivet and drill a 1/8-in. pilot hole into it. Follow this with a 7/32-in. drill to remove the remainder of the rivet, then tap it with a 1/4-20 tap. Reinstall the baffle using a 1/4-20 x 3/8-in. bolt and lock washer.

Before setting the manifold aside, check the water-pump bypass for corrosion. If the the tube is badly corroded or pitted, replace it. Remove the old tube by twisting it out with a pair of Vise-Grips. Put silicone sealer on the new tube before driving it in.

CHAPTER 7
Reassembly

Most popular big-block Ford, the 390 was produced from 1961 until 1976. Engine was available with two-, four- or six-barrel carburetion. This is a 1963 390. Photo courtesy of Ford Motor Co.

If you haven't already done so, clean the water pump, fuel pump, timing cover, vibration damper and pulleys, brackets, distributor and oil pan. Clean these as a group so you won't have to clean your hands more than once. With all parts machined or replaced and everthing clean, you are ready to assemble the engine.

It's not only more pleasant to work with clean parts, it's the simplest way to keep dirt out of the engine. Consider dirt your number-1 enemy when assembling the engine.

TOOLS, SEALERS
AND LUBRICANTS
To assemble the engine you will need some special equipment. If you have it ready before starting, it will simplify and speed up the assembly.
Tools—The most important assembly tool is a torque wrench. If you don't have one, buy or rent one. Two major types are available—*breakaway* and *pointer*. Although either type will do, I prefer the pointer type.

The advantage of the pointer type is that you can watch the progression while torquing a bolt. This enables you to observe whether a bolt is tightening properly as it turns.

If you reach a certain value while torquing a bolt, and the value remains there as you turn it, the bolt may be stretching, the threads stripping or the hole may not be clean. If you have chased and cleaned the threads, suspect that the bolt is stretching. If this happens, stop tightening the bolt. Remove and inspect it. Also, check its threads and those in the bolt hole for stripping.

A breakaway or *click* torque wrench has a dial so you can preset the desired value. When that value is reached, the wrench *clicks*.

The next unusual tool is a cam-bearing remover/installer. If you had the engine hot tanked, you should have removed the cam bearings. The caustic solution in the hot tank destroys bearing material. It's best to have the machine shop install the cam bearings. But, if you plan to install them yourself, you *must* have this tool.

There are two different kinds of installers. One pulls the bearings into place with a threaded rod. The other drives them into place. If you rent either tool be sure you get the collet that matches the bearing ID to the sleeve. I cover cam-bearing installation later in the chapter.
Sealers—To prevent water, oil and air leaks you need sealers. All engines will need sealer on the intake, oil pan, timing cover and most of the gaskets. A silicone sealer works well.

If you have a 427 or are using shim-type head gaskets, use a head-gasket sealer. Do not use silicone here; a light coat of non-hardening gasket cement works best.

The important thing to remember when using sealants is that they should not be used in areas where they can do harm. You don't want sealant in oil galleries, on bearing surfaces or circulating with the oil. Most adhesives do not dissolve in oil and the excess will circulate until it blocks an oil passage. Use care—don't use too much or get sloppy.
Lubricants—There are three lubricants I use for assembly; Lubriplate, Ford's Oil Conditioner and molybdenum-disulfide grease, commonly called *moly grease*. Use moly grease for high-contact-pressure areas—cam lobes, lifter bases, push-rod ends, rocker arms, valve-stem tips and the fuel-pump eccentric. I use Lubriplate or Oil Conditioner for the remaining areas: bearing journals, timing chain, rocker shafts, wrist pins, lifter bores and the distributor-drive gears.

These lubricants are essential to good break-in. Moly grease helps to prevent camshaft wear on initial start-up. You only need a small amount.

Complete, high-quality gasket set simplifies assembly.

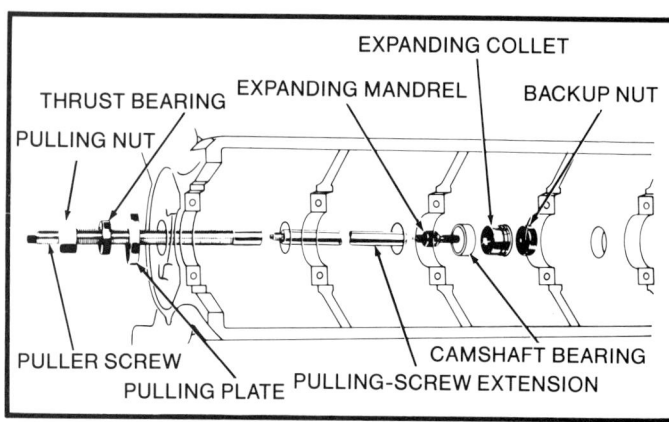

Though it looks complicated, puller-type cam-bearing installation tool is simple to use. Drawing courtesy of Ford Motor Co.

Most cam manufacturers include moly grease with their cams—enough for three or four rebuilds.

You will also need motor oil for assembling the engine. Figure on one quart for assembly, plus oil to fill the engine's crankcase. Use the same oil for assembly you plan to use in the engine. Put some in a squirt can to lube bolts and other areas.

Whatever brand you use, be sure it's an SF- or SE-grade oil. Besides being highly resistant to high-temperature breakdown, these oils have good cleaning qualities. If you are building a race engine with large bearing clearances—which includes stock clearances for the 427—use racing oil from the start. Choose a weight suited to the engine and climate.

Finally, keep the engine parts, oil containers and oil *clean*. If you get a part dirty, wash it and start over again. Never install a dirty part thinking everything will be OK; it won't be.

Plastigage is a *tool* for double-checking bearing clearances. I say "double-check" because if the crank and rod journals were ground correctly and the bearings match, the clearances should be exact. Even still, it's always possible to have a mismarked or out-of-spec bearing or journal.

Using Plastigage is somewhat time-consuming, but inexpensive. Use it. You need a package of green—0.001—0.003-in.—Plastigage. Make sure it's fresh, otherwise it will indicate more-than-actual clearances.

CAM-BEARING INSTALLATION

Cam-bearing installation is more time-consuming than difficult. Some people think anything that takes time is difficult, but if you made it this far you're not in that category.

You only need one collet or mandrel to install the cam bearings. The bearings have the same inside diameter. The thread-in or puller-type installer is the slower way, but you are less likely to damage a bearing or bearing bore.

Bearing Preparation—Before installing the cam bearings, chamfer their IDs. The chamfers will make it easier to install the camshaft. Chamfering both edges also ensures there are no ridges touching the bearing journal to cause extra wear.

Chamfer the bearing with a knife, conical or cylindrical fine-grained stone or *bearing scraper.* A bearing scraper has a triangular blade with hollow-ground sides. Remove only enough material to eliminate the sharp edge. The amount removed from all five bearings shouldn't register on a gram scale.

Hold the cutting tool at about a 45° angle to the bearing, and peel or scrape away a *small* amount of bearing. If you use a fine stone to remove the edge, rotate the stone as you work your way around the bearing. If you use a knife, hold it so that the blade won't dig in as you cut. The idea is to scrape, not whittle.

When you finish, check the edges for smoothness, especially the *leading edge*—the edge that will be installed toward the front of the block. If they're not smooth, go over them again—just don't remove a lot of material.

Next, put the bearings in order for installation. Although the ID of all the bearings is the same, the OD increases from rear to front. Also, each bearing has to match up with different oil holes within the block. There are different-width bearings for different blocks, so check them against the old bearings.

The bearing bores are numbered one through five, front to rear. Most bearings have a corresponding number on them, or the box has individually numbered slots. Occasionally the bearings are boxed separately, with numbers on each box.

In addition to the bearing numbers, each bearing insert has a front and rear. Most cam-bearing bores are drilled for an additional oil passage. Check the oil-system drawing for your particular engine, pages 29 and 31.

With the exception of number-3 and -5 on the side-oiler 427, all big-block cam bearings are drilled to channel oil to other parts of the engine. If a bearing is installed incorrectly, it may block the gallery and starve part of the engine.

Double-check each bearing *before* you install it. Match the holes in the cam bore with the holes in the bearing. The holes should match in location, width and diameter or shape.

Order of Installation—When a camshaft bearing is installed, it must align squarely with its bore in the block. To keep the bar, mandrel and bearing aligned, the bar must pilot off another cam-bearing bore. The farther away the two bores are from one another, the more-accurate the alignment.

As a result, when you install the cam bearings, its best to work from the ends of the block toward the center. So install bearing number-5, -4 and -3 from the front of the block and -1 and -2 from the rear. You can install all the bearings from the front of the block if you must, but it's easy

Before installing cam bearings, line them up in order of installation.

Cam-bearing oil holes must align with those in block. Check alignment before and after installation.

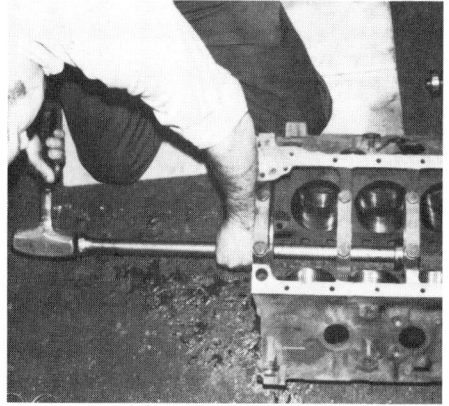
Driver-type cam-bearing installer is more common. Puller or driver type works well.

to misalign one of the front bearings.

Puller Type—Select the correct *expanding collet* or straight *mandrel*, depending on which type is available. Match the collet or mandrel to the backup nut. You also need the puller screw and plate, thrust bearing, nut and extension.

The mandrel fits the bearing ID. A shoulder or step butts against the bearing edge without interfering with the bearing bore in the block.

The expanding collet eliminates the need for dozens of different-size straight mandrels. The collet expands and fits the bearing ID to hold the bearing in alignment.

The mandrel or collet should be a close fit in the bearing, but not so tight that it distorts the bearing. The collet should be expanded so it contacts the bearing ID.

The straight mandrel should slide in easily but not allow the bearing to cock during installation. If it doesn't fit the bearing snugly, the mandrel can be shimmed by wrapping it with paper, masking tape or thin cardboard. Be sure the shoulder fits through the bearing bore in the block.

Oil the collet or mandrel to protect the bearing surface. Start with the rear, or number-5 bearing. Place the pulling plate against the front of the block, the thrust bearing over the top of that and then run the pulling nut down against the thrust bearing.

When tightening the pulling nut, use a wrench on the end of the bar to keep it from turning. If the bar turns, it will damage the bearing or turn it and misalign the oil holes.

Pull the bearing in until the edge is about 1/16-in. away from the front face of the bearing bore. On the center bearings be sure the oil holes line up, and the bearing is centered in the bore. Exact center isn't important, but the oil holes must align.

To install the number-1 bearing you need a feeler gage and a straightedge. This bearing should be 0.005—0.020-in. back from the front face of the block. Back the puller off periodically to check bearing position. Use feeler gages and a straightedge to check its depth.

You can also check the squareness of the front bearing using the feeler gage and straightedge. Check the depth of the bearing in several places around its front surface. It should be the same all around the front edge.

If you go too far, reverse the bar and pull the bearing back. Don't remove it and start over again unless you get a new bearing. If the mandrel contacts the plate before the bearing reaches the correct position, space the pulling plate away from the front of the block. Use a thick metal plate with a hole in it or a piece of bar stock under each side of the pulling plate to do the spacing.

Driver Type—The driver type of cam-bearing installer is the more common. You have to use the correct-size mandrel and make sure the oil holes are aligned.

To keep the driving bar aligned, a tapered guide pilots in the number-1 or -5 bearing bore. This centers the bar in the bore during installation.

Start with the number-5 bearing. Place the bearing onto the mandrel and position the mandrel in front of the bore inside the block. Align the oil holes. Place the bearing and mandrel squarely over the opening, with the driving bar centered in the bores.

Drive the bearing into place until the bearing is about 1/16-in. behind the bore edge. Once the bearing is installed, double-check oil-hole alignment. If it's OK, install the other bearings.

Install the number-1 bearing from the rear by driving the bearing from the inside of the block. Be careful not to damage the other bearings with the driving bar as you slide it through. Protect them by wrapping rags around the rod.

Again, position the number-1 cam bearing 0.005—0.020-in. behind the front surface of the block. Move the bearing if it is not the right depth; don't remove it. Simply install the mandrel and tap the bearing to reposition it.

Cam Plug—To install the camshaft plug you need a hammer, a driving tool, sealant and the plug. Use a driver that will bear evenly against the plug near the edge. Hammering on its center will distort the plug.

A special tool is available for this, but a pipe reducer or length of 2-in.-diameter pipe works well. A pipe reducer works best because of its smooth, wide face. File the end smooth, particularly at the mold joints. If you want to be fancy, put the reducer on one end of a short length of pipe and a cap on the opposite end to make a professional-style driver.

The cam plug is an expansion-type plug. It is installed with the cup flange toward the block. Before installing the plug, seal the edges. Use a small amount of silicone sealer or weatherstrip adhesive. Don't use too much—it may get on the cam bearing.

Line the plug up squarely with the block and drive it in until it bottoms in the bore, page 144. Don't keep hitting the plug once it bottoms. This will distort it so it may not seal. Once

119

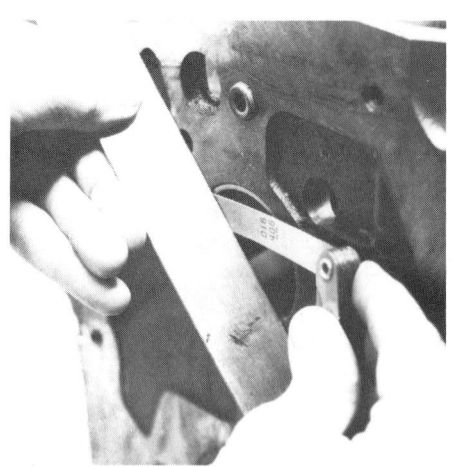

Front cam bearing should be set back 0.005—0.020-in. from front face of block. This ensures adequate lubrication for cam-sprocket and thrust plate.

Apply silicone sealer to outer edge of core plug and drive plug into place. Use large-diameter punch to prevent distortion of plug.

Drive plug no farther than 1/8-in. past chamfer.

it is installed you can add some sealer around the periphery of the plug for insurance. Wipe or trim the sealer to make it flush with the block. Remove any sealer from the rear cam bearing.

Core Plugs—Most big-block water-jacket core plugs are the cup type, though some engines use screw-in plugs. To install either type, be sure each opening is free from foreign material: metal shavings, paint, dirt or grease. Spread sealer around the outer edge of the plug.

To install a cup plug you need a driver and hammer. The driver can be the back of a socket or a piece of round stock. It should fit inside the plug without actually touching the edge. Unlike the cam plug, these are installed concave or "open" side out.

Drive the plug into the hole squarely until the plug is about 1/8-in. past the outside edge of the chamfer. If you drive the plug in too far, remove it and replace it. The same holds true if the plug is crooked or distorted. Wipe any excess sealer from the hole and you can install the next plug.

If the engine has screw-in core plugs use a 1-in. hex-drive adapter. The plug should be screwed in until it is "snug." You don't need to over tighten a screw-in plug.

Screw-in core plugs—and screw-in oil-gallery plugs—can be sealed with either silicone or Teflon tape. I prefer Teflon tape. It not only seals well but reduces drag during installation.

Wrap the tape around the threads and trim the end flush with the bottom thread. Wrap the tape so the end of the tape is *trailing* as the plug threads in. This means the tape should be wrapped opposite the way the bolt threads in, or counterclockwise, viewed from the outer end of the plug.

Oil-Gallery Plugs—Like the other core plugs, oil-gallery plugs can be either screw-in Allen-head or drive-in cup type.

You need a driver to install the cup plugs. Use a short length of 1/4-in.-diameter round stock or a long 1/4-in. bolt with the threaded end shaped to fit squarely against the cup plug.

Some cups are flat and others have rounded bottoms. The best ones have a slightly raised bottom that forces the outer edges of the cup out when the plug seats. You can feel this dimpling action as the cup seats.

Use sealer with the cup plugs: silicone or weatherstrip adhesive. Drive the plug *squarely* into place until it bottoms. If the plug goes in crooked, remove it and install another.

The screw-in type is a 5/16-in. Allen-head plug. It can be sealed with either silicone sealer or Teflon tape. Simply screw the plug in until it bottoms.

No matter which type of plug or sealer—tape or silicone—you use, make sure the sealer doesn't get into the galleries. Oil-gallery blockages are hard to find and oil-starved bearings don't last very long!

There are a lot of plugs to install. The standard hydraulic-lifter block has 11 plugs. The first solid-lifter blocks have six as did the early top-oiler 427s; the solid-lifter side-oilers have seven; the hydraulic-lifter side-oilers 15. The two screws for the camshaft thrust plate also plug the lifter

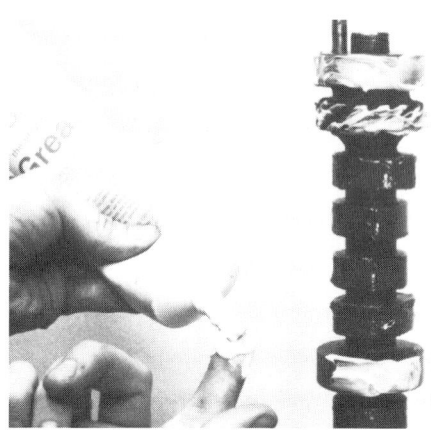

Apply moly grease to cam lobes and oil or white grease to bearing journals.

galleries. Blocks with thrust-button-type camshafts use cup plugs.

Side-oiler blocks also have a bolt or plug in the front at the lower right of the timing cover, viewed from the front. This seals the main oil gallery. All but two of the side-oiler plugs are screw-in. Cup plugs are used in the gallery above the cam on the hydraulic-lifter 427 and at the front of the left lifter gallery.

If you are installing a plug with a *jiggle-pin*, use a hollow, 1/4-in.-diameter tube to install it. Otherwise you may damage the opening or pin when driving the plug into place.

A *jiggle-pin* is only used with hydraulic lifters. It bleeds air from the oil before it reaches the lifters. It is not absolutely necessary—many engines have been run successfully without it.

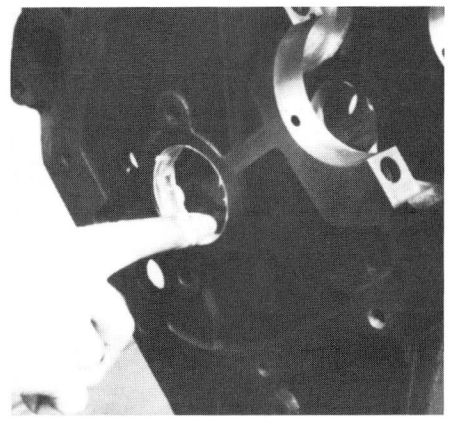
Lubricate cam bearings. Be sure fingers are *clean* when doing so. One advantage to white grease over other assembly lubricants is that any dirt shows up right now! A disadvantage is it doesn't dissolve with the engine oil, so don't overuse it. It'll end up as a huge gob in the oil pan.

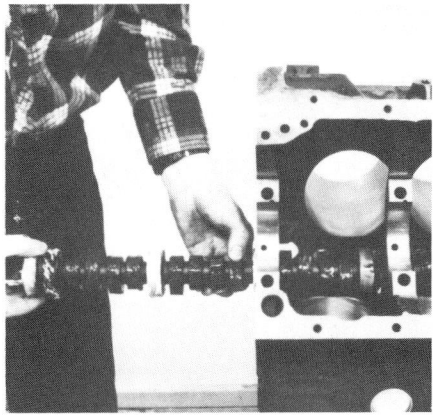
Carefully feed cam into place. Do not bump lobes against cam bearings.

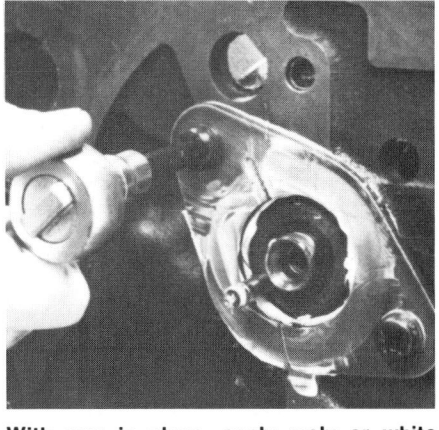
With cam in place, apply moly or white grease to back of thrust plate. Install screws with lock washers and torque 12—15 ft-lb.

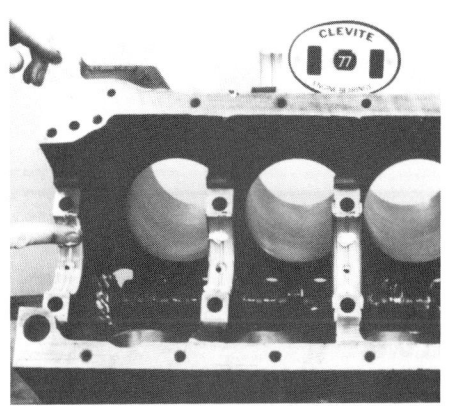
Main-bearing inserts are pushed into place and coated with moly grease or Lubriplate.

CAMSHAFT INSTALLATION

Proper installation is very critical to camshaft life. A camshaft isn't the easiest thing to remove with the engine installed, so be very careful during its installation.

Camshafts are normally *Parkerized* when new or reground. Parkerizing is a chemical surface-treatment process that applies a manganese-phosphate solution to the cam lobes.

Parkerizing serves a similar function as honing marks in cylinder walls. Just as honing retains oil and helps seat the rings, Parkerizing retains oil and helps each lifter mate with its cam lobe. A Parkerized surface also retains dirt or metal particles, so clean the cam throughly before installing it. Also clean the cam bearings.

If you are installing a new or reground camshaft, install the drive pin before lubing the cam. Remove the pin from the old cam with a pair of Vise-Grips or clamp the pin in a vise and pull it out of the cam. If this burrs the surface, take advantage of it and insert the burred end into the cam. The "knurl" will help hold it in place.

Occasionally, the fuel-pump eccentric will loosen and chafe the drive pin. If the pin is badly worn, replace it. Check its length—thrust-button cams use a 7/8-in.-long pin. Thrust-plate cams use a longer 1-1/2-in. pin.

Apply moly grease to the lobes and Lubriplate or oil to the journals and cam bearings. Each lubricant has its purpose, just as the journals have one function and the lobes another. The moly coating helps cam-and-lifter break-in and protects them until they receive oil.

With the cam cleaned and lubed, loosely install the camshaft spocket as a handle. *Carefully* guide the cam through the cam bearings. You don't want to nick one now with one of the sharp cam lobes. Reach inside the block to guide the cam. Before you reposition your hands, slip one journal into a bearing to support it.

Binding Problem?—Once installed, the cam should turn freely. If it doesn't, remove it and check for dirt or varnish on the journals or bearings. Also check bearing and journal diameters, particularly on a new or reground cam. All journals should be 2.1238—2.1248 in.; bearings should be 2.1258—2.1268 in.

If the journals are OK, or if you are using the original cam, the binding could be caused by incorrectly sized bearings, improper installation or a bent cam. Most likely, it's damaged bearings or a bent camshaft. Recheck camshaft runout, page 84.

If the journals are clean and correctly sized, check the bearings. Clean them with laquer thinner and apply layout dye—usually Dykem, or engineer's blue—to the bearing ID. Reinstall the cam, turn it a few times and remove it. Wipe the bearing surfaces clean and check them. Areas where the camshaft binds will be wiped clean of the dye.

If only one or two bearings are binding, the points of interference can be relieved with a bearing scraper. Don't scrape too much before rechecking. It's easy to remove bearing material, but once it's gone you can't put it back without replacing the bearing. If there is serious binding on three or more bearings, replace them.

Thrust Plate—If everything is ready, remove the cam sprocket and install the thrust plate and screws. The thrust plate should be installed with the oil slot to the outside, facing up. You need that #4 Phillips-head bit again. Loctite the threads and torque the screws 12—15 ft-lb.

Camshaft End Play—You must install the camshaft sprocket to measure camshaft end play. Temporarily install the spacer—if needed—cam sprocket, fuel-pump eccentric, flat washer, lock washer and bolt. Loctite the bolt threads at final installation.

Torque the bolt 40—45 ft-lb. To keep the camshaft from turning while tightening the bolt, insert a piece of round stock or a screwdriver through the cam sprocket and into the hole in

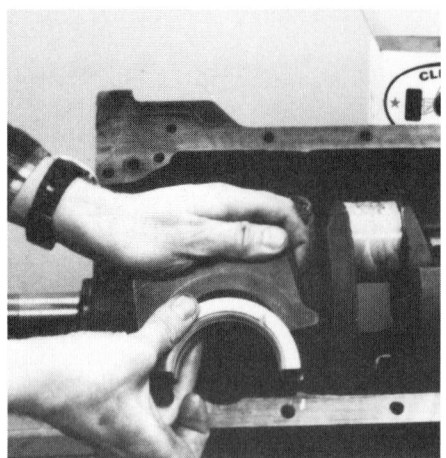
Make sure you have correct thrust bearing. Bearing for '64 and later engines won't fit '58—'63 blocks as later thrust-face OD is larger.

Be sure bearing inserts stay in position as caps are installed. Caps should seat firmly in their registers before bolts are torqued. Tap them into place with a soft mallet.

Install rear-main lip seal with lip facing toward the front. Offset ends about 3/8 in. from bearing cap-to-block parting faces.

front of the distributor drive.

Once the sprocket is installed, measure camshaft end play. There are two ways to measure end play: with a dial indicator or with feeler gages.

With a dial indicator, mount the indicator with the plunger perpendicular to the sprocket and the tip against the nose of the cam. Push the cam to the rear of the block and zero the indicator. Pull the camshaft forward and read end play directly.

To check end play with feeler gages, simply pull the camshaft forward. Use feeler gages to measure the gap between the camshaft sprocket and the thrust plate.

End play should be 0.001—0.007 in. for FE engines and 0.003—0.009 in. for FT and race engines. Maximum allowable end play is 0.012 in. for FE and 0.010 in. for FT and race engines.

If you want to reuse the old camshaft sprocket, replace the thrust plate and recheck end play. If you replaced the camshaft sprocket the only other major wear point is the thrust plate. Replacing the plate should correct end play.

Another possible wear point is the thrust surface on the front of the number-1 cam journal. It is rare that this would be worn enough to require camshaft replacement.

Install Main Bearings—Clean the bearing bores in the block and main caps. Wash them with a paper towel and a *little* laquer thinner or other fast-drying solvent. Dry the bores with a clean paper towel and make sure there are no particles of dirt or paper left.

Dirt or other foreign particles between the bearing half and its bore will distort the bearing, causing insufficient bearing-to-journal clearance and possibly cause the bearing to overheat.

Dirt trapped between the bearing and journal will *embed* into the bearing surface. Most modern bearings are designed with some *embedability*—the material is capable of absorbing some small particles to prevent bearing-journal damage. Still, they can only absorb so much before the dirt ruins the bearing and journal.

Each bearing comes in two halves, upper and lower. In most cases, the upper—block—half is identified by a groove. The lower—cap—half has no groove. High performance cranks with grooved main journals use grooved bearing inserts in the block *and* cap.

The center bearing—number-3—acts as the thrust bearing. Number-3 insert halves have wide flanges.

Make sure the bearings are correct. Double-check the size to make sure it is what the box markings indicate. Install the upper halves into their bores to check them.

To install the upper bearing half in the block, insert the bearing tab into its register on one side. Hold it down with a thumb or finger and push the opposite end down flush.

All bearing halves should go in easily except for the thrust bearing. Because of its thrust flanges, it will require considerably more force. **Don't hammer on the bearing!** Push *gently* on the *end* of the bearing with a plastic mallet or the end of a hammer handle.

You will notice the oil holes in the bearing don't align with those in the block. This is due to the machining of the block. Except on the side-oiler blocks, the oil holes are drilled through the main-bearing bores to the top oil gallery. Bearing bores for the crank and cam journals, however, are not in perfect alignment. Consequently, the oiling holes are slightly off-center in the bearings.

There is a way to be sure the restriction is correct. Measure the oil holes, with the shank end of drill bits. The following chart lists the drill bit that should pass through the bearing into the oil hole:

Number-1	7/64 in.
Number-2	9/64 in.
Number-3	9/32 in.
Number-4	5/32 in.
Number-5	9/32 in.

If the opening is not large enough, remove the bearing to keep any metal out of the galleries. Carefully enlarge the bearing hole to the correct diameter. Chamfer the hole to remove any rough edges or burrs.

Before installing the bearings in the caps, *lightly* file the block mating surfaces on each cap to ensure correct fitting. Select a smooth, flat or mill file, lay it flat on the bench and set the cap, mating surface down, on the file. Draw the cap back and forth a few times.

Do not remove any more metal than burrs or nicks—anything that projects above the mating surface. After filing the caps, clean the bearing inserts and install them in the caps.

Special tool used to install rope-type rear-main seal. Seal is driven into place and carefully trimmed against handy knife stops.

The same thing can be accomplished with large socket or piece of pipe.
Note additional core plugs on back of this '58 cylinder block. Also note casting-number location on top left rear of block.

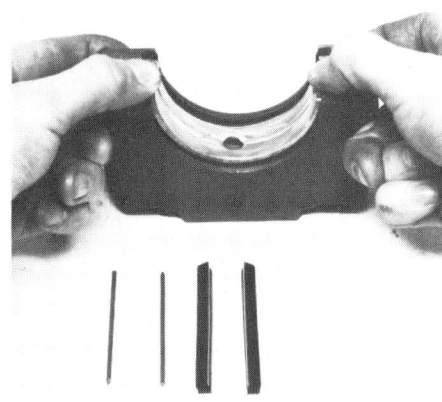

Snap bearing into place, oil lip-seal half and install it. Offset seal ends to mate with seal half in block.

REAR-MAIN SEAL

There are two types of rear-main seals. The original seal is the rope type. The later seal is neoprene. Both are two-piece seals and work well when installed correctly, but the similarity ends there.

The neoprene lip-type seal is far easier to install. It lasts longer and creates less drag on the crankshaft. Most gasket sets include the lip-type seal. If your gasket set includes only a rope seal, spend a couple of extra bucks for a lip seal.

If you plan to Plastigage the bearings, do it now, before installing the rear-main seal.

Plastigaging Bearings—Clean the main-bearing journals thoroughly and carefully lay the crank into the block. The bearings and crank journals are dry, so don't rotate the crank. Install all the main-bearing caps except the one you're checking. Torque the cap bolts 95—105 ft-lb. Don't bother to install the cross bolts on a cross-bolted 406 or 427.

Lay a piece of Plastigage on the crank journal in line with the crank and install the cap and bearing. Thread in the bolts and torque them 95—105 ft-lb. Do not rotate the crank or the gage will smear and you'll have to start over.

Remove the cap and use the scale on the paper sleeve to measure the Plastigage width. Compare the width with the printed scale to find bearing-to-journal clearance. Clearance should be 0.0005—0.0025 in., with 0.0005—0.0015 in. desired.

If the clearance is out of specification there is a problem. Remove the crank and recheck the bearings and crank journals. Again, the time to correct the problem is now.

Lip-Seal Installation—If the engine came equipped with a rope seal from the factory—most did—there may be a pin in the cap. The pin keeps the rope seal from turning in its groove and eventually tearing, leaking or bunching up. You must remove the pin to install a lip seal.

Drive the pin out with a small punch; knock it out from the back of the cap. This will keep you from inadvertently hitting the bearing-bore surface and distorting it. Seal the pin hole with silicone sealer and wipe away the excess.

Oil the bearing inserts and install the seal in the block; lip toward the front of the engine. One end should protrude 3/8-in. above the cap mating surface.

Install the cap half of the seal facing the same direction. Offset the opposite end of the seal 3/8-in. above the mating surface. When the cap is installed, the seal mating surfaces won't align with the cap-to-block mating surfaces. This lessens the chance of a leak.

Rope-Seal Installation—If you are installing a rope seal, leave the pin in the cap. Squirt oil on the seal halves before installing them; this will help them expand. Lay the rope in the groove of the block or cap and start it into the groove with your fingers.

With a hammer and a piece of round stock or a large socket, tap the seal into place. Don't damage the bearing-bore surfaces while installing the seal. Roll the socket or piece of stock back and forth while you hammer on the socket.

After the seal is seated, trim both ends. Hold the seal in place with your thumb and cut the seal flush with the mating halves of the cap surfaces. Use a sharp knife or razor blade and cut away from the bearing bore. Do this on both seal halves.

CRANKSHAFT INSTALLATION

With the main seal and bearings installed, you are ready to install the crankshaft. The crankshaft should be freshly polished and cleaned. Make sure there is no dirt in any of the crank's oil passages. Double-check the rod journal and bearing sizes.

The rear seal should already be installed unless you are checking the crankshaft for runout, page 82. Lubricate all the crank journals, and bearings. Carefully lay the crank on the main bearings. Do not bump the journals or bearings.

Install all the main-bearing caps except for the rear one—number-5. Make sure the caps are in order and facing the right way. On the 406 and 427, install the cross bolts with their spacers and washers but don't tighten them. Each spacer and washer should

Lubricate side seals and check length against main-bearing cap.

Install cap, side seals and nails. Be sure cap is fully seated before driving in the nails.

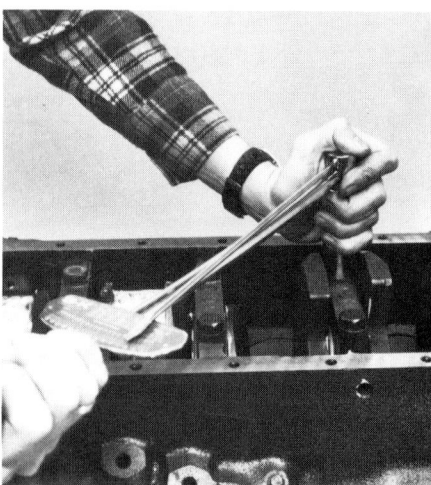
With all caps in place, torque cap-1, -2, -4 and -5 to specifications.

be marked for main-cap number and location, R or L for right or left. Be sure they are correctly installed.

If you have trouble getting the caps in position, wiggle them back and forth, front to rear, while pushing down on them. If the caps are really stubborn use a plastic or rubber mallet to *tap* them into place. Lube and install the main-bearing bolts finger tight as guides so the caps don't go off-center and hit the crank or bind in their registers. Be sure to hold the bearing insert in place with your fingertips while you're installing each cap.

Installing Rear-Main Side Seals— Before installing the rear main-bearing cap, place one of the side seals against either side of the cap. Check how much of the side seal will project above the cap when the seal is fully seated in its register.

Install the main-bearing cap and bolts, but leave the cap about 1/8-in. off its register. Oil the side seals and slide them into place, with the wide side of the seals away from the cap. Be sure the seals bottom against the block. Push the cap all the way down and snug the bolts down by hand.

Ford recommends torquing the main-bearing cap before installing the side seals, but I've found it's common for the seals to bunch up or tear as they are installed. Though leaks are common in this area, I've never had a rear-main seal leak when the side seals were installed *before* torquing the main-bearing cap.

Before installing the *nails,* check their points. Each should be bent slightly at the point—if not, bend them. The bend keeps the point from digging into the seal and pushing it down. All you have to do is install each nail with the bend in the right direction.

Insert the two nails. Install each nail on the crank side of the seal with the point *away from the seal.* Drive the nail down until it is flush with the top of the cap. Leave any remaining seal sticking out until you install the oil pan. The distance the seals stick out should conform to your previous check.

Torque the Main Caps— When torquing the main-bearing bolts, snug all the bolts about 5 ft-lb, except those on the center—number-3—main-bearing cap. The thrust-bearing surfaces must be aligned before the center cap is tightened. Leave these bolts finger tight.

Torque the bolts on cap-1, -2, -4 and -5 to 95—105 ft-lb. Torque the bolts in stages, alternating between bolts on the same cap about every 20—25 ft-lb. You don't want one bolt at 105 ft-lb and the other at 30 ft-lb.

Once the other main-bearing caps are torqued to spec, align the thrust surfaces on the number-3 main bearing. Place a large screwdriver between the front of one of the main-bearing caps—not the center main!—and one of the crank counterweights. Pry the crankshaft forward in the block.

Hold the crankshaft in this position. Slip another screwdriver between the front of the center main-bearing cap and the crankshaft. Pry the bearing cap toward the *rear* of the block.

Remove the screwdriver you used to pry the bearing back, but maintain pressure against the one holding the crank forward. Torque the center main-bearing bolts the same as the others. Release the screwdriver once the bolts are tight.

On the cross-bolted main bearings, torque the cross bolts next. Be sure each spacer and washer is correctly installed. Torque the cross bolts in two stages, 20 ft-lb, then 40 ft-lb. Follow the sequence shown at right.

Crankshaft End Play— With the center main cap tightened, check crankshaft end play. Use either a dial indicator or feeler gages.

With a dial indicator, position the indicator at the crank nose or the flywheel or flexplate crank. Set the tip of the indicator against the flange. Pry the crankshaft away from the indicator and zero the dial. Pry the crankshaft toward indicator and read end play.

With the feeler-gage method, pry the crankshaft forward or back. Measure the gap between the center bearing and crank thrust surface with feeler gages.

Crankshaft end play should be 0.004—0.010 in., with 0.014 in. maximum. In this instance, less clearance is desired, especially for racing engines. For racing use a tolerance of 0.004—0.008 in.

If end play is *less than* 0.004 in., loosen the center main-bearing cap and realign the thrust faces. Recheck end play.

If there is still too-little end play, remove the crank and center bearing. If necessary, thin the front thrust-bearing flange by laying a piece of 320-grit sandpaper on a flat surface and *lapping* the bearing to bring end

Before torquing center—number-3—main-bearing cap, force crank back and forth to seat thrust bearing. Torque center main cap while holding crank forward.

If the engine has them, install cross bolts next. Spacers are marked for bearing-cap number, 2—4, and position, R for *right* or L for *left*. Left is the oil-pump side. Spacers should be a light press fit, 0.000—0.001-in. clearance.

Torque cross bolts in two stages following sequence shown below. Remember: Main-cap bolts must be torqued before cross bolts. First stage is 20 ft-lb, second 40 ft-lb.

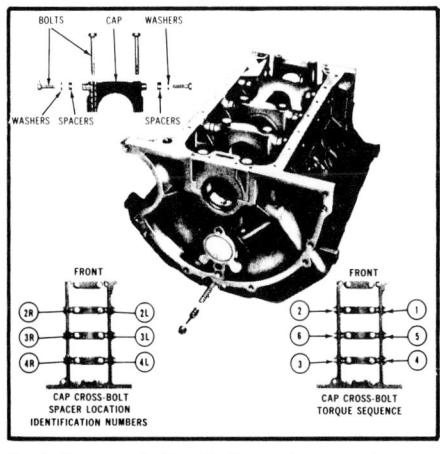
Install cross bolts, their washers and spacers as shown. Install and torque cross bolts 20 ft-lb, then 40 ft-lb in sequence *after* torquing vertical main-cap bolts. Photo courtesy of Ford Motor Co.

play within tolerance. Measure the front flange with a micrometer. Continue to check it until you've removed enough material to bring the end play to at least 0.004-in.

If you find there's too much end play, trade the crank and bearings for a *crankshaft kit*. This is very unlikely if you checked the crank carefully, but it can happen. So bite the bullet and make the correction now.

If the end play is acceptable, check crankshaft *breakaway torque*—force required to start the crank turning. The only significant resistance the crank should have is rear-seal drag.

If you installed a lip seal, the crank should turn by hand. With a rope seal, breakaway torque should be less than

Once mains are torqued in place, check crankshaft end play. Dial indicator is used here. It can also be checked by finding distance between crank thrust face and center main-bearing flange with feeler gages.

15 ft-lb. Recheck runout if the torque is excessive, page 82.

PISTON AND CONNECTING-ROD INSTALLATION

The pistons and connecting rods should already be assembled at this point. If not, refer to page 93.

To install the pistons and rods you need the following: a ring expander, ring compressor, fine-tooth file, rod-bolt protectors, a large can for oil and a hammer with a rubber-covered or wood handle. You also need one quart of oil. Special rod-bolt protectors can be picked up at the parts store, but two short pieces of 3/8-in.-ID flexible fuel line will work.

Fitting the Rings—Standard and most oversize piston rings usually have the end gaps preset. End gap should still be checked. Racing and some other rings may need the end gaps adjusted before installation.

To check ring end gap, insert each ring 1/2-in. into a cylinder bore. Square the ring in the bore by pushing it in with a piston. Measure the end gap with feeler gages and compare it to the chart, page 126.

You are matching each ring to a particular cylinder, so once you have measured or corrected the end gap, keep that ring with its cylinder or connecting rod and piston.

Note: If the cylinders *have not been rebored*, push the ring *all the way to the bottom* of the cylinder. End gap will be less at the bottom of a tapered cylinder. Minimum piston-ring end gap ensures the ends of the rings don't butt at maximum operating temperature.

Although the above dimensions are acceptable, best oil control and cylinder sealing is ensured by keeping end gap as small as possible. For the 330 FT and standard FE engines, compression-ring end gap should be 0.010—0.020 in. Oil-ring side rails should be 0.010—0.035 in.

Rule-of-thumb for compression-ring end gap is to multiply the bore diameter by 0.004 in. The reason end gap depends on bore diameter is simple.

As bore diameter increases, so does ring circumference, or "length." The

Correct ring installation requires care. Keep everything organized. Ring expander saves wear and tear on fingers, prevents broken rings.

Even though piston rings are sized, check end gap. Place ring in bore and push it in with piston.

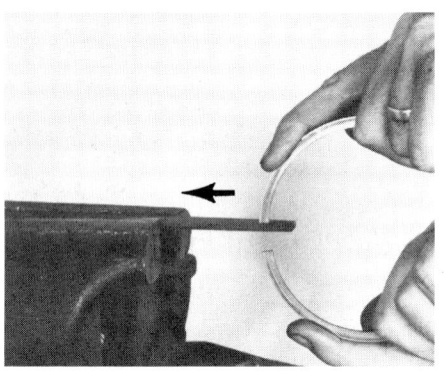

If there is insufficient end gap, ring end must be filed. Place fine-tooth file in vise. File only one end of ring and in direction shown. Pinch ends of ring together occasionally to be sure you are keeping end square. Give filed end light cleanup to remove any burrs or sharp edges.

RING END GAP (inches)			
	Top Compression (1 ring)	Bottom Compression (1 or 2 rings)	Oil Control (steel rails)
FE exc. 427	0.010–0.031	0.010–0.020	0.015–0.066
FE 427	0.018–0.028	0.015–0.025	0.015–0.055
FT 330	0.010–0.015	0.010–0.015	0.010–0.035
FT 361, 391	0.015–0.023	0.010–0.020	0.015–0.025

larger ring has more circumference, so it expands more when heated.

Heat expansion is also the reason *minimum* end gaps are larger in heavy-duty and racing applications—these engines run hotter. The 330 FT engine can use the smaller FE end gaps because of its relatively small—3.87-in.—bore.

Oil-ring end-gap range is wider simply because an oil ring runs cooler, does not seal combustion pressures and two rings are used in an oil ring assembly. Consequently, end gap is much less critical.

If you are setting end gap on an engine for light service, use 0.004-in. per inch of bore. If the engine is going to receive hard use—heavy-duty trucks or racing—increase the end gap. Exactly how far depends on use. Follow the ring-manufacturer's recommendations.

Heavy trucks will normally be worked hard all the time, while a "race" engine may be driven sedately on the street. On the 427, using the 0.004-in. factor gives an end gap of about 0.017 in. This is the *minimum* for *street* driving. When Ford raced this engine, compression-ring end gap was set at 0.028 in. to accommodate the high heat and resulting ring expansion.

Again, if you are considering the "racing" specifications, be honest with yourself and realistic concerning your use of the engine. If the engine is not kept hot enough, it will have more oil-contaminating blowby and will lose compression past the ends of the rings.

Hot in this sense doesn't mean coolant temperature. It refers to high combustion-chamber temperatures that come from *sustained* wide-open-throttle operation. The result is hotter valves, pistons and *piston rings*. Consequently, more ring end gap is needed to allow for increased ring *growth*.

If you find gaps are too *large*, return the rings for new ones. If the gaps are too *small*, you have some additional work to do. You must adjust the end gap by filing *one* end of the ring.

To avoid breaking or chipping the rings, use a fine-tooth file. Place the *file*—not the ring—in a vise. Clamp it lightly between wood blocks. File only *one* end of the ring to correct the gap—use the other end as a reference to make sure the filed end is square. Also, hold the ring end *flat* and *perpendicular* to the file surface. If you are working with moly rings, file from the outside edge to the inside to avoid chipping the moly coating.

Stop and recheck the end gap often. Be sure to check it in the same bore in which it will be installed. *Clean* the ring before you check it so no metal filings get into the block.

When you finish gapping each ring, carefully remove any burrs or sharp edges with an abrasive stone (whetstone) or *small* file. The outside edge will have the smallest burr, the inside edge the largest and the sides somewhere in between. Once you finish all the rings for that bore, keep them with their piston and move on to the next bore.

Install the Rings—With the rings fitted and deburred, install them on the pistons. Rings can be installed by hand. But as a precaution against twisting or breaking the rings or

Piston squares ring in bore so end-gap measurement is accurate. If cylinders have not been bored, push ring to bottom of bore to measure end gap.

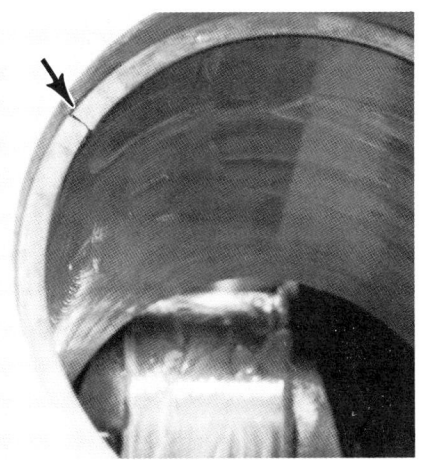
Note, ends of ring overlap! Ring set was meant to be sized. This one definitely doesn't have correct end gap.

Check end gap with feeler gages.

Begin ring installation with oil ring. Slip expander/spacer into place. *Expander/spacer ends butt together—they do not overlap.* Sealed Power expander/spacer has foolproof plastic inserts to help prevent ends from overlapping.

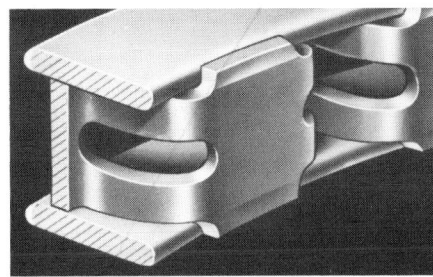
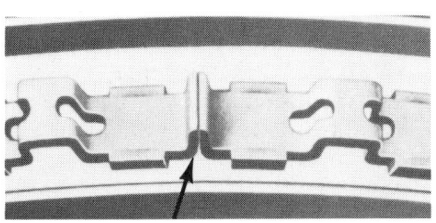
Section of oil-ring assembly. Top drawing shows how expander/spacer fits behind two side rails to spring load them against cylinder wall. Bottom drawing shows correct relationship of expander/spacer ends. Drawings courtesy of Sealed Power Corporation.

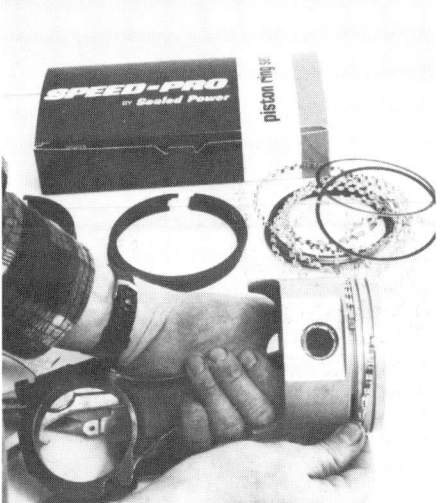
Slip first side-rail end into oil-ring groove above expander/spacer. Guide free end over piston top to prevent scratches. Recheck expander/spacer ends. Second side rail installs below expander/spacer.

scratching the pistons, use a *ring expander*. A ring expander will allow you to spread the rings while holding them flat. This lessens the chance of breaking or twisting a ring.

Before installing the rings, make sure the rings and pistons are clean. Lay out each complete ring set. Remember, if the end gaps are matched to the bores, put the rings with the correct piston.

To install rings, I kneel on the floor, place the rod between my knees and install the rings with a ring expander. Another way is to clamp the rod between two blocks of wood in a vise. Rest the piston on top of the vise to keep it from rocking back and forth.

Oil Rings—Oil rings come in three pieces: an expander/spacer and two side rails. Install the expander/spacer first. Stretch it over the top of the piston and slide it into the oil-ring groove.

Make sure the ends of the expander/spacer *butt* together—*they must not overlap*. The rings I used have plastic blocks to prevent them from overlapping. Rings without these ends tend to overlap.

Install the side rails next. Install the upper side rail first to keep the expander/spacer in position. Slip one end into the oil groove, above the expander/spacer. Spiral the rail into place by pressing it into position with your thumb. The rail should seat against the back of the expander. Check the expander/spacer ends.

Next, spiral the bottom rail down over the expander/spacer and seat it in the groove. Double-check the expander/spacer ends to be sure they have not overlapped. Also be sure the side rail is not wedged between the *back* of the expander and the top or bottom of the ring land. The expander/spacer and both side rails should rotate freely as a unit in their groove when you finish.

Next align the ring end gaps. The butted ends of the expander/spacer should be 180° from the front of the piston as it is installed. The side-rail

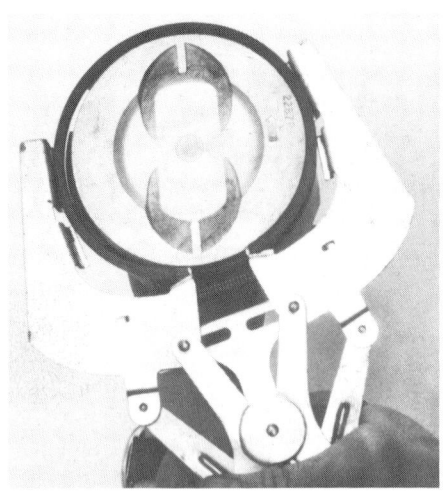

Compression rings are installed with ring expander. Compression rings are very brittle; ring expander minimizes chance of breakage. Install them in the right grooves and be sure their top sides are up. Once rings have been installed, align end gaps according to chart.

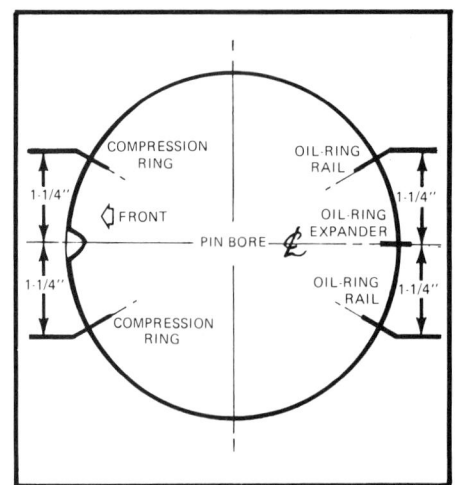

Unless piston-ring manufacturer specifies otherwise, align ring end gaps according to above drawing.

Equipment for piston installation. Don't skimp on oil quality or quantity. Thoroughly lubricate everything.

Push bearing into place and lubricate.

end gaps should be offset about 1-in. on opposite sides of the expander/spacer gap.

Compression Rings—There are a number of things to watch for when installing compression rings. Compression rings are much stiffer than the oil rings, and much easier to break. That's why I recommend using a ring expander.

Compression rings also have a *top* side. Each compression ring should have a little mark—usually a dot—to indicate the top side. If the rings don't have that mark, look carefully at the package. There should be an illustration showing the orientation of the rings.

Compression rings are designed to *twist* slightly when they are in the cylinder bore. This twist improves cylinder sealing and can be either up or down depending on ring design. One direction offers slightly improved blowby control, the other direction improved oil control.

This does not mean that you can reverse the ring depending on what you want from your engine. The twist direction is part of the ring *design,* not a function of installation. Compression rings must always be installed according to the manufacturer's recommendations.

Install the bottom compression ring first from the *top* of the piston. Spread the ring only far enough to clear the circumference of the piston. Lower the ring to the second groove and ease the ring into its groove as you release the expander or your thumbs. Install the top compression ring in the same way.

Set the ring gaps in the correct position. There are a number of ways to position the gaps. The above illustration shows one way. If the ring manufacturer has a specific recommendation, use it. The major consideration is to make sure the gaps are not in line, which can cause excessive blowby.

PISTON AND CONNECTING-ROD INSTALLATION

To install the piston-and-rod assemblies, you will need a ring compressor and hammer, rod-bolt protectors, and oil in a large coffee can or equivalent to dip the pistons. You will also need a torque wrench.

Cleaning the Bores—The bores, rods and bearing inserts should all be cleaned. Wipe them off with some good-quality paper towels. A paper towel run through the bores should come out clean. If it doesn't, clean the bores one last time. This is your last chance.

Use paper towels rather than cloth rags so no damaging lint, fuzz or residue is left behind. Smear the cylinder walls with *clean* oil from a freshly opened can prior to installing the piston-and-rod assembly. Empty the remaining oil into the large can.

Position the Block and Crank—Position the cylinder block so you can reach inside to guide each rod onto its crank journal. Also install the front crank bolt and washer so you can use a wrench to position the crank throws as you work. Each rod journal should be at BDC as you install its piston-and-rod assembly. This will give you room to reach underneath to guide the rod onto the crank journal and install the rod bolts or nuts.

Piston and Rod Preparation—Place the rod-bolt protectors over the bolts. Install the bearing insert in the rod. Smear moly grease on the bearing surface, then oil it. Recheck piston-ring gap one last time.

Dip the piston assembly, dome first, into the can of engine oil. Let the oil go over the wrist pin and smear it over the skirt. To avoid a mess, keep oil from running into the bottom of the piston. You are now ready to in-

Install thread protectors.

Dunk piston in oil to lubricate rings, piston and wrist pin.

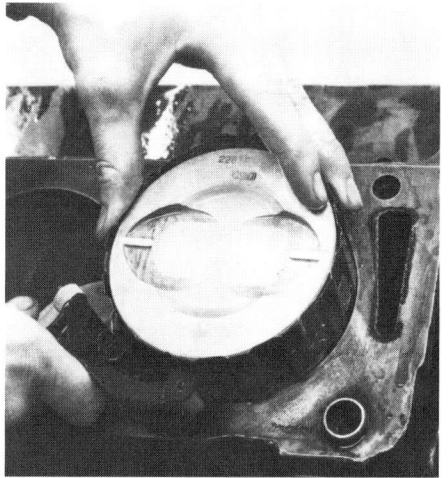
Position crank so piston will be at BDC. Be sure piston faces correct direction, with notch or arrow toward front. Double-check ring end-gap positions. Install ring compressor with bottom edge about 1/4-in. below oil ring.

Push compressor down square against deck and tap piston into bore. If piston refuses to enter bore, STOP. One of the rings may have snagged on edge of bore. Forcing it may break ring or ring land. Remove piston, reinstall ring compressor and try again. Once rings enter bore, remove compressor. Use free hand to guide rod onto journal.

stall the ring compressor.

Ring Compressors—There are many different kinds and qualities of ring compressors, but all serve the same purpose—to compress the rings flush with the outside of the piston. This allows them to enter the cylinder bore with the piston without hanging up.

One of the oldest and usually least-expensive types has a metal band with a strap around the outside. The band is tightened by an over-center screw or a ratcheting handle.

Another type uses a *corrugated*—rippled—band. The corrugated band produces less friction on the piston and rings. The corrugations also keep the compressor from entering the bore and provide a more stable base at the block deck.

There are also several arrangements used to compress the band. The first I mentioned is with an over-center screw or ratcheting handle. The second way is with a set of handles like a pair of pliers. Some handles lock in position and others don't. I suggest that you get the corrugated-band type with the locking handles.

Fit the ring compressor loosely over the rings. If there is an arrow on the band, be sure it points *toward* the piston skirt. With the plier handles, clamp down until the band is snug around the piston. With the wrench type, rotate the ratchet mechanism to tighten the band.

Installation—Make sure you have the correct piston for the bore and the arrow or notch on the piston points forward. Hold the piston and ring compressor and ease the connecting rod and piston skirt into the bore.

Put one hand inside the block to guide the rod. Slide the piston down until the compressor is flush with the block deck. With a block of wood or hammer handle, *tap or push* the piston into the bore.

If the piston hangs up before it enters the bore, **STOP!** One of the rings has probably slipped over the edge of the block deck. Forcing it into the bore can bend or crack a ring or ring land. Remove the piston, reinstall the compressor and try again.

If the piston refuses to enter the bore, take a close look at the piston rings to see if one is snagging or hanging up in its groove. *If you didn't check the ring end gap, remove the rings and do it now.* It's possible the ring ends are butting together and not seating fully in the grooves. This will cause *severe* damage to the piston, rings and bore if not corrected.

Once you are sure the rings are OK, install the ring compressor and try again. When the top ring enters the bore, the compressor will relax. Set it aside. Tap or push the piston down in the bore. Use your free hand to guide the rod over the rod journal until it seats.

> **DOUBLE-CHECK**
>
> If the rod and piston are correctly assembled, the oiling notch on the rod should point toward the opposite cylinder when the arrow or notch on the piston faces forward. At the same time, each rod cap must be installed in its original direction—with the numbers on the rod and cap facing out, or towards its cylinder bank. If you must reverse the cap to get the oiling notch in the correct direction, *STOP!* The piston-and-rod assembly has been assembled 180° out and must be disassembled and turned around. *Never* reverse a cap on its rod.

Rod-Bearing Cap—Apply moly grease and oil to the rod-cap bearing insert. Remove the protectors from the rod bolts and install the cap. Fit

Install rod-cap insert and lubricate.

Remove bolt protectors and install rod cap. Install nuts or bolts finger tight.

Torque rod-cap nuts or bolts after all piston-and-rod assemblies are in place. Though I'm not doing it here, it's a good idea to support rod caps with feeler gages slipped between them.

the rod cap to *its* connecting rod. Oil the bolt threads.

When installing the cap *be sure* it is the right one for the rod—double-check the numbers. They should match. Make sure the numbers align and that the oiling slot points in, toward the "V" of the engine. The rod-bearing tangs should be on the same side as the oiling slot.

You can wait for final torquing until all the pistons and rods are installed. Just make sure the caps are snug so the bearing insert will stay in place as the crank is turned.

The advantage of waiting is twofold. You save time doing them all at once. And you can support each rod while torquing the nuts or bolts.

Once all the rods are in place, insert feeler gages between each pair while torquing to take up the side clearance. This lets the big end of the rod support the load rather than the bearings. This is particularly important with the higher torque on high-performance rod bolts or nuts.

Torque standard big-block rod nuts 40—45 ft-lb. Rods using 13/32-in. bolts and nuts, such as those used with cast-iron-crankshaft 427s, all 406s, 428CJs and high-performance and Police Interceptor engines, should be torqued 53—58 ft-lb. LeMans and NASCAR rods have cap bolts. Torque these 65 ft-lb.

Connecting-Rod Side Clearance— With the rods installed, check the side clearance. Measure with a feeler gage between the rod faces.

Standard FE and 330 FT engines should have 0.010—0.020-in. clearance, with a maximum of 0.023 in. Racing FE engines should be

Both cam dowel and crank key should be toward top of engine (arrows). Before installing timing-chain assembly, apply Lubriplate to crank snout, thrust plate and back of sprockets.

Slide sprocket onto crank about 1-in. from where it seats. Note *pip* mark on face of crank sprocket. There is a matching mark on cam sprocket. When timing chain is installed, both of these marks should be in line with the cam and crank centers.

0.018—0.028 in., with 0.025 in. recommended. FT engines, except the 330, should be 0.010—0.030 in.

These clearances should have been checked before you removed the rods during teardown, so this should be just a final check. If the clearance is not correct, remove the piston and rod assembly. Have the face of the rod resurfaced if the clearance is too small. If the clearance is too large you'll have to install new rods.

TIMING CHAIN AND SPROCKETS

If it's on an engine stand, position the engine upright, as it would be when installed. Rotate the crankshaft until the rearmost keyway is straight up. With the engine on a bench, turn the crank so the keyway is in line with the crank and camshaft.

If you removed the Woodruff key, check the slot to be sure it is not burred. Remove any burrs with a fine-tooth file. Install the key by using a brass punch and hammer.

Remove the bolt and washer from the end of the crank—leaving the Woodruff key pointing straight up. The best way to do this is to turn the crank *slightly* clockwise past where you want it. Hit the breaker bar or wrench sharply with your hand or mallet to break the bolt loose.

This will usually rotate the crank

Replacement timing-chain sets. At left is original-equipment type for FE and medium-duty FT engines. At right is Cloyes True Roller replacement set for heavy-duty and racing applications. FT engines with forged cranks are originally equipped with roller chain and larger-ID crank sprocket.

Loop chain over cam sprocket. Align timing marks, then loop chain over crank sprocket. Marks should align when chain is pulled taut.

Slide cam sprocket into engagement with cam dowel. Slight adjustment of cam or crankshaft may be necessary before sprocket engages dowel. If timing marks are aligned, alternately push cam and crank sprockets into place.

Install fuel-pump eccentric, bolt and washers. Torque bolt 35—45 ft-lb. Lubricate chain and sprockets with oil. Moly grease goes on the fuel-pump eccentric. Before you forget, install oil slinger with concave side facing forward.

Before installing timing cover, install new front-main seal. Lay cover face up and knock old seal out with punch and hammer.

just enough to bring the key into alignment. It may take a few times to do it this way, but it helps to have the key *straight* up.

Next, rotate the camshaft until the dowel is on top, relative to the engine.

Install the Chain and Sprockets— Slip the crankshaft sprocket onto the crank with the timing mark—dot or arrow—facing you. Slide the sprocket on so it is about 1 in. away from the installed position.

Next, lay the timing chain over the camshaft sprocket. Turn the sprocket so the dot or arrow is facing you and straight down or in line with the crank. Loop the chain under the crank sprocket and pull the chain taut.

If the marks don't line up, disengage the chain from the crank sprocket and jump the chain one or two teeth to align the marks. Once you have the timing marks correctly positioned, place the cam sprocket over the dowel on the camshaft. Push both sprockets on simultaneously.

To assist sprocket installation, reinstall the vibration-damper bolt and washer. Turn the crank slowly while pushing on the cam sprocket. Once the timing chain and sprockets are in place, coat them liberally with Lubriplate or moly grease.

With the sprockets seated against the crank and camshaft, install the fuel-pump eccentric, flat washer, lock washer and cam bolt. Loctite the threads. Torque the bolt 40—45 ft-lb. Install the oil slinger on the crankshaft, with the raised center portion next to the crank sprocket.

TIMING-CHAIN COVER

If you haven't cleaned the timing cover yet, do it now. Remove the old front seal with a punch. Support the cover—most are cast aluminum and break easily. Drive against the edge of the seal with the punch and knock the seal out the back of the cover.

Install the new seal with the flat metal edge toward the front of the cover and the seal lip toward the rear. Lay the cover on a piece of wood, face down, and lay the seal over the opening with the flat metal part facing down.

To drive in the seal, either use a special seal tool, a large piece of round stock or a socket. If you are very careful, you can hammer the seal into place by tapping around the edge. I find that hammering lightly around the edge of the seal works fine, as long as I keep the seal going in equally all around.

Drive in the seal in until it bottoms in the cover. Lubricate the seal lip with oil or white or moly grease. Check that the garter spring around the inside of the lip is still in position.

With the seal installed, apply non-hardening cement, such as silicone, to the timing cover and install the gasket. Before installing the cover, be sure you have the camshaft-sprocket bolt torqued, the timing-chain assem-

Work new seal into place with either careful use of hammer or with large-diameter socket and hammer. With non-hardening sealer on both sides of gasket, install cover.

Center front cover with damper spacer. With cover centered, tighten and torque front-cover bolts 12—15 ft-lb.

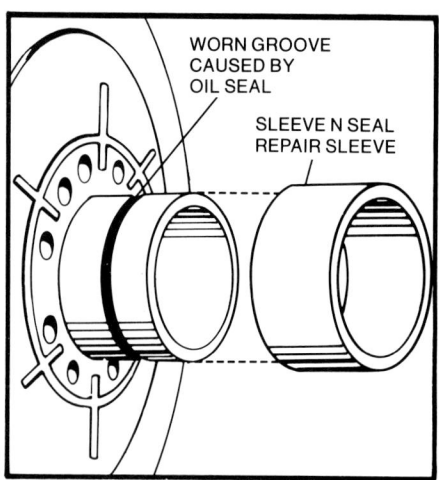
If spacer seal surface is severely grooved, this sleeve will restore the sealing surface. Although spacers are relatively inexpensive, sleeve is even less. Drawing courtesy Fel-Pro Inc.

Shim-type head gaskets (center and bottom) must be sealed before installation. Apply sealer around beading—embossed pattern—on both sides of gasket. If you are using Fel-Pro Permatorque Blue head gaskets, use no sealer. The gasket coating seals when heads are torqued in place.

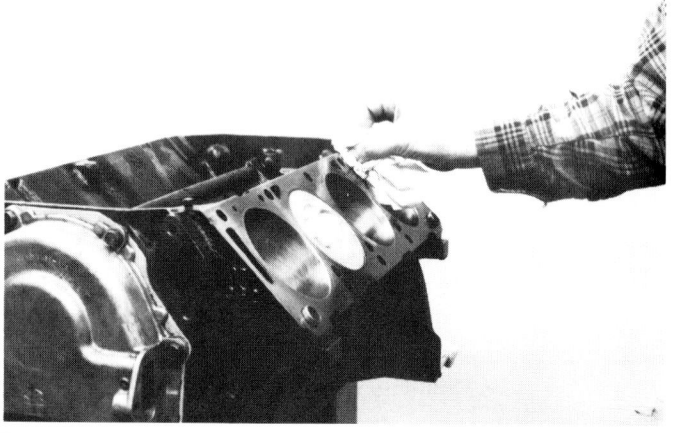
Thoroughly clean deck surface with paper towel and solvent or laquer thinner. Be sure hollow dowels are inserted in counterbores in two lower end head-bolt holes. Dowels align cylinder head and gasket on deck. Head gaskets are marked FRONT, but check water passages to be sure of alignment. Passage at rear of deck surface should be open and front one blocked.

bly lubed and the oil slinger in place. Install the eight cover bolts loosely.

Align the cover before you tighten the bolts. The cover must be aligned to center the front seal on the damper-spacer sleeve. If it's not centered, uneven loading will cause the seal to leak.

Use the spacer sleeve to align the cover. Simply slip it over the crank snout and through the front seal. This will locate the cover so you can tighten the bolts. Or, you can use a special tool designed to align front covers.

The special tool *positively* locates the cover to the crankshaft. If you use the damper spacer, it will probably be close—a lot closer than if you installed the cover without considering its alignment. With the cover positioned, torque the bolts 12—15 ft-lb.

Check the condition of the spacer next. Check for a deep groove or grooves where the seal rides on the spacer. If the groove is shallow you can smooth it out with emery cloth. Otherwise you have two options; replace the spacer or install a sleeve over the seal surface.

Damper spacers are relatively inexpensive but a repair is even cheaper. To install the sleeve, spread a thin film of silicone sealer on the inside surface of the sleeve. Drive the sleeve over the seal surface of the spacer. The sleeve is only about 0.030-in. thick, so a stock seal can be used.

Apply a light coat of Lubriplate, moly or oil to the spacer. On the inside of the spacer and the crankshaft nose, apply a waterproof lithium grease or anti-seize compound. This keeps the crank nose from rusting and the damper from sticking to the crank. Slide the spacer against the oil slinger.

CYLINDER HEADS

I prefer to install the cylinder heads next. You can move to the bottom end of the engine or start installing the fuel pump or oil-filter adapter if you wish. I had to choose some order for this book, so I used my preference

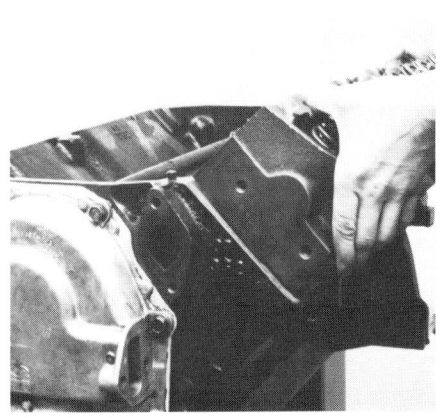
When setting head on block, get a good grip and let head contact dowels first, rather than the gasket. Move head around until holes engage dowels, then lower head onto gasket. Be careful at this point to avoid scars or other damage to head gasket—very important with shim-type gaskets.

Don't leave head sitting on block. Thread in one bolt as a precaution against the head falling.

Give bolt threads light coating of oil or anti-seize compound. Thread bolts into place and torque in sequence to specification.

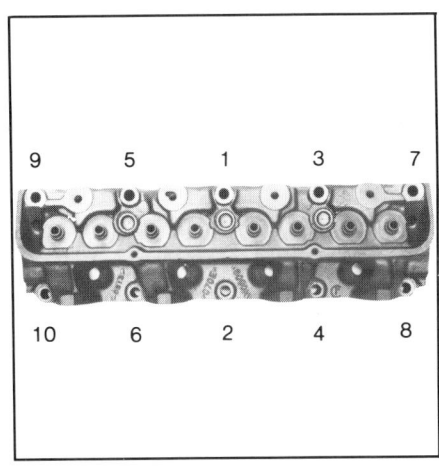
Use this sequence when torquing head bolts. Torque bolts in three steps as indicated.

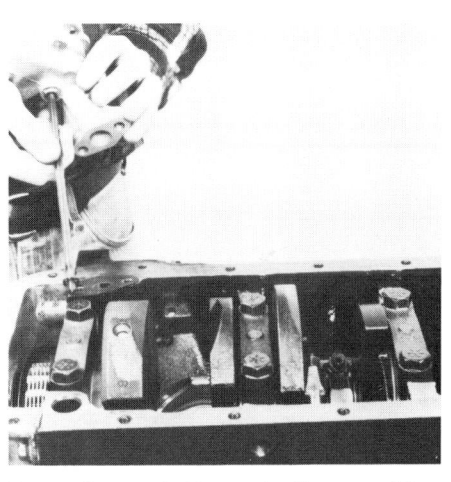
Note clip on bottom of oil-pump drive shaft. It is important that this clip be adjusted so shaft will remain in pump when distributor is removed, see text.

and installed the heads next.

Make sure the engine is supported so it wont fall or roll over as the heads are installed. If you are working on an engine stand, tighten or pin it so it will not let the engine rotate when you install the first cylinder head. If you are working on a bench or the floor, make sure the engine will not roll over. It must be well supported.

You will need the cylinder heads, gaskets, and 20 head bolts—10 short and 10 long. The short bolts are 2-13/16-in. long for all '58—'60 heads. The remaining FE and FT heads through 1975 use 2-7/8-in.-long "short" bolts. 1976—'78 FT heads return to 2-13/16-in. bolts. Short bolts can replace the slightly longer ones, but the long ones cannot be used in 1958—'60 engines.

Most big-blocks use a carburetor-heat exhaust crossover that runs diagonally through the intake manifold. On heavy-duty-truck engines the exhaust crossover runs straight through the intake manifold. See the photo, page 44. Heavy-duty-truck heads use a 3-1/2-in.-long bolt in place of one of the short head bolts. This bolt threads into the center hole on the exhaust-manifold side.

The 1958—'60 long bolts measure 4-7/32 in. except for the 352HP engine. The 352HP and '61—'78 engines use 4-29/64-in.-long bolts. Again, short bolts can be used in place of long ones but longer bolts cannot be used in older engines. Long bolts will bottom out in the bolt hole.

Clean the Gasket Surfaces—Clean both the block deck and the head-gasket surfaces with laquer thinner or solvent and paper towels. I prefer thinner because it leaves no residue. If you use sealant on the head gaskets, don't use solvent.

Before installing the gaskets, make sure there are two locating dowels on each block deck surface. These hollow dowels are installed in counterbores at the bottom corner head-bolt holes. If they are missing, buy a set and install them—they locate each head and gasket to the block.

Gaskets requiring sealant are metal shim-type, which I recommend on high-compression—11:1 or more—engines. An engine with a compression ratio this high should not be used on the street due to the lack of high-octane gasoline. But if you are building a car for racing, or restoring a high-performance engine and want to keep the original compression ratio, then use shim-type gaskets.

With lower-compression engines, I recommend Fel-Pro Permatorque Blue gaskets. These Teflon-coated gaskets do not need retorquing. You should *not* use a sealant with them as they are self-sealing. When pressure is applied by torquing the bolts the blue coating seals to the metal surfaces.

If you use the shim-type metal gas-

Torque pump and pickup bolts 12—15 ft-lb.

Trim timing-cover gasket flush with block and cover. Side seals for rear main bearing cap should protrude slightly—no more than 1/16 in.

Apply non-hardening sealer to rear-main-cap parting lines and edge of front-cover gasket. In most cases, I let gaskets do their job and leave sealer in tube.

kets you *must* use sealer. See photo, page 133. Apply sealer to both sides of the gasket. Spray Copper Coat works well, as do some other sealers.

Be sure to install the gasket in the right direction—there is a *front* to each gasket. The front of the big-block head gasket has no water passage. The rear has a tear-drop-shape water passage about 2-1/2-in. long.

Install the Heads—With the gaskets in place you can install the heads. Be sure you have a firm grip when you lift a head to set it in place. I find the easiest way is to stretch my fingers across the ends of the head and just *clamp on* and set the head in place. Or place a couple of fingers in an intake port at one end and an exhaust port at the other and lift.

As you lower the head, let the bottom of the head contact the *dowels* first—not the *gasket*. Slide the head around until its counterbores engage the dowels. Thread in a couple of bolts to hold the head in place. Make sure the engine is stable, then install the other cylinder head.

Coat the bolt threads with oil or anti-seize compound and thread them in. Torque the bolts to specification in three stages. Use the sequence shown in the photo, page 133.

In the first step, follow the sequence shown and torque the head bolts 70 ft-lb. Repeat the sequence, torquing the bolts 80 ft-lb. Finally, go through the sequence a third time, torquing the bolts 80—90 ft-lb on all engines except 1963—'67 427s. On these engines, torque the bolts 100—110 ft-lb.

Again, if any bolt seems to reach a certain torque value and stay there

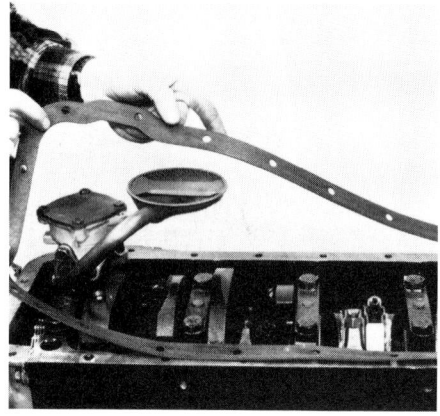
Lay pan gasket into place. Install oil pan and bolts.

when tightened further, remove the bolt. Recheck the threads on the bolt and in the block. If these seem clean and undamaged, the bolt is probably stretching—replace it.

Most cylinder-head gaskets do not require retorquing the head bolts after initial engine run-in. Ford recommends not retorquing head bolts with the steel-shim gaskets used on the 427. I usually do not even check the composition gaskets, but I have retorqued with the shim type with no problems.

OIL PUMP, PICKUP & PAN

As I noted, you could install these items before the heads. The decision is yours, but I always prefer to wait to seal off areas of the engine until all major components are installed.

You should have inspected and cleaned the oil pump or replaced it. If you didn't inspect the pump, do it

Most gasket sets include new nylon or fiber washer for drain plug—use it.

now, page 96. You should also have a *new* oil-pump drive shaft, a clean oil pan, pickup assembly, oil-pan gasket and non-hardening sealer.

Oil-Pump Drive Shaft—The oil-pump drive shaft comes with a Tinnerman retainer on the distributor end. The retainer keeps the shaft in the oil pump when the distributor is removed. *Be sure this retainer is in place.* The only way to retrieve and reinstall the shaft is by removing the oil pan and pump.

Confirm that the shaft will stay in place when the distributor is removed. Place the shaft in the oil pump and set the pump on its mounting surface. Slide the shaft away from the pump until the Tinnerman retainer hits the bottom of the distributor-bore boss. The shaft should still be firmly engaged in the oil pump.

If it pulled out of the pump, slide the retainer toward the distributor-end of the shaft. Recheck pump and shaft

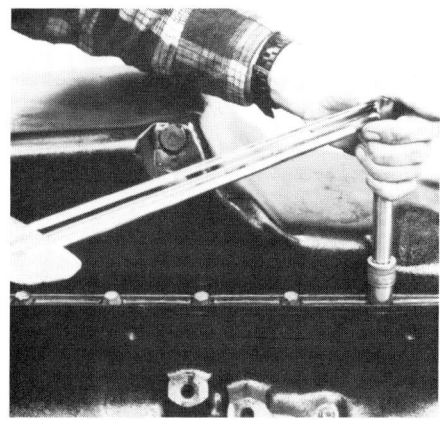

Torque oil-pan bolts 9—11 ft-lb and drain plug 15—20 ft-lb. Any tigher on the drain plug will split the washer and guarantee a leak.

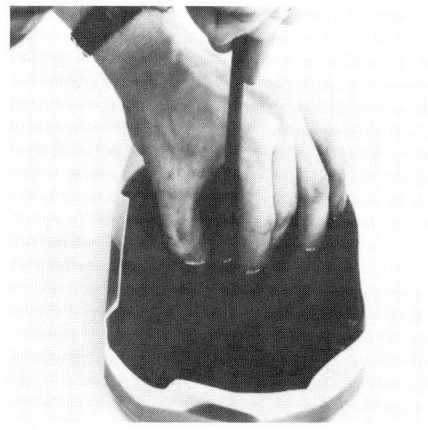

If you don't want to hear five minutes of lifter clatter on start-up, prime the lifters. Soaking lifters will not work. Submerge lifter in oil. To prime, work plunger down in lifter several times with pushrod.

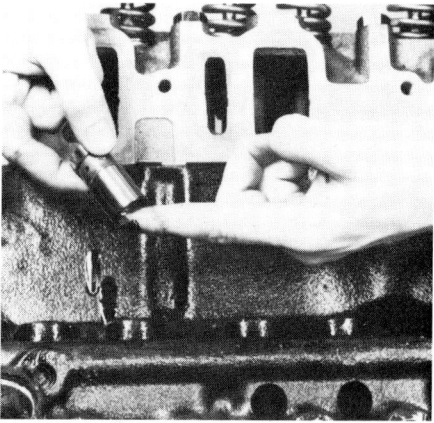

Apply moly grease to lifter foot. If you are reusing old lifters, be sure they are installed in their original positions. *Never install lifters out of order, or install old lifters on a new cam.*

Slide lifter into position. Lifters should slide in by their own weight.

Intake-manifold installation requires special attention. I have seen many manifolds leak after installation, but are usually leak-free if correct installation procedure is followed and machine work is correct. Use high-quality gasket set and sealer. Apply non-hardening sealer around front and rear water passage.

in block.

Install the Pump—Do not use sealer on any oil-pump gaskets—some will end up in the oil galleries. Lay the oil-pump gasket on the block and position the pump on top with the drive shaft in place. Install the clean and lightly oiled oil-pump bolts. Torque them 12—15 ft-lb.

Be sure the pickup screen is clean. If it isn't, have the pickup hot tanked. If you clean it at home, soak it overnight in a covered container of caustic cleaner. If the screen is still not clean, buy a new one, unless it is one of the deep-sump pickups. If so, keep cleaning it—they are rare.

Install the *clean* pickup tube and gasket. Use no sealer, although you can use a small amount of grease to hold the gasket in place. Torque the bolts 12—15 ft-lb.

Oil Pan—Make sure the oil pan, mating surfaces on the block, rear main-bearing cap and timing cover are clean. If the rear-main side seals project more than 1/16 in. from the bottom of the block, trim them to just above the surface of the block.

Use a thin coat of silicone sealer around the oil pan and install the gasket. Apply an extra dab of sealant across the joints between the timing cover and block, the rear-main cap and block and at the side-seal ends.

Install the oil pan and bolts. The longer bolts—either two or four—go into the timing cover. Torque all bolts 9—11 ft-lb.

Install the drain plug and a new gasket. There should be a nylon plug gasket in the gasket kit. Torque the plug 15—20 ft-lb—no more. Any tighter will distort the pan and split the gasket.

LIFTER INSTALLATION

If you are installing the old cam and lifters, be sure to install the lifters in their original bores. If you mixed them up or forgot where they go—buy new ones. *Never mix used lifters or install used lifters on a new camshaft.*

Before you buy new lifters, see the pushrod chart in Chapter 3, page 57. There are some differences in plunger or lifter heights, as well as some other valve-train differences.

Installing lifters is easy, but some preparation is required. Hydraulic lifters must be *primed*. Priming can be done in one of two ways. You can simply pump oil into the hole at the side of the lifter with an oil can. The other way is to place each lifter in a shallow container of oil and work the plunger up and down with a pushrod. This draws oil into the lifter.

With the lifters primed, wipe the bases of the lifters dry and dab moly on them. Spread more oil on the sides of the lifters and inside the lifter bores in the block. Finally, push the lifters into their bores—they should go in with little effort.

INTAKE MANIFOLD

Install the intake manifold with non-hardening gasket sealer and silicone sealer. You also need some dexterity, strength and your torque wrench.

Manifold Baffle—With the lifters in place, install the baffle under the intake manifold. As you can see by the photos, I installed the manifold gaskets first. The order doesn't make much difference, but be careful not to

Set manifold-to-head gasket in place. Make sure locating tabs and opening for heat-riser passage are correctly positioned.

Apply silicone sealer to front and rear manifold-to-block surfaces, with an extra dab in each corner.

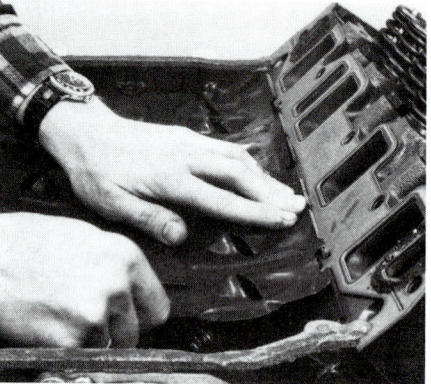

Install gaskets, lift them back up to allow sealer to become tacky and then press back into place.

Apply extra dab of sealer in each corner on top of gasket.

Last application of sealer is around water passage on gasket. Install oil baffle.

knock the gaskets loose if you install them first.

The baffle keeps hot oil off the underside of the intake manifold. This keeps the intake charge cooler and reduces oil baking under the manifold exhaust-gas passage.

Make sure the baffle is clean and straight. Line up the tabs on the baffle with those on the head gaskets—see nearby photo. Press down on the center of the baffle while working the tangs into place.

Gasket Types—Even though FE engines use several different intake-manifold-to-head gaskets, most use one style. So you'll probably have no problems unless yours is one of the high-performance big-blocks.

The confusion starts with the 427—no surprise, right? There are several manifold-to-head gaskets for the 427. The first is for the High-Riser, which has taller intake ports and no carburetor heat passage. There are two versions for the Medium-Riser, one with a heat passage and one without. The Tunnel-Port uses one style, with round ports and no heat passage.

If you have a choice, use a gasket with the open heat passage for a street-driven car. The heat passage provides better fuel atomization, which improves low-speed driveability and warmup.

Early high-performance engines—352HP, '61–'62 390HP and '62–'63 406 and Low-Riser 427—with taller intake passages also used a different intake-manifold gasket. Low-Riser gaskets will work for all of these. The Low-Riser gaskets are available with blocked heat passages, but remember, use the open passages on street engines.

FT engines have one set of intake-manifold gaskets, which is not interchangeable with any FE type. FT engines have heat passages in the center of the cylinder heads, FE heads have them between cylinders -1 and -2, and -7 and -8.

Gasket Installation—Check the alignment dowel at the front of the block to make sure it is not mushroomed at top. Pull the dowel out with Vise-Grips and dress the edge with a file. If the edge is severely mushroomed, replace it.

Run non-hardening sealer around the four water passages—two in each cylinder head, one at each end. Lay the manifold-to-head gaskets into place, locking the two tabs with the head gasket.

Run a 1/8-in.-wide bead of non-hardening gasket cement along the front and rear block surfaces. Dab a little extra in the corners near the heads. Press the front and rear manifold-to-block gaskets into place. Lift them back up gently several times to help the cement tack. When it starts to tack, stick the gaskets down permanently.

Now add some silicone sealer on top of the gaskets where the front, rear and side gaskets meet—make it a big dab. Use the non-hardening gasket cement again and smear some around the water passages on the manifold side of the gasket.

Make sure the intake-manifold sur-

Don't be embarrassed about asking for help; manifold is awkward to handle and heavy. Set it down squarely on the block so you don't disturb gaskets. Double-check front and rear manifold-to-block gaskets to make sure they didn't get pushed out of position.

Remember there are two lengths of manifold bolts—two short bolts go at rear on standard manifolds, four short ones go at corners on Tunnel-Ports. Oil threads and install all bolts two or three turns. Apply sealer under bolt heads. Gasket or head can shift as bolts are tightened, so install all bolts at once to ensure gasket or manifold isn't pushed over any bolt hole. Possible exception is end holes. These may not align until other bolts are partially tightened.

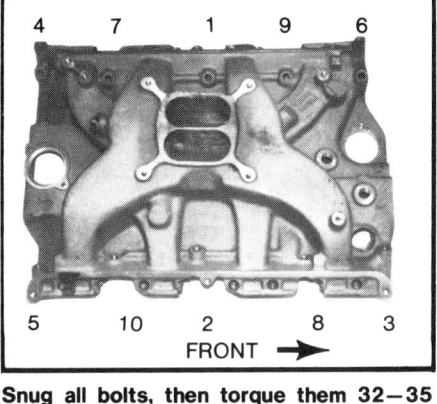

Snug all bolts, then torque them 32—35 ft-lb in sequence. Use this sequence on 10-bolt manifolds.

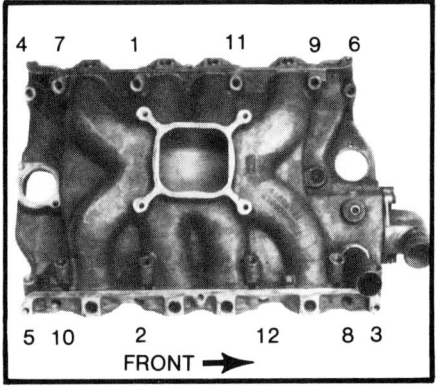

Tunnel-Port intake manifolds use 12 bolts. Put sealer on bolt threads. Torque manifold 32—35 ft-lb in sequence.

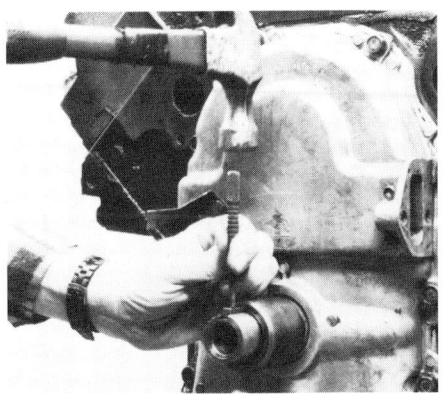

Install timing pointer and front crankshaft key. Original timing pointer was riveted—this one has been drilled and tapped for bolts. Inspect key and keyway for burrs before installing, and for tight fit after installed. Tap Woodruff key into place.

faces are clean. Lift the manifold and carefully position it on the engine. Again, the big-block intake manifold is either heavy or extremely heavy, depending on whether it's aluminum or cast iron. Either way, don't be embarrassed about asking for some help or using a lift—that's a lot better than ruining your back or risking a leak.

When setting the intake in place, be careful not to disturb the gaskets. The front and rear manifold-to-block gaskets are easily knocked out of place. Check the gaskets to make sure they are flush with the outside edges at the front and rear, then insert the bolts.

The normal bolt set includes 10 3/8-16 bolts, two 1-1/2-in. long and eight 2-5/8-in. long. The short bolts go in the rear bolt holes.

Some intake manifolds have four short and six long bolts. Tunnel-Port manifolds have four short and eight long bolts. On these manifolds, the short bolts go at the corners.

Apply sealer under the bolt heads or where they contact the manifold.

Torque the Manifold Bolts—Make sure you have all the bolts threaded at least two full turns before running any of them down. This will prevent the manifold from shifting out of position, making it impossible to start the remaining bolts.

The exception is the end bolts; these may not thread in right away. If you have trouble with any of them, thread in all the other bolts and begin the torque sequence without the end bolts. Install the end bolts as soon as you can and bring them into the torque sequence shown nearby.

CRANKSHAFT DAMPER

Clean the damper and inspect the rubber cushion between the hub and outer ring. Surface cracks in the rubber are normal, but they shouldn't extend under the metal between the hub and the weight.

If the cracks are deep or the rubber is badly deteriorated, pick up a new damper from your Ford dealer or find a good one at the junkyard. Before you buy another damper, check the discussion on types of engine balance in Chapter 3.

Apply a light coat of moly grease to the inside of the hub. Make sure the Woodruff key and spacer are installed. The damper is next.

I use a large rubber mallet to drive the damper onto the crank nose. Whatever you use, make sure you hit the center and not the outer ring. If you need to, use a spacer—a large socket is good—over the hub to help drive on the damper.

The damper can be pulled on by installing the center bolt. On the 428 in the photos, the outer pulley had to be installed first, then the bolt and washer. Hold the crankshaft while torquing the damper bolt. I did this by installing one of the flywheel bolts and resting a wrench against the engine stand.

If you haven't installed the oil pan, hold the crankshaft by slipping a

Apply moly grease or anti-seize compound to inner hub of balancer. This makes installation—and removal, if necessary—easier.

If you drive damper on, use rubber mallet on center hub only.

Some engines must have pulley installed before center balancer bolt. Tang on pulley aligns with keyway in balancer.

Tighten pulley bolts before installing harmonic-balancer bolt.

Install harmonic-balancer bolt and torque it 70—90 ft-lb.

screwdriver, pry bar or wood block between a crank counterweight and the right side of the block. Just don't slip and injure yourself or damage the engine. Torque the damper bolt 70—90 ft-lb.

OIL-FILTER ADAPTER

Time to install the oil-filter adapter. If you haven't already done so, clean the adapter. Squirt some oil down the right passage, which feeds oil from the pump to the filter. This will help prime the pump. Fill the adapter to the top.

Apply gasket cement to the gasket side of the adapter. Only apply enough to coat the surface, but not so much that it will squeeze out into the passages when the adapter is installed. Place the adapter against the block and torque the bolts 12—15 ft-lb.

Standard bolts are 5/16-18, 1-1/8-in. long. Longer bolts are used with spacer blocks on trucks. Shorter bolts are used to clear the adapter on engines that use an oil cooler. The 352HP engine, the first big-block with an oil cooler, had 1-1/4-in.-long bolts.

Install the oil-pressure sending unit if you removed it. Some '63—'64 427 engines also used an oil-temperature sending unit, which also mounts to the adapter. Use Teflon tape to seal the threads.

FUEL PUMP

Clean the fuel pump. Apply gasket cement to the pump housing and install the gasket. Smear some moly grease on the contact surface of the actuating arm.

Look through the hole in the timing cover. Turn the engine over until the low point on the fuel-pump eccentric is down. This will give the most clearance and least pressure on the actuating arm to ease installing the fuel pump. Install the pump and torque the bolts 20—25 ft-lb.

Fit oil baffle into place. Be sure it doesn't contact valve springs or retainers.

WATER PUMP

Make sure the water-pump bypass tube and the tube on the intake manifold are clean and in good condition. If the tubes are badly corroded, re-

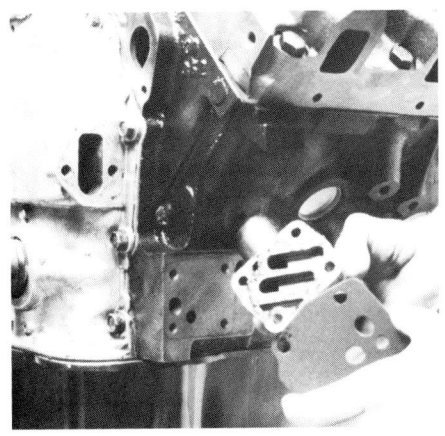
Apply light coat of sealer to oil-filter adapter and press gasket into place. Be sure gasket holes align with those in block. Torque adapter bolts 12—15 ft-lb. Clip for automatic-transmission-cooler lines bolts to adapter in some cases. Location of these clips varies.

Apply sealer to pump and press gasket into place. Apply coat of moly to arm contact surface. Install pump and torque bolts 20—25 ft-lb.

Bypass hose is installed with soapy water inside to ease pump installation.

Apply sealer to both sides of water-pump gaskets and slip pump into place. Torque bolts 20—25 ft-lb and tighten bypass-hose clamps.

Install pushrods. Be sure they are seated in lifters. Apply moly grease to pushrod ends and valve-stem tips.

Before installing rocker-shaft assembly, make sure bolts and stands are correctly positioned. Necked-down bolt at left installs in bolt hole with oil gallery (arrow). Bolt is also slightly longer than other rocker-shaft bolts. Cast-iron rocker stands are grooved on one side only. Groove must align with oil gallery in head.

Recheck baffle for adequate clearance around valve springs. Bend baffle if necessary to prevent chafing.

place them. If you need to replace them, twist the old tubes out with Vise-Grips. It's a lot easier now than after the water pump is installed and the engine is back in the car.

Clean the passages, put sealer on the new tubes and drive them in. Cut a 2-3/4-in. length of 5/8-in.-ID heater hose and slip it over the bypass tube in the intake manifold.

Place two hose clamps over the hose, but leave them loose. Position the clamp screws so you can get at the heads with the overflow tank and distributor installed. See the nearby photo.

Next apply gasket cement to the water-pump outlets. Place the gaskets on the water pump and smear a small amount of gasket cement on the gasket surface.

Have the bolts ready before you install the water pump. Slide the water-pump tube into the bypass hose and place the water pump against the block. Make sure the gaskets did not slip out of place. Install the bolts.

You will now appreciate those pictures or notes you took when you removed or disassembled the engine. Most water-pump bolts are 3/8-16 X 1-1/4 in. Some are shorter and some longer. Others are double-ended so the small thread serves as a stud for mounting other accessories. Torque the water-pump bolts 20—25 ft-lb, then tighten the bypass-tube hose clamps.

ROCKER-SHAFT INSTALLATION

Rocker-shaft installation is fairly simple, but there are things to watch for. I cover these as they come up.

Alternate between bolts to bring torque up evenly to 40—45 ft-lb.

If engine has solid lifters, adjust valve lash now. Use procedure outlined in text.

Lightly oil threads and thread in rocker-shaft bolts. Capture pushrods by rotating rocker arms over pushrod ends as shown.

Pushrods—Install the pushrods first. If your engine has adjustable rocker arms, the socket end of the pushrod goes up. Non-adjustable rockers use a pushrod with a ball at both ends. Apply moly to both ends of each pushrod before inserting it.

If you are installing used pushrods with non-adjustable rockers keep them in order. The smaller wear pattern goes into the lifter. However, if you're using used pushrods with new rocker arms, turn them end for end: big wear pattern at the lifter. Order is not important if this is the case.

Make sure the pushrod *seats* in the *center* of the lifter. If you push down on the pushrod and move it from side to side, it should feel as if it is pivoting from one point. The bottom end shouldn't slide back and forth.

Install the sheet-metal oil baffle next. The curved side fits next to the springs and the flat side against the head.

Rocker-Arm Assembly—With the pushrods in place, turn the crankshaft clockwise until number-1 cylinder is at TDC at the end of the *compression* stroke, then rotate the crank an additional 45° past TDC. This sets the cam so none of the lobes are at full lift for installing the rocker-arm assembly. This minimizes stress on the rocker shaft as you bolt it into place.

To find TDC, watch the intake pushrod for number-1 cylinder while rotating the crank. When the pushrod falls back or goes down you have it, the next time TDC lines up at the damper pointer! You can also check for camshaft position with the rocker-arm assembly installed and the bolts finger tight. It's a bit easier to see the pushrods move.

The additional 45° is marked XX on some dampers or pulleys. If the damper is not marked, rotate the damper or pulley about 2-in. past TDC on the damper. That's close enough.

Recheck that all the pushrods are still seated in their lifters. Apply moly grease to the valve-stem tips and set the rocker-arm assembly on the engine. Rotate the rockers so the pushrods seat.

On each rocker shaft there are four 3/8-16 bolts: one 3-19/64-in. long and three 2-31/32-in. long. The long bolt is for the oil-feed hole in the cylinder head. The shank part of the bolt is narrower—necked down—to allow for oil flow.

Make sure the two long bolts are in the right position—one for each shaft. The long bolt goes in the second hole from the front on the left shaft and in the second hole from the back on the right shaft. As a double-check, look at the holes in the head. The long bolt fits into the counterbored hole with an additional hole for the oil. See nearby photo.

There is *one* exception. For High-Riser heads, all four bolts are the same size, 2-1/2-in. long. High-Riser heads do not have a counterbore for the rocker-stand bolts and the bolts are not necked down. These heads have the oil hole, but the stands have an oil-passage slot for oil flow.

Thread the bolts in until they all meet resistance from the valve springs. Make sure all the pushrods line up and are seated in their sockets or adjusting screws. Be sure the oil baffle is not binding on the valve springs, rockers or head.

Start at one end of the shaft and turn each bolt two turns. Continue to the next bolt until the bolts are torqued 40—45 ft-lb. Install the rocker-arm-and-shaft assembly on the other head using the same procedure.

ADJUST VALVE CLEARANCE

Solid Lifters—If the engine has solid lifters, adjust the valve clearance or *lash*. Correct lash is 0.025—0.027 in. *cold*.

You only need to turn the crankshaft twice and set the corresponding valves for each setting. You can set half the valves with the crankshaft at TDC on the compression stroke for number-1. The other half can be set at TDC on compression stroke for number-6.

To make sure it is the compression stroke, watch the intake-valve pushrod for the cylinder you are setting. The first time TDC comes around after the intake valve closes is top dead center for that cylinder. The valves to set with the two different crank positions are as follows:

#1 CYLINDER TDC ON COMPRESSION STROKE	
Number-1	Intake & Exhaust
Number-3	Intake
Number-4	Exhaust
Number-5	Exhaust
Number-7	Intake
Number-8	Intake & Exhaust

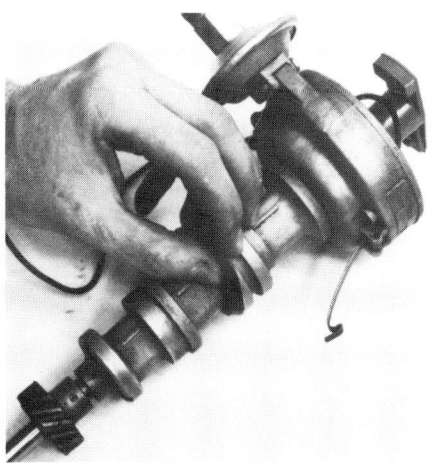

Install new distributor-base gasket or O-ring.

Majority of big-blocks use conventional, point-type ignition. Install new points, condenser and rotor.

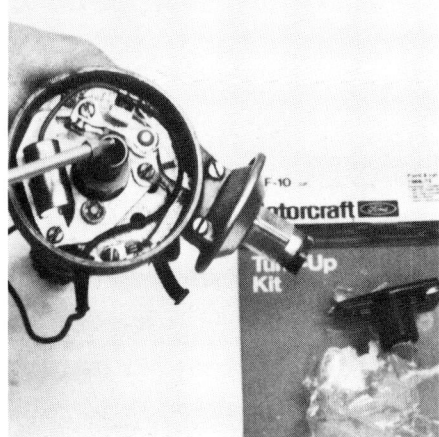

Felt wick in distributor shaft is often forgotten. Shot of oil will help keep the mechanical-advance mechanism working freely.

#6 CYLINDER TDC ON COMPRESSION STROKE	
Number-2	Intake & Exhaust
Number-3	Exhaust
Number-4	Intake
Number-5	Intake
Number-6	Intake & Exhaust
Number-7	Exhaust

On long-duration cams, turn the crank to TDC for each cylinder to the adjust its valves. These cams are used in the 427 with a stock High-Riser or two 4-barrel carburetors. You can also use this method with a standard cam. The order of the valves, front to rear or rear to front, on each side is EIEIIEIE.

To set the valves for individual cylinders, go through the engine's firing order. Turn the engine to TDC for number-1. Set its valves. Turn the engine clockwise 90° and set the valves for number-5. Turn the engine 90° again and set the valves for number-4 and so on. Turn the engine 90° each time and set the valves for the next cylinder in the firing order.

Firing order for the big-block is 15426378, with the left cylinders numbered 5—8 and the right cylinders 1—4, front to rear.

Hydraulic Lifters—Rocker arms used with hydraulic lifters have no adjustment provisions because they rarely need adjusting. However, conditions do arise that restrict lifter-plunger travel at one end or the other.

On a rebuilt engine, the major causes are incorrect parts, or machining on the cylinder head, block deck, valves or valve seats. If you have reused some of the original parts, it's possible to overlook a worn valve-stem tip or collapsed lifter, bad cam lobe or bent pushrod.

It sounds like there is a lot that can go wrong, but the chances of any of these items causing problems is *slim*.

To check the "clearance" on hydraulic lifters, you must bleed the lifters. There is a special tool for this, which locks onto the rocker arm. You simply lever down against the pushrod until the lifter plunger bottoms.

You can use a pair of *long-jaw* 10-in. Channellock, or water-pump pliers. Open the jaws to their widest position. Place the top jaw under the valve end of the rocker arm and the bottom jaw on top of the pushrod end. Carefully lever the rocker over to compress the hydraulic-lifter plunger.

Be patient, don't try to force the lifter down immediately. A lifter can take a minute or more to bleed down—especially if it is new! Once the plunger bottoms in the lifter, hold it there. Measure the clearance between the rocker arm and valve stem.

Desired clearance is 0.10—0.20 in. Adjust the clearance by installing a longer or shorter pushrod. Turn to the chart on page 57 for available pushrod lengths.

If only one or two valves are not within spec, recheck the lifter, cam lobe, pushrod and rocker before installing a different-length pushrod, as listed in the chart. Pay particular attention to used parts, especially the lifter and pushrod.

Remember, when replacing pushrods, loosen the rocker-shaft bolts one or two turns at a time until the shaft is unloaded. Loosen each bolt two turns at a time so the load is removed evenly.

DISTRIBUTOR INSTALLATION

Turn the crank until piston number-1 is at TDC on the compression stroke. Make sure the distributor is clean. Install a new distributor-base O-ring. Apply moly grease or Lubriplate to the distributor gear and a light coat of grease to the O-ring.

Align the distributor housing with the mark you made on the manifold. Check the distributor cap to see where the number-1 cap tower is in relation to the distributor housing. Put a mark on the housing to correspond to the number-1 cap tower.

As originally installed, the tower for number-1 is on the side opposite the vacuum-advance diaphragm at the clockwise edge of the retaining clip. Check the nearby photo. If the number-1 plug wire is not in this position, put it there. Rearrange the plug wires to follow the firing order when you install the distributor cap.

Turn the distributor rotor to line up with the number-1 cap tower. Next, turn the rotor an additional 45° counterclockwise. This is done because the distributor drive gear has helical-cut teeth and will rotate as it engages the drive gear on the cam.

Ease the distributor into place. Move the rotor back and forth slightly to engage the teeth on the two gears. When you feel the gears engage, push the distributor the rest of the way down. If you did it right, the rotor will rotate back in line with the number-1

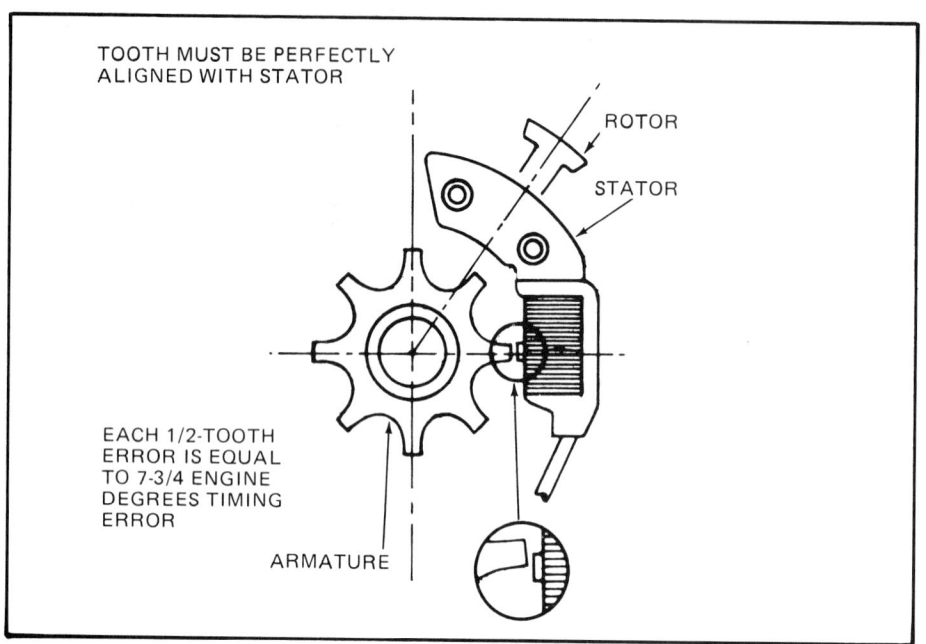

With number-1 piston at TDC on its compression stroke, align armature teeth as shown. This gives approximately correct timing for electronic ignition. Reset timing with timing light during engine run-in.

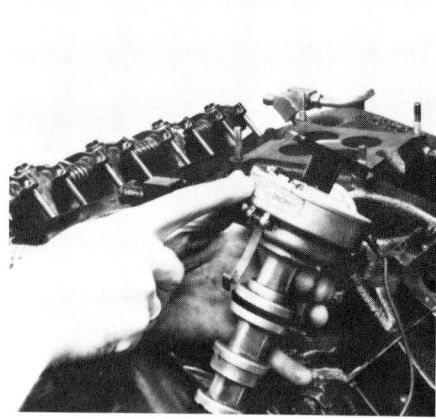

Note cast-in 12127 on distributor housing. Rotor should point toward number with distributor installed and number-1 piston at TDC on compression stroke.

position when the distributor is seated.

If the distributor won't seat fully after the gears engage, the oil-pump drive shaft is not lined up. Pull the distributor up slightly, but leave it engaged with the drive gear. Rotate the crank while occasionally pushing the distributor down until the oil-pump drive shaft engages.

To check that the distributor is in the right position, rotate the crank one full turn: The rotor will rotate 180°. The next time the damper reaches 0, the rotor should point to the number-6 position on the distributor-cap tower—check it with the cap. Crank the engine around to 0 a second time and it should be back to number-1.

With the distributor installed, you should be able to rotate the *housing* about 25° clockwise and about 15° counterclockwise. This will ensure enough room to rotate the distributor to set ignition timing.

If you pull the distributor out after it has been fully seated, be sure the oil-pump drive shaft stays in place. You should be able to see it—the shaft should be about centered in the hole.

When the distributor is seated, the housing will project about 1/16-in. above the surrounding surface of the intake manifold. Place the distributor hold-down against the distributor housing. The hold-down fingers point

Slide distributor into place while turning rotor. If you have trouble engaging oil-pump drive shaft, lift distributor slightly and rotate crankshaft a few degrees and try again. Repeat process until distributor base is flush with manifold. Install hold-down clamp and bolt.

down. Torque the bolt 12 ft-lb, unless it's in an aluminum intake manifold—then make it 10 ft-lb.

VALVE COVERS

Before installing the valve covers, be sure they are clean and the mounting flanges are straight. Check the mounting-bolt holes. If they are dimpled, set the bolt flanges on the edges of a workbench or vise and hammer them flat. Also check the grommet for the PCV valve. Replace the grommet if it is brittle or rotted.

Apply non-hardening sealer to valve cover and press gasket in place. Torque valve-cover bolts 4—7 ft-lb.

Apply non-hardening cement or silicone sealer to the valve-cover flanges and install the gaskets. Install the valve covers. In most cases, the PCV valve goes to the rear on the right side; the oil-filler neck goes to the left front. Check your photos or notes to see if yours is different. Torque the valve-cover bolts 4—7 ft-lb. Do not over-tighten these bolts or you'll distort the valve-cover flanges. Leaks will result.

ENGINE MOUNTS

If the engine mounts are near 10-years old, buy new ones. If the mounts are less than 10-years old, clean them up and check them for cracks. Also be sure the rubber is not starting to separate from the metal

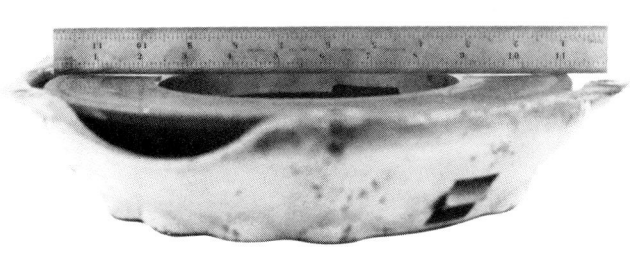

Two pressure plates damaged by overheating. One is warped, as indicated by gap between straightedge and pressure-ring surface. Heat checks and chatter marks on other pressure-plate surface indicate it has been extremely hot. If pressure plate or flywheel is suffering from either of these symptoms, replace it. Photos by Tom Monroe.

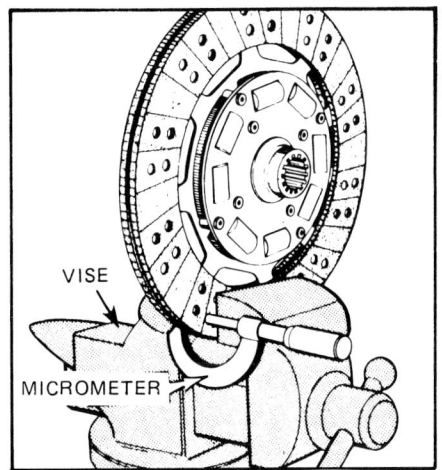

To measure clutch-disc wear accurately, you must compress the marcel spring that is between the two faces. Measure disc close to vise jaws. Drawing courtesy of Schiefer.

compressing the wave spring flat. Measure disc thickness close to the vise or clamp.

A new disc measures about 0.330 in. At about 0.280 in. the rivets are flush with the worn friction surface and will groove the flywheel and pressure plate as the disc wears more.

Sorting this out, there's about 0.050 in. of friction-material thickness to play with. If you know the mileage on the clutch disc, a few calculations will give you an idea of how many miles it has left.

For instance: If the disc measures 0.300 in. and it's been in the car for 40,000 miles, you have used 0.030 in. to go 40,000 miles. Simple extrapolation indicates it has another 25,000—30,000 miles left.

Use your own judgement on whether to replace the disc now or later. There is one advantage to replacing it now—it's easy.

Pressure Plate—The typical approach to pressure-plate inspection goes something like this: If it was OK when it came out, it'll be OK when it goes back in! This rule of thumb is not too far off.

If the plate looks good—bright and shiny, free of rivet grooves, heat checks or cracks, it's probably OK. But if it has *hot spots*—black or blue spots on the face—the plate has been overheated. It very likely has been overstressed and should be replaced.

Also, if the release levers are badly worn where they contact the release bearing, replace the pressure plate.

No matter how good it looks, if the pressure plate has serviced more than two clutch discs, replace it. The bushings in the release mechanism are probably worn and the springs fatigued. This reduces clutch capacity. The newly rebuilt engine will develop more torque than the old one, so it will need a clutch which has full torque capacity.

Flywheel Inspection—What happens to a pressure plate also happens to the flywheel, but the flywheel's larger mass means it suffers less. Heat that would warp a pressure plate can be absorbed by the flywheel without harm.

The flywheel is *not,* however, immune to grooving by rivets. Grooving, hot spots and maybe heat checks can be removed by resurfacing—another job for the machine shop. If they can't do the job, they'll be able to direct you to a shop that can.

A note of caution: Be sure the flywheel is *ground*, not turned on a lathe. Turning on a lathe produces *bumps* on the surface where the hot spots have hardened the metal. A ground flywheel surface will be flat.

There is one problem left. *If heat checks have developed into long radial cracks toward the center, the flywheel must be replaced!* These are stress cracks. Even at low rpm, a stress-cracked flywheel can *explode* like a grenade. More than one person has lost lower appendages to an exploding flywheel because someone ignored the cracks.

Be sure to clean the flywheel. Grease or oil on any clutch friction surface will cause chattering or grabbing. Deposits may be removed with 400-grit sandpaper. This also gives the flywheel a little *tooth*.

Follow up with a thorough cleaning with lacquer thinner, CRC Brakleen or alcohol. Avoid petroleum-base cleaners such as kerosene or cleaning solvent. They leave an oily film. Treat the pressure plate to the same lacquer-thinner bath and sanding.

Clutch-Release Bearing—Replace it. Don't check it out, don't fool around with it, just replace it. It's not expensive and it's a lot easier to replace now rather than after the engine is installed.

INSTALLATION

Both manual and automatic transmissions have an *engine plate*. This plate serves as a "gasket" and as the bellhousing cover.

If you have installed an engine with one of these plates before, you may have noticed it tends to fall off its dowels. Eliminate the problem by gluing the plate to the back of the engine. Weatherstrip adhesive works best, but you can try a small amount of grease first.

Make sure the plate is installed in the right direction. If you replaced it, be sure the new one fits—check *all* the bolt holes. If the engine uses screw-in oil-gallery plugs, you may need to drill holes in the plate. Unless the gallery plugs are flush with the block, they may butt against the plate and distort it as the transmission is bolted to the engine.

Flywheel or Flexplate—Inspect the crank flange. Be sure it is not nicked, scratched or full of grease and dirt; otherwise the flywheel or flexplate may not seat squarely. If the flywheel or flexplate doesn't seat squarely, it may eventually loosen.

The flywheel or flexplate bolt pattern is the same for all big-block Fords. Both index with the crankshaft flange in only one position. Be sure all the bolt holes align before installing any bolts.

Most flexplates use a load-spreader ring between the bolt heads and flexplate surface. Make sure you install it if the engine came with one.

Hold the crank with the vibration-damper bolt and torque the flywheel bolts to 75—85 ft-lb. On FT engines, make sure the bolts engage all of threads in the crankshaft flange, because not all flywheels have the same thickness. Torque the bolts in a crisscross pattern.

With a flexplate, rotate the crank so the mark you made on one of the converter drive-stud holes is at the bottom.

If you're working with a manual transmission, be sure to install a new pilot bushing or bearing, page 82. Smear a *light* coat of moly grease on the bushing or bearing ID. Install the clutch and pressure-plate assembly next.

Clutch Installation—Spend some time in this area: If the clutch is installed wrong you may have to pull the transmission or engine again to repair it.

Keep the friction surfaces clean from dirt, grease and oil. Wash your hands before handling the clutch disc and pressure plate. If you happen to get grease or oil on either part, clean it again with alcohol or lacquer thinner.

Clutch alignment is best done with a clutch-aligning tool. You can also use a dowel that fits the pilot bearing and clutch-spline ID or an old input shaft from another transmission.

Align flywheel or flexplate holes with those in the crankshaft. Be sure to install load-spreader ring for flexplate. Install bolts.

Torque bolts 75-80 ft-lb in crisscross pattern. Flywheel turner is handy to keep crank from turning when torquing bolts. Screwdriver through flywheel or helping hand holding damper bolt with breaker bar and socket also works.

Make sure the clutch is assembled correctly. The disc should be marked *flywheel side* or *pressure-plate side*. Place the clutch disc against the flywheel, with the hub side toward the transmission. The hub is made up of the forged center, damper springs and sheet-metal retaining ring.

Place the pressure plate against the flywheel and clutch disc. Reach through the pressure plate to position the clutch disc. Install all the lock washers and pressure-plate bolts—turn them in only a few threads.

To use an alignment tool, place the tool through the splines on the clutch disc and into the pilot. Its nose should be snug in the pilot bushing or bearing. Slide the correct-diameter sleeve or cone-shaped adapter onto the alignment tool to center the clutch disc on the flywheel. Tighten the six bolts one or two turns at a time and torque them 12—20 ft-lb. Remove the alignment tool.

INSTALL THE ENGINE

With the exception of some high-performance engines with stock cast-iron headers, the engine should have the exhaust manifolds installed before the engine is installed. For installing an engine with factory headers, read the sidebar below.

Factory Cast-Iron Headers—In full-size cars with the late 406 or 427, the right header must be installed *during* engine installation. In some Comets and Fairlanes, the left header must also be installed at this time. Engines with these headers will have manual transmissions. Usually there are problems only with the right header.

With the engine and transmission bolted together but off the engine mounts, install the starter motor, page 150. Make sure the wire terminals are clean.

Lower the right exhaust header into place. Most headers must go in from the front of the engine until the collector is below the cylinder head. Lower the engine until it is just above the engine mounts. Install the header bolts and torque them 18—24 ft-lb.

Note: The two bottom holes in the center mounting flanges are slotted. Insert these bolts in the cylinder head and lower the header into place. Install the six remaining bolts. The engine can now be lowered onto its mounts if the left header is installed.

The only case where the left header may create problems is a Fairlane or Comet; the clutch-equalizer bar may get in the way. Usually, leaving the header bolts loose will give the clearance needed. The '66—'67 427 Comet and Fairlane have holes through the shock towers to allow access for tightening the bolts through the wheel wells.

While lowering engine, be sure wires or linkages don't become damaged. If you discover engine is hanging at angle you can't work with, pull it back out and readjust chain position on engine or hoist.

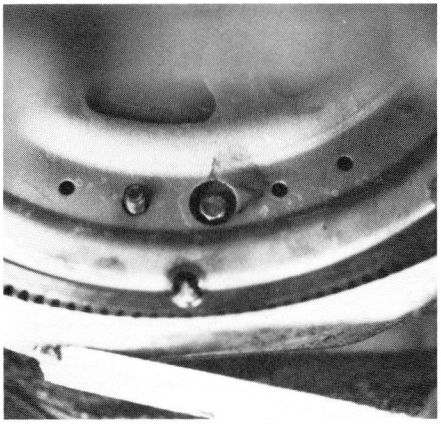

Be sure converter studs align with holes in flexplate, then push engine back into engagement with the transmission.

Install bellhousing-to-engine bolts and torque then 40—50 ft-lb. Make sure engine and bellhousing are together before tightening bolts or you will break the bellhousing.

Pick it up—Bolt the chain to one end of each cylinder head, running from left front to right rear. The chain can also be hooked or bolted to the exhaust-manifold brackets, if you reinstalled them. The brackets make it easier to move the engine around, especially when tilting it.

Roll the jack under the transmission. With the transmission on the jack, remove any transmission supports. Raise the transmission so it lines up with the engine as the engine is lowered into the compartment.

Check that the engine compartment is clear of obstructions. If anything isn't positioned correctly, the engine may damage it. Make sure wires are out of the way. Install new exhaust-manifold doughnuts or gaskets. A dab of grease will hold them in place.

Jack the front of the transmission as high as it will go. This makes it easier to engage the engine and transmission before the engine mounts contact their supports. On unit-body cars, ease starter-motor installation by resting starter on the center steering linkage at right.

Raise the engine enough to clear the vehicle. Position it over the engine compartment, then slowly lower it and guide it into place.

Manual Transmission—Be sure the clutch-release bearing is installed. Place the transmission in gear and guide the clutch over the input-shaft splines. Once the clutch hub contacts the splines, rock the engine left and right while pushing back on it. Keep engine and transmission aligned.

The input-shaft nose will center in the clutch disc before the disc and input-shaft splines engage. Once the nose of the input shaft is centered in the clutch disc, check the gap between the bellhousing and engine. It should be even from top to bottom and side to side.

If you have difficulties, don't panic. Take a good look at the situation. Reinstall the clutch-aligning tool; it should center in the clutch disc and pilot bearing without binding, and be perpendicular to the flywheel and clutch.

If you are sure the clutch is aligned, rotate the transmission input shaft. Take the transmission out of gear and rotate the input shaft *slightly*.

You can disconnect the drive shaft at the rear axle and rotate the shaft while the transmission is in gear and while someone else pushes on the engine. Or simply jack one rear tire off the ground and turn it. Use extreme caution when pushing the engine around so you don't knock the car off its supports.

Once the transmission splines and clutch engage, push the engine back into place while guiding the engine dowels into the bellhousing. Install two bottom bolts and one top bolt finger tight.

Automatic Transmission—Align the flexplate with the torque-converter stud. Lower the engine until the bottom converter stud aligns with the bottom flexplate hole. Position the engine so it is slightly farther away from the top of the bellhousing than it is at the bottom. Move the engine back until the alignment dowels on the engine engage the bellhousing.

Square the engine with the converter housing. Once you have an equal gap all around, push the engine back. You may have to lift and twist on the engine. The converter hub pilots into the back of the crankshaft. At the same time, the engine dowels must fit into the bellhousing *and* the converter studs must engage the flexplate.

With the engine and transmission engaged, the converter should rotate with the crankshaft without dragging or scraping. Install two bottom and one top bellhousing bolts. Thread them in finger tight.

Install the remaining bellhousing bolts. Be sure to install the automatic-transmission filler-tube bracket and any wire and vacuum-line brackets under the bolts. Torque the bolts 40—50 ft-lb.

Engine Mounts—Lower the engine into place. If the engine has stud-type mounts—most do—the studs should slip through the slots in the supports. If not, guide the studs with a screwdriver. Install the nuts and torque them 45—60 ft-lb. Through-bolt mounts are used in Mustangs, Cougars, early Fairlanes and Comets. Torque the through bolts 35—50 ft-lb.

TORQUE-CONVERTER NUTS

The torque-converter nuts are self-locking, with either nylon inserts or oval-shaped openings. Unless the old torque-converter nuts have a really good grip, replace them.

Line up engine mounts and guide engine into position while lowering it. Torque nuts on stud-type mounts 45—60 ft-lb, through bolts 35—50 ft-lb.

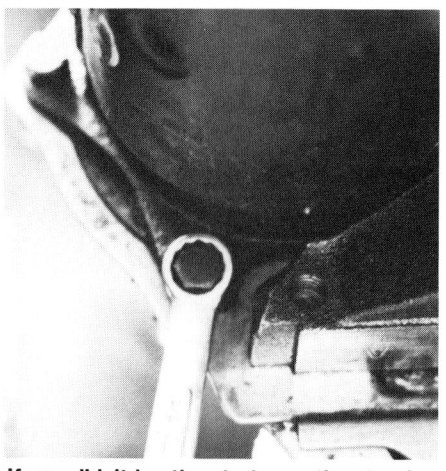
If you didn't lay the starter on the steering linkage or brace before installing the engine, thread the starter into engagement with the ring gear. You may have to remove a support brace or move the steering linkage by turning the steering wheel. Install *all three* bolts—I know they're tough. Starter flange may crack if any are left out. Torque bolts 15 ft-lb in an aluminum bellhousing, 20 ft-lb if bellhousing is cast-iron.

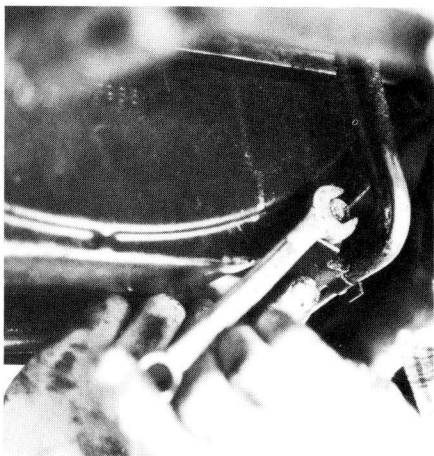
Tighten torque-converter nuts 20—30 ft-lb. Install inspection plate and support brackets.

Install fuel-tank-to-pump hose after checking it for cracks. Tighten clamps securely.

Before installing metal fuel-pump-to-carburetor line, install any brackets or components that line may have to be routed around. Remember you must be able to turn distributor to set timing, so don't route fuel line too close to vacuum-advance unit.

Use flare-nut wrench to install line to pump.

Install the torque-converter-to-flexplate nuts. Keep the crankshaft from turning with a breaker bar and socket on the vibration-damper bolt. Snug *all* the nuts, then torque them 20—30 ft-lb.

After one nut is tightened, use the wrench to turn the flexplate to the next bolt. Don't try this with the sparkplugs installed. The additional resistance from engine compression may damage a torque-converter stud.

If you don't want to remove the sparkplugs, wait until the starter is installed to turn the crank, or turn it with the vibration-damper bolt.

STARTER MOTOR & INSPECTION PLATE

The '58—'64 Fords use a different-style starter than the '65-and-later models. Later Fords use a conventional starter, where the drive gear is pushed toward the transmission and into engagement with the ring gear. The late starter can be used with the early big-blocks with no problem.

On the early engines, the starter gear is pulled into engagement with the ring gear. These starters have a much longer nose, which fits into a recess in the bellhousing. The '65-and-later bellhousings do not have this recess, so early starters won't fit.

A rubber gasket/spacer on the early starter positions the drive gear the correct distance from the flywheel. The spacer must be used on the old-style starter. Without it the starter drive won't engage the flywheel. Buy a new spacer if the old one is cracked or rotted.

Starter Installation—On some cars, such as the Thunderbird shown, the easiest way to install the starter is to remove a cross brace to install starter motor. On Comet/Fairlane, Cougar/Mustang and '58—'64 full-size cars, turning the steering wheel will provide clearance to install the starter.

On some installations you must unbolt the idler arm at the frame to get the needed clearance. Once unbolted, pull the steering linkage down for clearance. On pickup trucks you won't have any problem. On heavy-duty trucks you can just about stand up while installing the starter.

Hold the starter up and thread in the lower bolt far enough so it won't come out. Push the starter into place and insert the top and side bolts. Torque them about 20 ft-lb with your calibrated hand. Use about 15 ft-lb with an aluminum-bellhousing.

Before connecting the starter cable, make sure the cable connector and surfaces of the nuts are clean. If not, clean them with a wire brush and a file or sandpaper.

Fasten the starter wire to the single stud on the starter. Most engines have one or two clips that fasten to the side of the block to keep the starter cable from moving around. Route the cable through the clips.

Install the inspection cover to protect the clutch or torque converter from snow, dirt and road hazards.

EXHAUST PIPE/S

Use new gaskets where the pipe or heat riser connects to the exhaust manifolds. If the pipes are full of carbon, soot or oil, the new engine will appear to smoke for a long time afterward. If this bothers you and you can afford it, consider installing a new exhaust system.

It can be difficult to hold the exhaust pipes in place while trying to run the nut up the stud. Once I have them in position, I wedge a board between the ground and the pipes to hold them. This leaves my hands free to start the nut and hold the clamp or collar in place. The best exhaust nuts are the double-length brass ones. Torque them 25—30 ft-lb.

OIL FILTER

If you didn't already install it, fill the oil filter with oil and smear a light coat on the rubber seal. Turn the filter until it contacts the adapter, and then turn it another 3/4-turn. *Add oil before you forget*—the usual amount is five quarts, but add four quarts for now. Recheck oil level after the engine has been run the first time.

LACING IT UP

The way you finish the installation

Power-steering pump, air-conditioning compressor, alternator, air pump, brackets and other components are installed in various orders depending on each application. If you remember the order you took them off, reverse order should suffice. Otherwise it's trial and error. When installing accessories be sure pulley grooves align.

job has a lot to do with the look of the finished product. In some cases, it can also affect engine performance. It's important at this point to take your time to do the job right the first time.

Carefully inspect any parts you plan to reuse, particularly the little things: hoses, belts, plug wires and hose clamps. Replace any questionable parts. It would be a real heartbreaker to spend this much time and money rebuilding the engine and have it overheat and score a piston because a 50¢ hose clamp failed.

Fuel Lines—Hook up the flexible fuel-tank-to-pump hose. Be sure the hose is not cracked, rotted, too short or twisted. Use hose clamps at both ends of the hose.

Next, install the pump-to-carburetor line. This line is usually metal. Be sure it's not crimped, cracked or twisted and the fittings at both ends are in good condition.

If the old metal fuel line is in bad shape, don't replace it with flexible hose. Pick up a new metal one at the parts store. Take in the old line to match the size and length, then bend the new one to follow the same routing.

Install the metal line, but leave the fittings loose. Check the routing of the line, particularly for clearance around brackets, coil and distributor. Tighten the fittings after you install the carburetor.

Alternator, Power-Steering Pump and Air Pump—The order you use to

Install wire/s to oil-pressure and temperature sending unit/s.

install these components depends on the bracket and component arrangement. This can get very confusing, especially if the engine is loaded with options and is in a cramped compartment. If you took pictures or notes use them.

Install the alternator wires, then the alternator and its brackets. *If you connected the battery for any reason, disconnect it before installing the alternator.* The power lead for the alternator is *hot* and unfused. Shorting the alternator leads may damage the diodes in the alternator or burn out its wiring harness or *fusible link*.

An air-injection pump—smog pump—usually mounts on the right side of the engine above the alternator. Install the air pump.

Install the power-steering pump next. In some cases the pressure line must go on after the main bracket surrounding the pump. Otherwise install any lines before placing the pump back onto its brackets. Be sure to install the line to the hydraulic wipers if you have a Thunderbird so equipped.

Install all the spacers and brackets so the pulleys line up and the belts can be tightened. This is an area where photos and notes will serve you well. Sight down the pulley grooves to double-check pulley alignment.

Air-Conditioning Compressor—Install the air-conditioning compressor and brackets. Be sure to install all A/C-compressor supports. The compressor usually has an extra brace to the intake manifold.

If there is an idler pulley, spin the pulley before you install it. If the bearing clicks or feels rough, replace the pulley.

Hood is installed before job is completed due to fast-approaching sunset. Line up hood and hinges to reference marks and tighten bolts. Carefully lower hood to check alignment and clearance to cowl and fenders.

Install throttle and transmission-kickdown linkages.

Emission-control and vacuum-hose hook-ups can be very confusing. Use your photos or notes to sort things out.

Install carburetor spacer if used, gasket and carburetor. Torque bolts gradually to prevent carburetor base from being distorted. Check throttle positions—open *and* closed—when finished.

Install throttle-return springs. Recheck throttle travel. Above all, be sure springs pull throttle closed.

This is most-common type of throttle-linkage retaining clip. Be sure it snaps firmly into place over rod.

Ignition Coil & Wiring—Install the coil and wiring harness on each side of the engine. Route each wire so it will clear moving parts and hot exhaust manifolds.

Connect the wires to the oil-pressure and water-temperature sending units, coil, A/C-compressor clutch, throttle-stop solenoid and choke heater. Again, refer to your photos or notes for the correct installation.

If you saved the clips for securing the wires, use them. They protect the wires and keep them from chafing on metal parts.

Carburetor & Linkage—Install the carburetor with the appropriate spacer and gaskets. Make sure the throttle opens all the way and, more importantly, *closes* all the way. Recheck it after installing the linkage.

Use a tubing—*flare-nut*—wrench on the fuel line and support the fitting on the carburetor or fuel pump with an open-end wrench. Also make sure the line doesn't twist as you tighten each nut. If it does, spray the nuts with penetrating oil and loosen them on the line before installing them.

Install the stabilizer bracket, if used, on the intake manifold or fire wall, then hook up the throttle linkage. Recheck the throttle and choke for full open and close.

Attach the transmission-kickdown rod and check for correct action of the rod. The rod should engage at about 3/4—7/8 throttle.

Next, install the return springs. If they weren't correctly hooked up in the first place, have the parts man at the dealer show you the illustrations in a parts book—make sure you ask

nicely and buy parts from him. After the linkage and springs, install the speed—cruise—control if the vehicle has it.

With all the linkage hooked up, pull the throttle wide open and release it. *Make sure* the throttle closes all the way. With the springs installed you shouldn't have to push it closed. Check it with the choke open and closed. If the engine has multiple carburetors, be sure *all* carbs return to the idle position.

Vacuum Lines—Install all the vacuum lines to the carb, PCV valve, distributor and any other accessories. Make sure all the connections are tight. Check that the hoses are not split or rotted, which will cause air—"vacuum"—leaks and poor performance. Remember to connect the vacuum line/s to the automatic-

Be sure you hook up vacuum-advance line. On Holley carburetors, line connects to nipple screwed into main metering block.

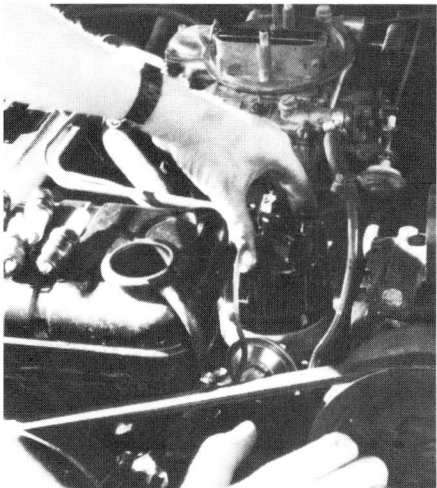
Install rotor, cap and wires. Firing order for all big-blocks is 15426378.

Take care when installing thermostat and its housing or expansion tank. If thermostat is out of position, it may crack thermostat housing when bolts are tightened. Note that thermostat spring goes into intake manifold.

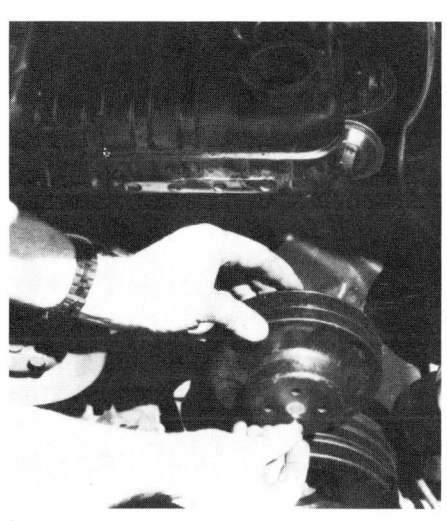
Fit pulley on water pump and thread in one bolt a turn or two to hold it in place. This allows you to install all belts without having the fan in the way.

Air-injector pump and alternator share brackets on this installation. Installing these components correctly requires a good memory or notes, especially when spacers are involved. I hope you took good notes.

transmission modulator. If you leave this one off, the transmission will shift late and very hard.

Sparkplugs & Leads—All big-block Fords have the same firing order: 15426378. Number the wires now before installing the distributor cap. This makes it easier to route them to the plugs.

All the wires can be left in order on their valve-cover clips except for number-7 and -8. Move number-7 to the front position, then -5, -6 and -8. The reason for this is *induced firing*, explained on page 10.

Make sure the number-7 and -8 wires don't run *parallel* to each other. You can leave one of the wires out of the valve-cover clip, or clip one of the wires at a 90° angle to the others on the loom.

Water-Pump Pulley, Fan & Belts—Install the water-pump pulley and fan as an assembly. If the water-pump hub has studs, simply install the pulley, then any spacers and the fan. Torque the nuts 10—15 ft-lb.

The bolt-type water-pump hub isn't quite as easy. Run one bolt through the fan and the pulley. Place fan and pulley over the water-pump hub as an assembly and thread one bolt in. Install the remaining bolts before threading any of the bolts in more than a couple of turns.

With a fluid-clutch fan, torque the bolts after the belts are installed and tightened. Otherwise there's no way to keep the hub from turning.

Be sure the belts are not cracked or split. The best way to check for cracks is to bend the belt inside-out.

Install the belts. When tightening belts, be careful. Do not pry on the P/S pump, valve cover, fuel line or some other damageable item. Most belts can be tightened by pulling on the accessory and tightening the bolt while holding the tension. This works best with two people.

There is a special tension gage for accessory belts, but except for a few specialty shops and the military, I have seen very few used. If you have the gage, use it. Tension on a new belt should be 140 lb, a used belt 100 lb. *A used belt is one that has been run on the*

Set radiator on supports, install top shroud and bolt them securely in place. Don't forget to install automatic-transmission-cooler lines.

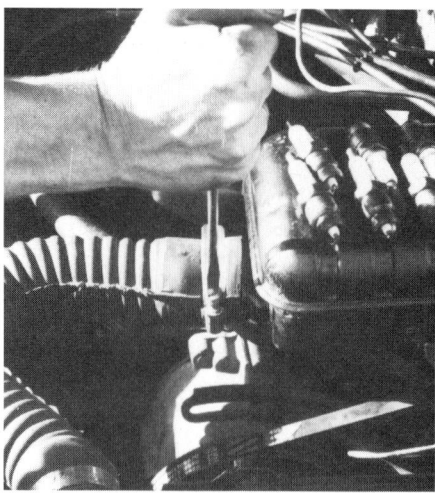
Coming into the home stretch. Install upper radiator hose. New hose clamps, by the way, is a good idea.

Rotate hose clamps so you'll have access to their bolt or screw heads after everything is installed.

Install hoses and clamps so there is room for other components and access for future removal.

Single bolt is removed, fan installed and bolts tightened. The reason belts were installed first should now be obvious. You can use belts to keep pulley from turning when tightening fan-clutch bolts.

Add engine oil. Don't skimp on quality. Use same grade and quality you plan to run in engine.

engine more than 10 minutes.

The simplest way to check belt tension is by deflection—pushing on the belt. Simply measure how far the belt deflects when you push on it with your thumb midway between two pulleys. The belt should not deflect more than 1/2 in. in a 12-in. length between pulleys.

To check that you haven't overtightened the belt, see if you can twist the belt 90° from its resting position. If you can, the belt should be OK. Belts that are too tight cause excessive wear on the water pump, alternator and front main bearing. Loose belts don't always squeal, but they can cause undercharging, jerky power steering or overheating.

COOLING SYSTEM

Before you install the radiator and heater hoses, make sure they are in good shape. If they are more than two-years old, replace them anyway. On a Comet/Fairlane or Mustang/Cougar, you will save a lot of grief if you fit the lower radiator hose to the water pump before installing the radiator. Align the hose so the radiator end points toward the front of the car. Tighten the clamp at the water pump.

Fan Shroud, A/C Condenser & Radiator—If the engine has a full-circle fan shroud, put it in place and lay it back over the fan.

Install the air-conditioning condenser if you removed it. Set it on its brackets or the brackets shared with the radiator.

Install the radiator. Take care not to damage the fins or cut yourself when lowering it into place. Protect both with a sheet of cardboard taped over the core. The radiator is held in one of two ways. Some have saddles on the bottom and and bolts in the upper shroud. Others are retained by side mounts.

Saddle-mount radiators are simply set on their saddles and bolted in. To install side-mount radiators, run a small Phillips-head screwdriver through the radiator flange and a radiator-support nut. Align the holes with the screwdriver and install the bolts.

Once the radiator is bolted in, move

the fan shroud forward and install it. There should be at least 1/2-in. clearance all the way around between the fan-blade tips and shroud. If you can't obtain clearance by adjusting the shroud, loosen the radiator and adjust it.

Hook up the lower radiator hose. Make sure it is not crimped or twisted before you tighten the clamp.

Automatic-Transmission Cooler—If the vehicle is equipped with an automatic transmission, connect the cooler lines. Again, use a tubing, or flare-nut, wrench for the flare nut and an open-end wrench to support the fitting at the radiator.

Be careful not to twist the line or fitting on the radiator and be sure the line routes straight into the fitting. Leave clearance between the lines and the fan.

Upper Hose & Expansion Tank or Thermostat Housing—Install the thermostat in its housing or expansion tank. The spring should point toward the intake manifold.

Smear silicone sealer on the flange around the thermostat and place the gasket onto the flange. Apply sealer to the gasket surface.

Place the expansion tank or thermostat housing on the manifold and install the bolts. Torque the bolts 10 ft-lb—no more, especially with aluminum manifolds. Next, install the upper radiator hose—routing it so it won't contact the fan or the hood when closed.

Heater Hoses—Install the heater hoses—cut off the ends if they look badly flared. It's a good idea to replace the heater hoses. If you think they are OK, bend each hose sharply. If cracks appear at the outside of the bend—you can see the reinforcing cord—replace the hose.

Install the hoses so they flow coolant in the correct direction for heating. The lower heater-core outlet connects to the intake manifold; the upper goes to the water pump.

BATTERY

Put on your safety glasses and install the battery and cables. Hook up the positive cable first. If you see a light spark when you hook up the ground, it's probably current from the clock or dome light.

If you have more than that—there's a loud "pop" when you hook up the cable—investigate! Start by disconnecting the voltage regulator, genera-

Air cleaner *with new air filter* is installed for pre-oiling and initial fire up. Air filter also protects against backfire and possible underhood fire. Remember to crank engine over with plugs out and ignition system disabled until oil pressure builds up.

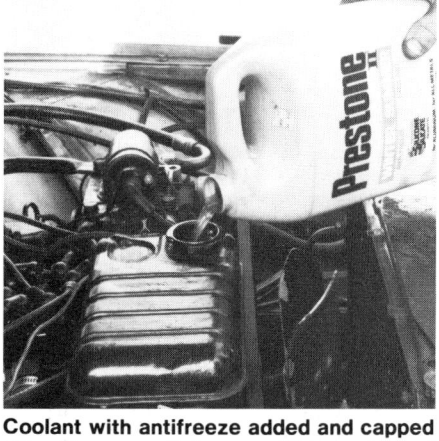

Coolant with antifreeze added and capped off with new radiator cap. Old cap leaked through center rivet.

tor or alternator. If there is still a short, make sure you haven't pinched any wires when you installed the engine. Also check that the hot wire on the solenoid isn't touching the fender.

Engine Ground Strap—Don't forget to hook up the ground strap at the fire wall. On most engines, it bolts to the rear of the right cylinder head.

PREPARATION FOR STARTING

You'll need a few things before starting the engine for the first time. With all the time you've spent, don't take any shortcuts now.

Move the vehicle off any ramps or stands. If you drained the torque converter, refill through the transmission filler tube immediately after the engine starts. Have at least five quarts of automatic-transmission fluid ready.

Hook up the timing light before you crank the engine. Be sure the timing marks are legible. Loosen the distributor hold-down bolt enough that you can turn the distributor by hand, but not so it is loose and will turn with the engine running.

Fill With Coolant—The new engine is going to be extremely *tight* until the rings seat, which will make the engine run considerably hotter than normal. Add to this the fact that during initial run-in the car will be stationary, so there is only fan-induced airflow through the radiator. These two factors add up to the possibility of the engine overheating.

To be sure the engine receives enough cooling, fill the cooling system with water before you start the engine. Leave the garden hose running slowly in the filler neck and the radiator petcock open. Balance the flow so the water doesn't flow over the top of the neck but keeps a constant level in the radiator. Once the engine starts, adjust the water flow to keep the radiator filled.

Pre-Oiling—Remove the sparkplugs and disable the ignition system. Crank the engine over until oil pressure is indicated. The oil-pressure light should go out or the gage should show a reading within 15 seconds.

You can repeat this with the sparkplugs installed. Now, reinstall the sparkplugs and wires, leaving the ignition disabled. Torque the plugs no more than 20 ft-lb. Crank the engine again. You should see pressure fairly soon because the oil galleries will be partially filled. As a bonus, cranking the engine also fills the float bowls in the carburetor/s.

Now reconnect the ignition system. Give everything one last check. Make sure you have plenty of oil in the pan. See that everything is clear of the fan and the vehicle is in **PARK** or **NEUTRAL** with the parking brake applied and the wheels chocked.

Air Cleaner—I prefer to do the initial run-in with the air cleaner on, then remove it to adjust linkage and the carburetor. So install the air cleaner, along with any emission-control plumbing.

Be sure to clean the housing and use new gaskets around the air-cleaner base and the oil-filler necks. Don't install a dirty air-filter element or your fresh engine will draw in old dirt. The same goes for the PCV valve, hoses and vacuum lines. They should all be cleaned or replaced.

STARTING ENGINE

You'll have a lot to do once the engine starts, so it is safer and easier to have a friend's help at this point. The engine will belch smoke when it first starts, and hydraulic lifters may clatter for several minutes before they prime and quiet down. Both of these are normal. Solid lifters won't be as noisy at first.

Once the engine starts, hold engine speed to about 1200—1500 rpm. If you have any difficulty with the engine during the first 1/2 hour, shut it off, *do not let it idle*. Check the oil-pressure gage or light as soon as the engine starts. Add four quarts of transmission fluid immediately in case of an automatic.

If everything seems OK, check the automatic-transmission fluid. Make sure the transmission is in **PARK**, the vehicle is on flat ground and the parking brake applied. Also, the engine must be running to make an accurate check.

If the engine is shut off, the torque converter drains into the transmission oil pan, giving a false over-full dipstick reading. On a cold automatic transmission—one that has not been driven 5—10 miles—the correct level is the **ADD** line, not the **FULL** line. Recheck the level after you've driven the car a few miles.

Check the engine for leaks, and be sure the oil pressure and coolant level are staying up. The coolant level will probably drop when the thermostat opens, so readjust the water flow. The engine will be running on the warm side and generating a lot of heat at first—this is normal. The temperature gage should still be well within limits and not indicate **HOT**!

Set the timing as soon as you can. Watch the timing light around the pulleys and *stay out of line of the fan!* Keep the engine at about 1000 rpm and plug the vacuum-advance hose. Turn the distributor *clockwise to advance* timing and *counterclockwise to retard* it.

Unless you have a distributor wrench and can stay far away from the fan, finish the run-in of the engine before tightening the distributor hold-down bolt. After you shut off the engine, tighten the distributor hold-down bolt and hook up the vacuum line to the distirbutor-advance diaphragm.

When the engine is running smoothly, run the automatic transmission through the gears. This will circulate the fluid and stabilize the fluid level. Recheck the fluid level afterwards.

Post Checks—When you shut the engine off, drain the coolant far enough to allow room to add rust-inhibitor or antifreeze. From this time on, never, under any circumstances, run 100% water in the cooling system. Make sure coolant is at the correct level.

Check the engine for leaks or loose bolts. As soon as the engine has been run and then cooled, the gaskets will relax and parts loosen up. Check the valve covers, timing cover, oil pan, exhaust manifolds, intake manifold, carburetor mounting, hose clamps, thermostat housing and exhaust-pipe connections.

Hood—Align the hood to your marks and reinstall; round up some help to make it easier. Use short 2x4s at the back of the hood to protect the body and windshield when installing the hood. Close the hood slowly at first to make sure you have it aligned. The gaps should be even all the way around, just as it came from the factory.

TEST RUN

Before driving off, be sure you have all your tools. Remove the blocks from the wheels. Check the radiator-coolant level again. If you can, check it with the nose of the car facing uphill a little and the heater turned on to bleed out any air.

Take a short test drive. Fill the tank with gas and check under the hood and car for leaks. Be particularly leery of any strange noises. Recheck automatic-transmission-fluid level after a few miles.

When driving with a freshly rebuilt engine, take it easy. You can drive 55 mph with no problem; just don't accelerate hard up to that speed. With the rings seating, the cooling system has enough heat to dissipate without your causing more.

The first thing you should notice on the freshly rebuilt engine is that it is quieter and feels more solid. The engine will be stiff at first. It will crank hard when starting and gas mileage will be down until it is broken in and has loosened up a little.

After about 200—300 miles you should feel the engine loosen up substantially. As the engine gets more miles on it you can accelerate harder. This is also true for heavy loads; keep them light until the engine is broken in. The longer you wait until you *work* the engine, the more you increase its longevity.

Vary road speed when you are on the highway—don't maintain a constant speed until you have a few-thousand miles on the engine. A good break-in procedure isn't great for fuel economy but it pays off in the long run.

Ignition timing shouldn't vary during break-in, but you will constantly have to lower the idle speed as the rings and cylinder walls conform to each other. Keep a constant eye on everything—particularly oil level—even if everything seems fine.

SOLID-LIFTER ADJUSTMENT

Ford's recommendation for hot lash on solid lifters is the same as it is cold. Check the lash again using the same procedure you used during engine assembly, page 140.

You can adjust valve lash with the engine running, but if you use a heavy wrench you can damage the valve train. with a light wrench you might damage your hands. If you decide to adjust the valves with the engine running, I think a 7/16-in. offset box-end wrench works best.

When setting the lash with the engine hot, use the smaller tolerance of 0.025-in. A pair of old clean valve covers with the tops cut off are great for minimizing oil splash when setting the valves with the engine running. Wrap a thick rag around the wrench to absorb some of the vibration. Turn the adjusting screw until you feel a light drag on the feeler gage.

If you set the valves with the engine off, go through the firing order with each piston at TDC with a long-duration camshaft. Use this technique with the stock camshaft for the 427 with the High-Riser or two 4-barrels. If you adjust the valves while the engine is shut off, you may need to start it occasionally to keep the engine warm and adjustments equal.

CHAPTER 9
Tuneup

After the initial break-in, have the engine tuned for maximum performance. Performance includes three basic factors: power, economy and emissions.

If you are not experiencing any serious problems—detonation or overheating—run the engine for a few-hundred miles before making a final tune. Most engines will need constant idle adjustments in the first 200—300 miles as the engine loosens up.

As the rings seat they create less friction and they seal better. As a result, the power and economy of the engine will increase. It will take about 1000 miles before you will have an accurate idea of the power and fuel economy the engine can produce. The improvement will be dramatic for the first 200 miles, tapering off as miles accumulate. Most engines will deliver maximum performance when they have a few-thousand miles on them.

PERFORMANCE AND EMISSION TUNING

The performance of the newly rebuilt engine should be a dramatic improvement over the old. Still, don't be satisfied with the performance of the new engine until it has been thoroughly tuned. Just because it's better doesn't mean it's the best it can be.

You may be suprised at the hidden power or economy in the engine. Having the carburetor or distributor set just right can make the engine perform even better than new.

Depending on how much money you can spend there are several ways to tune the engine. The first method is the most common—doing it yourself with a timing light and dwell/tach meter. With this method you can check initial timing, vacuum and mechanical advance, idle settings and point dwell. This method works fine if you left the engine stock, but it's still not the best way.

An engine analyzer with an oscilloscope is the next step up. With this setup you can check air/fuel ratio,

Having your engine professionally tuned is the best way of getting the most out of it. If your engine was rebuilt to original factory specifications, an electronic engine analyzer like this one will do the job. This Sun analyzer even has a printout so you'll have a record of your engine functions after the tune-up. Photo by Tom Monroe.

point dwell at various rpm, power balance, engine emissions and the primary and secondary circuits of the ignition system. Again, with a stock engine this is a good way to tune the engine, and far better than the home tuneup.

The final method is to have your vehicle tuned and monitored on a chassis dynamometer. The chassis dynamometer measures rear-wheel horsepower through a large set of rollers set in the floor. Different loads are dialed into the rollers to simulate changing road conditions. The engine is also hooked to an electronic analyzer, which can monitor engine performance under road conditions.

A chassis-dynamometer tuneup costs a bit more, but it is worth it to check out the performance of the engine. This tool can help to adjust the engine for performance under most road conditions.

If the engine is in tune for power, fuel economy and low emissions usually follow. So it pays to have the engine tuned right.

INDEX

A
A/C compressor 17, 151
A/C condenser, install 154
A/C system, discharge 17
Accessory-drive belts 17, 153-154
Air cleaner 15, 155-156
Air conditioning, see A/C
Air pump 15, 18, 151
Air-compressor oil drain 32
Air-injection tube 100
Air/fuel mixture 5, 7, 157
Alternator 16, 18, 151
Automatic transmission 149
Automatic-transmission cooler lines 18, 19, 155
Automatic-transmission fluid 155, 156

B
Battery 16, 155
Bearing crush 91-92
Bearing-crush height 91-92
Bearing-journal finish 81
Bearing-to-journal clearance 4
Bellhousing 150
Bellhousing dowells 149
Bellhousing inspection cover 20
Block deck, mill 75, 102
Block
   330HD 32
   330MD 32
   361 32
   390HP 29
   391 32
   406 29
   427 29
   428 32
   industrial 32
   FE 26-32
   FT 32
   side-oiler 427 32
Blowby 4, 128, 129
Bore taper 73, 74
Bushing 93

C
Camshaft 67, 69, 121
   bearing journals 84
   design 83
   inspect 83
   spacer 121
   thrust button 120
Camshaft bearings 69, 118-120
Camshaft end play, measure 121-122
Camshaft plug, install 119, 144
Camshaft sprocket 121
   cast iron 96
   nylon 96
Camshaft thrust plate 68, 96, 120-121
Camshaft-bearing
   oil holes 119
   order 118
Camshaft-lobe
   base circle 83
   closing ramp 83
   duration 83
   heel 83
   lift 83
   opening ramp 83
   profile 83
   rake angle 83
   toe 83
   wear 6, 8, 12, 84
Camshaft-lobe lift 12, 84
Cape chisel 92
Carbon deposits 6, 7
Carburetor 16, 143-144, 152
   float level 9
Carburetor spacer 143-144

Cast-iron headers, install 148
Casting numbers 26-27
   see also tables
Catalytic converter 6
Center oiler, see top oiler
Change level 25
Cherry picker 13
Clutch 59, 148
Clutch bolts, torque 148
Clutch disc 146-147, 148
Clutch linkage, disconnect 19
Clutch pressure plate, inspect 147
Clutch-release bearing 146, 147, 149
Cobra 24, 25
Cobra Jet 32
Cobra Jet Mustang 34
Compression ratio 7, 75
Compression test 6, 10
Compression tester 10
Connecting rod 125-130
   bent 90, 91
   long 34, 36, 37
   short 34, 36, 37
   twisted 90, 91
Connecting rods 34, 36
Connecting-rod
   bearing 91
   bearing bore 92
Connecting-rod bolt
   inspect 92-93
   LeMans 92
   measure 92
   NASCAR 92
Connecting-rod type
   330HD 34
   330MD 34
   332 34
   352 34
   360 34
   361 34
   361 Edsel 34
   390 34
   391 34
   406 34
   410 34
   427 34, 36
   428 34
   428CJ 34
   428SCJ 36
   LeMans 33, 34, 36, 66, 95
   NASCAR 36, 66, 95
   Police Interceptor 34
Connecting-rod bushing 93, 93
Connecting-rod center-to-center length 34
Connecting-rod journals, hollow 34
Connecting-rod noise 8
Connecting-rod pin bore, check 87-88
Connecting-rod pin-to-bushing clearance 87
Connecting-rod side clearance 65, 130
Connecting-rod-bearing cap 66, 129-130
Connecting-rod-bearing-cap nuts, torque 130
Connecting-rod-bearing-cap bolts, torque 130
Coolant 58, 155
Core plug 69, 120
   cup type 120
   screw-in type 31, 120
Cranking vacuum test 9
Crankshaft 66-67, 123-125
   cast iron 33
   damage 79
   Detroit balance 33
   end play 124-125
   fillet 81, 95
   forged 33
   high performance 33
   inspect 79

   kit 81, 125
   out-of-round 79, 80
   runout 82-83
   thrust bearing 28
   thrust surfaces 28, 80
   thrust width 81
   zero balanced 33
Crankshaft damper, see Vibration damper
Crankshaft oil grooves, chamfer 81
Crankshaft oil slinger 61, 131
Crankshaft runout, measure 83
Crankshaft sprocket 65, 131
Crankshaft type
   330HD 33
   361 33
   390HP 33
   391 33
   406 33
   410 33
   427 33
   428 34
   428SCJ 33
   FE 32-34
   FT 32, 80
   LeMans 34
   NASCAR 34
Crankshaft-bearing bores 76
Crankshaft-bearing journals, cross drilled 33-35
Crankshaft-damper spacer 60-61, 132
Crankshaft-damper-spacer repair sleeve 132
Crankshaft-journal taper 80
Crankshaft-nose diameter 33
Cylinder block, align bore 76
Cylinder block, inspect 70, 72
Cylinder block, recondition 72
Cylinder-block type
   332 28
   352 26, 28
   352HP 28
   360 28-29
   361 28
   390 26-29
   390HP 29, 30
   406 26, 29
   410 28-29
   427 29-31
   428 32
   FE 26-32
   FT 32
Cylinder bore, chamfer 78
Cylinder head 63, 132-134
   assemble 112
   clean 100
   disassemble 98-100
   inspect 100
   mill 101-102
   pin and weld 101
Cylinder-head swapping 44-48
Cylinder-head type
   330 44
   332 39
   352 39
   352 HP 39
   360 39
   361 44
   361 Edsel
   390 39
   390HP 39
   391 44
   406 39-40
   410 39
   427 40-44
   FE 39-44
   FT 44
   Tunnel-Port 31, 101
Cylinder sleeve 30, 71

Cylinder-block crack 76
Cylinder-bore
   finish 76
   notch 78
Cylinder-head bolt
   torque 133, 134
Cylinder-head crack 11, 101
Cylinder-head dowel 137
Cylinder-head flatness 101
Cylinder-head warping 102
Cylinder-wall crack 11
Cylinder-wall failure 31

D
Daytona 3
Deck clearance 36
Decking, see Block deck
Detonation 7, 87, 157
Detroit balance 34, 35
Diagnosis 7
Dial bore gage 73, 92
Dipstick tube, install 143
Distributor 61, 141-142
Distributor-shaft bore 32
Diverter valve 18
Drain pan 13
Dynamometer 5, 157

E
Electronic ignition 10
Engine 13-23, 145-156
   decal 25
   hoist 13
   identification 24-26
Engine mounts 28, 63, 149
   four-bolt 28
   inspect 142
   two-bolt 28
Engine plate 60, 147-148
Engine tag 25
Engine-mount heat shield 142
Engineering number 27
Etching, see Notching
Exhaust manifold 14-15, 62, 142-143
Exhaust pipes 19, 151
Exhaust valve, burned 103
Exhaust, plugged 6
Exhaust manifolds 4-15, 143, 148
Exhaust-manifold 46-47
   doughnuts 149
   gasket 143
Expansion tank 17, 143, 155

F
Fan shroud 16, 154-155
Fan 15, 16, 153-154
Fender protector 13
Firing order 141, 153
Flare-nut wrench 18, 152
Flexplate 20-21, 33, 59-60, 148
Flywheel 20-21, 33, 59-60, 147, 148
Flywheel bolts, torque 148
Flywheel flange, inspect 147
Freon, bleed 17
Fuel lines 18, 151
Fuel pump 7-8, 61, 138
Fuel-pump eccentric 65, 121, 131

G
Gasket set 118
Generator mount 28
Glaze breaking 75
Ground strap 18, 155

H
H-pipe 6
Harmonic balancer, see Vibration damper
Head gasket
   blown 6
   composition type 101, 133
   shim type 101, 133, 134

Head-gasket
   leak 11
   sealer 133-134
Heat riser 6
Heater hoses 18, 155,
Heli-Coil 105
Holley carburetor 9
Holman and Moody 31
Hone 75
Hood 14, 156

I
Ignition coil, install 152
Improved-Combustion System (IMPCO) 15
Induced firing 10, 153
Inductance 10
Intake manifold 116
Intake-manifold type
   aluminum 62, 137
   cast iron 62, 137
   High-Riser 116, 136
   Low-Riser 427 136
   Medium-Riser 427 136, 137
   Tunnel-Port 427 62, 116, 136, 137
Intake manifold 62, 135-137
   mill 102
Intake-manifold, baffle 116, 135-136
Intake-manifold bolts
   torque 137
Intake-manifold gaskets 136-137
Intake-manifold heat passage 116, 136

J
Jiggle-pin 120

L
LN800 Line-hauler 24
LeMans 3
Leak-down test 6, 11
Lubrication, solid-lifter 29

M
Magnaflux 76, 79, 87, 92, 101
Main bearings, install 122
Main-bearing bores, align bore 67
Main-bearing
   clearance 8
   cross bolts 29, 31, 67
   edge ride 81
   noise 8
   oil hole 122
   webs 32
Main-bearing caps 66-67, 123-124
Main-bearing journals, grooved 33, 35
Main-bearing side seals
   install 124
   trim 135
Main-bearing-cap cross bolts, torque 77, 124
Main-bearing-cap bolts, torque 124-125
Manual transmission 149
Manual-transmission front seal 146

N
NASCAR 29
Notching 70, 75, 102

O
Octane rating 7
Oil baffle 140
Oil filter, install 144, 151
Oil pan 65, 134-135
Oil passages, clean 72
Oil pump 32, 64, 135
   assemble 97
   inspect 96
Oil 58, 154
Oil-filter adapter 63, 138
Oil-gallery plugs 69, 120, 144
Oil-passage crack 5
Oil pump 65, 134-135
   inspect 96-97
Oil-pump drive shaft 32, 97, 97, 134

Oil-pump drive shaft 65, 134-135
Oil-pump pickup 64, 135
Oil-pump pressure-relief valve 29, 96
Oil-pump rotors 96
Oil-pump-rotor end clearance 96
Oil-temperature sending unit, install 138
Oilite bushing 82
Oscilloscope 157

P
PCV valve 4, 5, 142, 156
Parkerize 121
Performance loss 5
Pilot bearing 82, 148
Pilot bushing 82, 148
Pilot shaft 82
Piston and connecting rod
   assemble 94-95
   disassemble 93
Piston
   bumpers 37
   compression height 36, 38
   dish 37
   dome 86
   eyebrows 37
   fitting 77
   inspect 86
   knurling 90
   pop-ups 37, 38
Piston rings 4, 86, 126-128
   fitting 125
Piston-ring type
   Dykes 37
   chrome 77
   moly 77, 126
   plain cast iron 77
Piston scuffing 87
Piston skirt 37, 86, 87
Piston slap 8, 87
Piston thrust surface 86, 90
Piston wrist-pin boss 88
Piston type
   427 36
   cast-aluminum 36
   FE 36
   forged aluminum 36
   four ring 36
   FT 36, 37
Piston-and-rod assemblies, 65-66, 128-130
Piston-ring compressor 125, 128
Piston-ring end gap 125-126
   compression ring 125
   oil ring 126
   position 127
Piston-ring expander 86, 125, 126
Piston-ring grooves, clean 88
Piston-ring
   noise 8
   twist 128
Piston-ring-groove cleaner 88
Piston-ring-groove width, measure 88-89
Pistons 36-38, 94-95
Piston-to-bore clearance 87, 89
Plastigage 118, 123
Power takeoff 33
Power-balance test 9-10
Power-steering pump 17, 151
Pre-oil 155
Preignition 7
Pushrod 8, 62, 140

R
Radiator hoses 15, 17, 154, 155
Radiator 16, 17, 154-155
Reading spark plugs 8, 9
Rear-main seal 123
Rear-main-seal type
   lip 123
   rope 123

Ridge 63-64
Ridge reamer 64
Road-draft tube 4
Rocker arm, adjustable 140
Rocker arms 61, 115-116
Rocker-arm assembly 61-62, 139-140
Rocker-arm
  oil galleries 76
  ratio 83
Rocker-arm screw 115
Rocker-arm shaft 113-115
  recondition 113-115
Rocker-arm stand 114-116
  aluminum 115
  cast iron 115
Rocker-arm-shaft
  notch 116
  oil holes 115, 116
Rocker-arm-to-shaft clearance, measure 115
Rod-bolt protector 66, 125, 128

S
SOHC 427 30
Service numbers 27
Shop manuals 160
Side-oiler
  427 30, 31, 37
  plug 120
Slide hammer 82
Sparkplug gap 144
Sparkplug wires, install 153
Sparkplugs 153, 155
Spotcheck 76, 87
Starter motor 21, 150-151
Starter-motor gasket/spacer 150
Stethoscope 7
Stress riser 81
Sunken float 9

T
Table
  Camshaft Lift 85
  Connecting Rods 54
  Cylinder Blocks 50-51
  Cylinder-Head-Bolt Torque 134
  Deck Clearance 38
  FE Crankshafts 52
  FT Crankshafts 53
  FT Cylinder Heads 56
  High-Performance Cylinder Heads 56
  Intake-Manifold vs Cylinder-Head Milling 102
  Main-Bearing Oil Hole 122
  Piston-Ring Widths 89
  Piston-to-Bore Clearance 89
  Ring End Gap 126
  Standard Cylinder Heads 55
  Standard Pushrods 57
  Valve Stem-to-Guide Clearance 103
  Valve-Seat Width 108
  Valve-Spring Specifications 110
  Valve-Stem Diameter 106
  Vehicle Identification Codes 48-50
  Wrist-Pin-to-Piston Clearance 88
  Wrist-Pin-to-Piston Clearance 88
  Wrist-Pin-to-Rod-Bushing Clearance 92
  valve-clearance adjustment order 140, 141
Taper 74
Taper tap 71
Thermostat housing, install 143, 155
Thermostat, install 143, 155
Thread chasing 71
Throttle linkage 18, 152
Timing chain 6, 64, 95-96, 130-131
  Morse type 96
  jumped 6
  roller type 96
Timing sprockets 6, 64, 95-96, 130-131
Timing-chain cover 61, 65, 131-132
Timing-chain-cover seal, replace 131
Top oiler 30
Torque converter 145-146, 149
Torque-converter nuts 20, 149-150
Transmission input shaft 82
Transmission-kickdown rod, 152
Transmission-pump seal, replace 145
Tuning 157

V
Vacuum fittings, install 143
Vacuum lines, install 152
Vacuum test 9
Valve, burned 6
Valve, grind 106-107
Valve clearance, adjust 7, 140-141, 156
Valve covers 61, 142
Valve float 108
Valve guides 5, 104-105
  bronze 101, 104-105
  cast iron 104-105
  inspect 101-104
  install 104-105
  recondition 104
Valve lapping 108
Valve lash, see Valve clearance
Valve leak 11
Valve leak test 108
Valve lift, measure 12
Valve lifters 62, 83-84, 86, 135
  inspect 83-84, 86
  replace 12
Valve margin 107
Valve pocket 102
Valve rotator 108, 114
Valve seat
  bottom cut 107-108
  top cut 107-108
Valve springs 98-100, 112-114
  inspect 110-112
Valve stem 5
  measure 105
  oversize 105
Valve tip, chamfer 106, 107
Valve-face angle 106, 107
Valve-guide insert 104-105
  aluminum/bronze 102
  bronze 105
  cast iron 103, 104-105
  thin-wall bronze 105
Valve-guide, knurling 104
Valve-keeper groove 105
Valve-lapping paste 108
Valve-leak test 108
Valve-lifter
  body 86
  bores 76, 135
  design 83
  foot 84, 86
  plunger 86
  prime 135
  pump-up 7
  wear 6, 8, 86
Valves 98-100, 112-114
  inspect 105
  recondition 105
Valve-seat
  angle 106-108
  bottom cut 107-108
  concentricity 108, 109
  reconditioning 107
  runout 108
  top cut 107-108
  width 108
Valve-seat insert, install 107
Valve-spring
  coil bind 110, 113
  free height 109, 110
  installed height 109, 112
  keepers 112, 114
  load 109-111
  rate 109
  retainer 98, 99, 112, 114
Valve-spring compressor
  C-clamp type 98-100
  hook type 98
Valve-spring retainer
  aluminum 114
  steel 114
  titanium 114
Valve-stem seal 5
  Teflon 104, 113
  umbrella 113
Valve-stem-tip hardness 106
Valve-stem-to-guide
  clearance 100, 102-104
  wiggle test 100
Valve-to-piston clearance 102
Valve-train noise 8
Vehicle identification number (VIN) 24, 25
Vibration damper 33, 35, 60-61, 137-138
  inspect 137

W
Water pump 60, 139
Water-pump bypass hose 139
Water-temperature sender, install 143
Wet test 11
Windage tray 35
Wrist-pin bore 90, 92
Wrist-pin bore, measure 88
Wrist-pin bushings 88, 93-94
Wrist-pin noise 8
  oil hole 95
  retaing-ring groove 87
  retaining ring 87, 93, 94, 95
Wrist-pin-to-rod-bushing clearance 92

Y
Y-block 3
Y-pipe 6

Z
Zero balance 34, 35

OFFICIAL SHOP MANUALS

For additional information concerning your car or truck you should obtain the official Ford shop manual. Single-volume manuals covering specific vehicles were printed for trucks prior to 1966 and for cars prior to 1970. These single-volume manuals were superseded by five-volume manuals which cover *almost* all car or truck lines. They are grouped according to the vehicle system—chassis, engine, electrical and body. The fifth volume covers general maintenance and lubrication. These manuals are available from Helm. To order, or for additional information, contact Helm, Inc., P.O. Box 07150, Detroit, Michigan 48207. Telephone: (313) 865-5000. Make sure you include the model and year of your vehicle when ordering.